Aeroelasticity

A V Balakrishnan

Aeroelasticity

The Continuum Theory

 Springer

A V Balakrishnan
Department of Electrical Engineering
Department of Mathematics
University of California
Westwood Blvd.
Los Angeles, California
USA

ISBN 978-1-4899-9499-8   ISBN 978-1-4614-3609-6 (eBook)
DOI 10.1007/978-1-4614-3609-6
Springer New York Heidelberg Dordrecht London

Printed on acid-free paper

Springer is part of Springer Science+Business Media (www.springer.com)

Foreword

The author is a noted applied mathematician and control theorist who has in recent years turned his attention to the subject of aeroelasticity. The present volume is the result of his research and provides an illuminating new view of the field. As he emphasizes in his introduction, his approach is one of continuum models of the aerodynamic flow interacting with a flexible structure whose behavior is governed by partial differential equations. Both linear and nonlinear models are considered although much of the book is concerned with the former while keeping the latter clearly in view and indeed a complete chapter is devoted to nonlinear theory. The author has provided new insights into the classical inviscid aerodynamics described by the celebrated Possio's equation and raises novel and interesting questions on fundamental issues that have too often been neglected or forgotten in the development of the early history of the subject. The author contrasts his approach with discrete models that have gained enormous popularity in the practice of aeroelasticity such as the doublet lattice model of Rodden et al. for unsteady aerodynamic flow and the finite element model for the structure. Much of aeroelasticity has been developed with applications firmly in mind because of its enormous consequences for the safety of aircraft. Aeroelastic instabilities such as divergence and flutter and aeroelastic responses to gusts can pose a significant hazard to the aircraft and affect its performance. Yet it is now recognized that there are many other physical phenomena that have similar characteristics ranging from flows around flexible tall buildings and long span bridges to flows internal to the human body such as blood flows through arteries and air flow over the tongue, and the author touches on these topics as well. For the theorist and applied mathematician who wishes an introduction to this fascinating subject as well as for the experienced aeroelastician who is open to new challenges and a fresh viewpoint, this book and its author have much to offer the reader.

Durham, USA   Earl Dowell

Preface

Aeroelasticity deals with the dynamics of an elastic structure in airflow with primary focus on the endemic instability of the structure called "flutter" that occurs at high enough speed. This book presents the "continuum theory" in contrast to extant literature that is largely computational; where typically one starts with the basic continuum model, a partial differential equation usually highly nonlinear but omitting the all important boundary conditions and disregarding the question of existence of solution; going immediately to the discretized approximation; presenting charts and figures for a confluence of numerical values for the parameters and conclusions drawn from them.

Here we stay with the basic continuum model theory until the very end, where constructive methods are developed for calculating physical quantities of interest, such as the flutter speed. Indeed this is considered "mission impossible" because it is nonlinear and complex.

As in any scientific discipline, continuum theory provides answers to "what if" questions which numerical codes cannot. It makes possible precise definitions—such as what is "flutter speed." Physical phenomena—such as transonic dip, for example—can be captured by simple closed-form formulae. And above all it can help develop intuition based on a better understanding of the phenomena of interest. As with any mathematical theory it enables a degree of generality and qualitative conclusions, increasing insight.

But the use of continuum models comes with a price: it requires a high level of abstract mathematics. For a precise statement of the problem, however, the language of modern analysis—developed in the latter half of the twentieth century—abstract functional analysis, in particular, the theory of boundary value problems of partial differential equations, is unavoidable. Indeed the aeroelastic problem, the structure dynamics in normal air flow-formulates as a nonlinear convolution/evolution equation in a Hilbert space.

On the other hand the numerical range of the physical parameters plays an important role in being able to generate constructive solutions otherwise impossible from the mathematics alone. What we do is indeed applied mathematics in the sense that we use mathematics to solve today's engineering problems addressed to engineers as well as mathematicians.

And now for some points of view, points of departure, of this book closer to the subject matter. Aeroelasticity is concerned with the stability of the structure in air flow. The air flow per se is of less interest. Thus we are not concerned, for example, whether there are shocks in the flow or not, in itself a controversial matter. The faith of the aeroelasticians in shocks, it turns out, is not substantiated by the mathematical theory (2D or 3D flow). It may be heresy to the clan but shocks may exist that do not affect the stability (or rather the instability) of the structure. Another and more significant view concerns the interaction between Lagrangian structure dynamics and Eulerian fluid dynamics, often the most mysterious part of computational work.

Here we take the simple engineering input–output point of view where the velocity of the structure is the input and the pressure jump across the structure is the output. The input–output relation is the integral equation of Possio that does not get any mention in as recent a work as [17] which features partial differential equations. The Possio Integral Equation can be looked as an illustration of the Duhamel principle and we make systematic use of it—linear and nonlinear—throughout the book. We show that flutter speed is simply the smallest speed at which the structure becomes unstable; it is a Hopf bifurcation point determined completely by the linearized model about the steady state. In turn this means incidentally that the control for extending the flutter speed need not be nonlinear, contrary to current wisdom.

The mathematical style of the book is largely imitated/borrowed from that of R.E. Mayer [14], and Chorin–Marsden [4] where they claim to "Present basic ideas in a mathematically attractive manner (which does not mean 'fully rigorous')." In this sense although we use abstract functional analysis, we try to reduce the abstraction and sacrifice mathematical generality, preferring to emphasize constructive solutions and basic ideas rather than get lost in Sobolev spaces and weak solutions. Quoting another pioneer in this style: "I shall not be guilty of artificially complicating simple matters. A phenomenon that sometimes occurs in mathematical writing." Tricomi in his book *Integral Equation*, 1957 [11].

We should caution that there are many problems that mathematical theory cannot currently answer especially in viscous flow and as a result also in aeroelasticity. We invoke the Prandtl boundary layer theory, for example, with this caveat.

We should also note a price to be paid for mixing the abstract with the concrete, saying too much or too little at either end.

Acknowledgments

It is a pleasure to acknowledge the generous help of my colleagues and my peers:

Earl Dowell (who was my chief mentor and also opened my eyes to the nonaircraft applications)
John Edwards
Oddvar Bendiksen
Dewey Hodges
Ciprian Preda
Marianna Shubov
Roberto Triggiani
Amjad Tuffaha

My former students:

Jason Lin
Oscar Alvarez Salazar
Irena Lasiecka (Post Doc)

And finally the NSF monitor Dr. Kishan Baheti for his unfailing belief in the value of the work and his help and encouragement throughout.

Los Angeles, California, USA   A V Balakrishnan

Contents

Chapter 1
Introduction

The subject matter of aeroelasticity is the dynamics of a structure in air flow, traditionally the wing of an aircraft in flight. There are important nonaircraft applications of more recent interest as well which we treat in Chap. 10. Thus it involves both structural dynamics and fluid dynamics without being either. In this sense it has an inherent identity complex but still has developed as a separate discipline over the many years since flight began. The central problem is "flutter," a potentially destructive instability that occurs at any altitude as the speed is increased to the "flutter speed," the "flutter boundary." Indeed the Federal Aviation Administration mandates that a margin of 15% be maintained in all flight. And thus "flutter analysis" is an ongoing activity in flight centers as well as in aircraft design.

Although in the aircraft context ultimately a flight test would be needed to verify the flutter phenomena, flight tests are costly and need to be kept at a minimum. Wind tunnel tests are much less expensive but paperwork is the least expensive. Hence the importance of analysis! The ultimate objective is of course to minimize the number of flight tests needed.

Here one has to distinguish between analytical (continuum) theory and (digital) computation (FEM/CFD) which is predominant in all the current work. Both structure and aerodynamic models originate as "continuum" models and digital computation needs perforce to approximate the continuum: convert the "distributed" parameter system to a "lumped" parameter system and replace the continuum partial differential equations by ordinary differential equations, however high the order. And of course digitization discretizes time and amplitude. Furthermore the system parameters must be specified numerically, numeric computation as opposed to symbolic computation. This restricts generality and can inhibit qualitative understanding of phenomena. Perhaps another way of putting it is to say that in this work we do not truncate the model at the beginning; we work with the continuum models or symbolic software and approximation as needed only eventually at the end-level calculation of specific functionals (such as flutter speed as a function of system parameters). We approximate the solution, rather than approximate the equations. Phenomena occur at the continuum level that do not occur in any solution of the discretized equations, however fine the discretization.

A V Balakrishnan, *Aeroelasticity: The Continuum Theory,*
DOI 10.1007/978-1-4614-3609-6_1, © Springer Science+Business Media, LLC 2012

In the computational approach the discretization is made without any concern about whether the continuum equations have a solution and whether are unique. Moreover there is no attempt to prove that there is a solution to the problem in the language of the starting continuum formulation. The digital computer solution is accepted as a good enough approximation to the solution. This can be particularly misleading when the continuum equations are shown to have more than one solution or no solution at all. Indeed it is typical in extant literature (see the List of Papers) in aeroelasticity where complex continuum structure models are introduced without any consideration of whether they have any solution and are immediately converted to discretized computational models (FEM). This is also true of the fluid dynamic models where the partial differential equations are introduced without any precise statement of the (fluid/structure) boundary conditions.

Thus we emphasize that a distinguishing feature of this treatment is that a precise statement of the aeroelastic problem is made especially at the Lagrangian–Euler fluid structure boundary; the dynamic equations are shown to have a unique solution and qualitative results are deduced and then only numerical computation is made for the functionals of interest, without discretizing the equations at any level.

The material has been distributed into 10 chapters in reasonable logical progression and are listed in the Table of Contents. Here we provide an overall preview of the various chapters and how they interlink.

We begin in Chap. 2 with the dynamics of structures, where the important notion of modes and mode shapes is described in the language of function spaces, specifically L_2-spaces. (It is interesting to note that the need for infinite-dimensional spaces is underscored by a recent publication "Dynamics of Very High Dimensional Systems" [24].) The primary reason for this level of abstraction is that a precise mathematical statement of the structure dynamics in air flow is not possible without it.

The first flexible wing model—the "Goland" model—was proposed by Goland [78] in 1945, a uniform rectangular beam of zero thickness with two degrees of freedom, plunge and pitch. Similar equations with higher degrees of freedom are employed in applications to spacecraft structures [54, 61]. It is a "fixed wing" model cantilevered to the fuselage with the wing tip (clamped/free end conditions). This is our "canonical" model; (all later versions still anchored essentially on this) including nonlinear models. A drawback is of course that the thickness or the "wing camber" is neglected. However, extending it to a plate model turns out to be too complex to handle and it is questionable whether the results justify the analytical complexity. There are no existence theorems, neither where the models include wing camber nor the wing section as a "tear drop" [18]. Another peculiarity is that the cantilever beam model is used only for the structure dynamics corresponding to the aerodynamic loads, and to calculate the latter one takes advantage of the "high aspect ratio" of the wing which must needs take into account the flexibility of the structure, "long slender wing." Some authors even further qualify it by "very flexible" wing [82]. The flexibility is only along the span; we have thus a "beam" instead of a "thin plate." Furthermore, this allows the enormous simplification that we may neglect the dependence on the span axis in describing the aerodynamic field equation. This is referred to as the "strip" theory or the "typical section" theory. We can also see this as a "first approximation" of the "finite plane" theory (cf. Chap. 3). In other words the structure "beam" model is consistent with the aerodynamic "strip" [6] theory.

The Hilbert space for structure state variables presented is standard and essential use is made of the notion of "elastic energy," including a characterization of the structure deflections for which the energy can be defined. This defines the modes and mode shapes as eigenvalues and eigenvectors respectively. In continuum theory the number of modes is infinite. In practice we know we will only deal with the "first few" modes, so that much of the mathematical theory of asymptotic modes is irrelevant. In the approximated computer model, the number of modes is indeed finite, but then we can always add one more! Indeed we construct a continuum model in which we match the "first few" modes of interest. This is because of the inherent damping of the structure that eventually increases as the mode number increases. But truncating the model at the beginning one loses physical phenomena that only the continuum model can capture.

Control theory is needed for our objective of stabilizing or enhancing the stability of the structure and here the truncated numerical model is useless. We need the continuum model to generate feedback control laws. One difficulty is in not knowing the inherent damping. The few models we have are not faithful enough. Hence in "robust" control we assume a model in which there is no damping (as in our models) but devise a feedback control which is guaranteed not to decrease the inherent damping.

Current control design for distributed structures uses colocated point controllers where the sensor and actuator are colocated. An inherent limitation here is that the damping will decrease (not necessarily monotonically) to zero as the mode number increases. An alternative is to use the "self-straining" actuators [70] which can ideally achieve "super stability." So they are attractive to try in the aeroelastic case.

Finally we consider nonlinear structure models (they may be considered extensions of the Goland model) still beam models with zero thickness. The models suggested [75, 76] unfortunately have not been analyzed in the continuum version unlike the linear case. There is no longer the notion of modes (eigenvalues) which makes the determination of stability difficult.

To make a mathematical statement of the structure dynamics we first need to invoke aerodynamics to calculate the aerodynamic loads. Chapter 3 thus begins with the aerodynamics part of aeroelasticity, the theory leading to the calculation of the aerodynamic lift and moment. We limit consideration to "nonviscous" flow where the viscosity of the air is neglected. Thermodynamics is involved and the assumption thruout is that the perfect gas law holds. In the absence of viscosity we need to assume that the entropy is a constant so that the heat exchange processes are adiabatic. The assumption of constant entropy enables us to relate the pressure to density:

$$p = A\rho^\gamma$$

and show that the flow is vortex-free and we have "potential" flow which can be then characterized by the Euler full potential equation. However, what distinguishes the aeroelastic problem from being simply fluid dynamics is the fluid–structure interaction, the boundary conditions.

The assumption here is "flow tangency," the wing velocity normal to the plane is equal to the fluid velocity normal to the wing plane. In addition we have to impose the conditions discovered by Joukowsky and the condition named after Kutta which is necessary for uniqueness of solution.

We are then able to come up with a complete statement of the aeroelastic equations. Here we also specialize to 2D flow, and zero angle of attack in which we ignore the dependence on the wing span (abandon the "Finite Plane" model) and consider the "typical section" approximation justified by the high–aspect ratio of the wing—a *raison d'être*—also for the flexible PDE model. Finally here we also extend the formulation to include nonlinear structure models.

The study of stability of the wing structure (which for us now depends on the assumed far-field speed that we consider a "parameter") requires that we specify stability about the state of the system. One usually thinks of the "equilibrium state" to which the system is "attracted" or returns in the absence of any disturbance. We need to determine the solution to the aeroelastic equation where we set all time derivatives to zero. We call this the "static solution". This is the objective in Chap. 4. The technique for solution is to exploit the analyticity in terms of the structure state variables so that we can make a Taylor series expansion about the zero or "rest" structure state. We show that the solution is nonunique for a sequence of values of the far-field speed parameter for which we have two different solutions, even if we may call one of them "trivial."

The main tool is the "static" version of the Possio integral equation which turns out to be the finite Hilbert transform solved by Tricomi, who showed the need for the Kutta condition for uniqueness of solution; see [11, and therein to the work of Sohngen, a German pioneer in the theory]. See also [31]. Here we are able to show the convergence of the power series pointwise. There are no discontinuities except on the boundary, for $z = 0$. Here we also derive a simple formula for the "Transonic Dip" for nonzero angle of attack which in numerical computation has to be extrapolated from a point on the graph. See [65] as well as the references therein. The nonzero static solution can be considered as an eigen value problem–albeit non–linear. The nonlinearity is primarily in the pitch or torsion variable.

The flow has no discontinuities in the field and the convergence is pointwise—in sharp contrast to the time varying case in Chap. 6.

We conclude the chapter with an example which turns out to be relevant to viscous flow treated in Chap. 7.

One of the major results of the continuum theory is the characterization of the smallest speed at which the wing is unstable—the "Flutter" speed can be characterized as a Hopf bifurcation point—an illustration of the Hopf bifurcation theory— the successor to the classical stability theory of Liapunov [42]. Here the stability is determined completely by the linear system obtained by linearizing the nonlinear system about the steady state. Naturally the bulk of the aeroelastic theory deals with the Linear case, and there is no exception here. By the linear case we mean the solution to the equations linearized about a steady state. The term "linear model" is often used in the aeroelastic literature without specifying the steady state about which it is linearized. Of course the natural steady state would be that of constant flow and zero structure state. The corresponding linear theory is treated in Chap. 5 the longest chapter in the book, with eight sections. This provides the bread-and-butter of aeroelasticity and the reason for its life in the sun. Here we study linear flutter analysis with continuum models without approximation of any kind. (No Pade approximation or the Rodden computer algorithm [17, 36]). The main concept is that of "aeroelastic modes" and the linear structure dynamics is formulated as a "convolution/evolution (semigroup) equation in a Hilbert space. What is also new here is the development of a state space theory which in particular shows that the aeroelastic modes are the eigenvalues of the state–space stability operator. We also show that the case $M = 0$ is a special one for which we have second degree noncirculatory terms not present for $M > 0$. Also we are able to characterize the mode shapes explicitly and their properties as eigenfunctions of the stability operator. The central theme is the linear Possio integral equation [56, 89] which connects the Lagrangian dynamics to the Eulerian. More specifically it provides the input–output relation between the structure velocity as the input and the pressure jump across the wing as the output, so that we can express the aerodynamic force and moment in terms of the structure state variables. The key concept is the "Mikhlin Multiplier" and the Balakrishnan formula based on it. The original version of the integral by Possio [89] was in the frequency domain and is transported here to the Laplace domain. This suffices for calculating the flutter mode and frequency. The aeroelastic modes are shown to be the zeros of the determinant of a 3 by 3 matrix function of a complex variable which is analytic except for the logarithmic singularity along the negative real axis for all values of M. The aeroelastic modes are shown to be in a bounded vertical strip. Zero is an aeroelastic mode for all $M < 1$. The associated flutter speed is known as the "divergence speed" and is shown to be finite. The latter result is essential in showing the existence of the flutter speed. It is discontinuous in M as a function of the angle of attack and an explicit formula exhibits a "transonic dip" for a high enough value of M, hitherto deduced from graphical computation. The modes are characterized as the eigenvalues of the stability operator in the state space theory. The uniqueness of solution of the Possio equation requires the Kutta condition and in fact uniqueness implies existence. Various expressions are deduced for the solution.

Key to the state space representation is the characterization of the aeroelastic structure equation as a linear convolution evolution equation in a Hilbert space. The convolution part plays the essential role in the flutter phenomena. We also derive an explicit time domain solution for

$$M = 0 \quad \text{and} \quad M = 1.$$

We note that all the results obtained in the classical treatise [6] based on the Theodorsen approximation can be deduced from the solution to the Possio equation as in fact we do here. The sonic case ($M = 1$) is interesting in that the role of the Theodorsen function is taken over by the error function.

Essential for the stability theory is the fact that the slope of the stability curve is negative for small speeds for all modes. The system becomes more stable initially with airflow and then becomes unstable at some point as we increase speed, which is the flutter speed. A relatively simple algorithm for it based on the continuum model is given in the appendix to the chapter. One of the interesting by-products of the continuum theory is the existence of "evanescent states" which decay too fast to be captured in the CFD computer calculations.

All of this is only a prelude to the stability theory for the nonlinear system developed in Chap. 6. The main result here is that the system is stable for all values of speed less than the flutter speed and at the flutter speed the asymptotic instability can be characterized as an LCO, limit cycle oscillation. It is consistent with the Hopf bifurcation theory but the details of the proof have had to be developed independently. The analysis turns out to be tedious and complicated involving many asymptotic estimates. We show that the limiting response is periodic with the period equal to the inverse of the linear flutter frequency with the preponderant harmonic being the third harmonic.

We also derive an expression for the amplitude of the LCO rather involved in terms of the various parameters. It is valid strictly speaking only for small initial amplitude and is shown to be proportional to it. Although probably of less importance in practice, the question of what happens for large initial amplitudes is left open here as in computation. Heavy use is made throughout of Volterra expansion. In fact in showing that the composite aeroelastic equation is a nonlinear convolution/evolution equation, the nonlinearity is expressed in terms of a Volterra series. And we note that in it the torsion variable is more dominant. We go with the linear structure model, the extension to the nonlinear case being straightforward based on the prequel. It is possible to examine the role of the structure in contrast to the aerodynamics although we don't go into it.

As noted, our primary interest is in the structure but we also obtain en route a revealing decomposition of the airflow—the velocity potential—into two parts. One part produces all the lift and can be linearized about the steady state whereas the second part is continuous across the wing and cannot be so linearized. Interpreted in the weak sense (which we do not attempt here), it may contain shocks, which is controversial.

In Chap. 7 we venture even more into controversial territory where we allow the flow to be viscous.

We assume small viscosity consistent with the fluid now being air. And we now have to refresh the conservation laws. The basic field equation now becomes the Navier–Stokes equation for incompressible flow. This is a much trodden area in fluid dynamics and yet still with many unresolved questions, which are then reflected in the structure dynamics as well. For example, there is no proof yet whether we return to the Euler equation as the viscosity becomes vanishingly small. The big question here concerns the presence of shocks. There is no mathematical proof yet of shocks in 2D flow. The boundary conditions now require that the structure velocity be completely equal to the fluid velocity. To make any progress at all we have to invoke the still unproven Prandtl boundary layer hypothesis, and even then our treatment is not as complete as we would like.

The question of whether we can altogether suppress flutter by means of control actuators (flutter suppression as it is called) is the theme of the eighth chapter. We do have a model, a state space model, that makes it possible for us to consider control design. Stabilizability is a central question in optimal control theory but the application to state spaces of nonfinite dimension essential to exhibit the flutter phenomena as here poses greater complexity. The main result here is that no actuator on the structure can stabilize the system; we simply cannot suppress flutter. It will always occur at some speed. However, we can design controls that enhance the stability in the sense that we can increase the speed at which flutter occurs. But this is no longer a crisp optimization problem and comparison of control performance is probably not possible. One fallout of our theory is that the control need not be nonlinear. Indeed as we have shown, the flutter speed is completely determined by the linearized system.

One of the important safety issues for aircraft in flight is the effect of wind gust which still continues to be of interest and is a familiar topic in flight dynamics [9,33]. Here we consider the problem with the 3D random field Kolmogorov model of air turbulence [35] as opposed to the deterministic gust models studied in [6, 84]. We calculate the spectral density of the structure response showing the role of the flutter speed. Illustrative examples are given for both bending flutter and torsion flutter.

Finally in Chap. 10 we provide an addendum on the latest area of research in flutter systems, nonaircraft and nonflight applications, for example to Piezoelectric power generation [106,109] and the biomedical problem of palatal flutter [102,105]. Here the main difference is in the nature of the air flow. It is no longer normal to the structure but is axial, axial flow. Our treatment is again based on continuum models as opposed to the regulation computational as in extant literature [50, 102]. This brings profound differences; the theory is more complicated and so this chapter is in the nature of an Addendum confined largely to problem formulation, emphasizing the difference from normal flow. It turns out that there is one simplification. In the application we need only consider incompressible flow which is then characterized simply by the linear Laplace equation. The nonlinearity is thus only on the boundary dynamics. We consider here again the Goland beam model, and in particular the symmetric case where the cg is on the elastic axis, and as a result the plunge and pitch dynamics decouple. However, the mathematics is more complex in that the Hilbert space has to be generalized to a Banach space: from an L_2 space to L_p, $1 < p < 2$. We are limited by space alone to problem formulation with details deferred to a forthcoming second volume.

Notes and Comments

It is interesting to note that in the beginning in aeroelasticity theory only continuum models were used by the German pioneers: Kushner, Schwarz, Sohngen, and Wagner among others. The convolution aspect was already shown by Wagner. But the Allies after WWII decreed a moratorium on any further research by Germany from 1945 until 1954. This was a death blow to German work in aeroelasticity from which it never really recovered. The advent of FEM/CFD can be traced to the late 1970s. A landmark is the appearance in 1982 of *Transonic Shock and Multidimensional Flows: Advances in Scientific Computing* edited by Meyer. It is ironic that this volume contains an article by *G* Moretti pleading for "Closer Cooperation between Theoretical and Numerical Analysis in Gas Dynamics." The Aeroelasticians combined the discrete approximation in structures FEM with CFD for the aerodynamics. But the integration of the two via the boundary conditions was always an art, often the more mysterious part. The battle between continuum theory and computation is still enjoined. And as late as 2010 we see one unresolved point: see [98] where the author writes: "verification of numerical solutions is a step in their scientific acceptability and a formal requirement of many engineering related enquiries... However, these equations are known to have nonunique solutions. If the true answer is not unique, what does correctness of approximation mean?" As we show in Chap. 7 that in viscous flow even existence of a solution is not proved compounding this difficulty, because now we can ask: approximation to what? So the conclusions from the computation have to be taken on faith.

It should also be noted that time is also discretized in the approximation but there are phenomena, for example, in stochastic process theory, Gaussianness of the innovation process in nonLinear filtering, which holds only in the continuous time model (Girsanov's theorem: see, for example, [26]).

Chapter 2
Dynamics of Wing Structures

2.1 Introduction

This chapter deals with the dynamics of elastic structures: unswept fixed-wing and free-free wings, modeled as beams of zero thickness, linear as well as nonlinear. The main concern is with spectral analysis, modes and mode shapes as a means to study stability. The language is that of abstract functional analysis: Hilbert spaces and semigroup theory of operators for time domain description.

2.2 The Goland Beam Model

This is the earliest continuum model introduced in aeroelasticity by Goland [78] in 1945. See Fig. 2.1. It is a uniform rectangular beam. The co-ordinate axes are chosen so that X-axis is the (rigid body) airplane axis also called the Chord-axis with the width or chord length $2b$:

$$-b < x < b.$$

The one-sided wing span is ℓ along the Y-axis:

$$0 < y < \ell$$

and it is assumed that the "aspect ratio"

$$\ell/b$$

is "high," justifying the flexible beam model. It is endowed with two degrees of freedom: bending displacement $h(\cdot)$ in a plane normal to the beam ("plunge" in aeroelastic parlance) and the torsion angle $\theta(\cdot)$ in radians ("pitch" in aeroelastic

A V Balakrishnan, *Aeroelasticity: The Continuum Theory*,
DOI 10.1007/978-1-4614-3609-6_2, © Springer Science+Business Media, LLC 2012

Fig. 2.1 Wing model

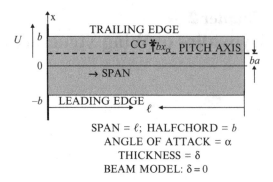

SPAN = ℓ; HALFCHORD = b
ANGLE OF ATTACK = α
THICKNESS = δ
BEAM MODEL: $\delta = 0$

parlance) about an axis parallel to the wing axis at distance bx_α from the cg as shown in Fig. 2.1. The bending dynamics is described by an Euler equation:

$$m\ddot{h}(t, y) + S\ddot{\theta}(t, y) + EIh''''(t, y) = L(t, y), \quad t > 0; \quad 0 < y < 1 < \infty \tag{2.1}$$

and the torsion by the (Saint Venant, string, or Timoshenko) equation:

$$I_\theta\,\ddot{\theta}(t, y) + S\ddot{h}(t, y) - GJ\theta''(t, y) = M(t, y), \qquad t > 0, \, 0 < y < \ell, \tag{2.2}$$

where the forcing terms:

$L(t, y)$ is the aerodynamic force per unit length (lift);
$M(t, y)$ is the applied aerodynamic.

The moment about the elastic axis per unit length, are both determined later from the airflow model.

m = Mass per unit length
I_θ = Crosssectional moment of inertia
EI = Bending stiffness
GJ = Torsional stiffness
S = Coupling constant = mbx_α, $|x_\alpha| < 1$,
see Fig. 2.1 for x_α; all constant, justifying the term "uniform."

See [5] for more details on the physical units.
Here x_α can be positive or negative, and it is required that $S^2 < mI_\theta$.
The convention throughout is that the superdots denote the partial derivatives with respect to time t and the superprimes the partial derivatives with respect to the spatial co-ordinate y. Note that the structure variables do not depend on the chord variable. To complete the dynamics we need to specify the end conditions.

End Conditions

1. For "fixed-wing" aircraft where the wing is attached to the fuselage (as in Fig. 2.1) we add the "Clamped-Free" or CF end conditions:

$$h(t,0) = 0; \quad h'(t,0) = 0; \quad h''(t,\ell) = 0; \quad h'''(t,\ell) = 0;$$

$$\theta(t,0) = 0; \quad \theta'(t,\ell) = 0 \quad t > 0. \tag{2.3}$$

2. FF Free-free condition
   This is typical of the recent UAV aircraft that have no fuselage and resemble a "flying wing." Here both ends are "free."

$$h''(t,0) = 0; \quad h'''(t,0) = 0; \quad h''(t,\ell) = 0; \quad h'''(t,\ell) = 0,$$

$$\theta'(t,0) = 0; \quad \theta'(t,\ell) = 0 \quad t > 0. \tag{2.4}$$

3. CC Clamped-clamped condition
   This is relevant to non-aircraft application–see Chap. 10

$$h(t,0) = 0; \; h'(t,0) = 0; \; \theta(t,0) = 0,$$

$$h(t,\ell) = 0; \; h'(t,\ell) = 0; \; \theta(t,\ell) = 0.$$

2.3 Time Domain Analysis

Our main concern is the stability of the structure. Stability of motion is a time domain concept even if all the techniques involve Laplace transformation, Laplace domain.

For an elementary treatment of vibration of beams reference may be made to the classic treatise of Timoshenko [32] and the textbook [34], and the more recent treatment relevant to aeroelasticity by Hodges and Pierce [5].

Here of course we use abstract functional analysis, albeit rudimentary, following [61] where a model with three degrees of freedom with damping is analyzed and [56] where specifically the Goland model is treated. Both of these follow the general Hilbert Space formulation in [16, 55]. Not only does it make the presentation compact and elegant, this level of abstraction is actually necessary for a precise mathematical formulation of the structure dynamics in airflow, the subject matter of this book.

Let H denote the Hilbert space:

$$L_2[0, \ell] \times L_2[0, \ell]$$

with elements

$$x = \begin{pmatrix} h(\cdot) \\ \theta(\cdot) \end{pmatrix}.$$

This choice of space is motivated by the fact that the elastic "potential energy" is given by

$$\frac{1}{2}\left(EI\int_0^\ell |h''(y)|^2 dy + GJ\int_0^\ell |\theta'(y)|^2 dy\right).\tag{2.5}$$

We begin with the abstract "differential" operator denoted A.

The Domain of A (denoted $\mathcal{D}(A)$): where we need to distinguish between CF "clamped–free", FF "free–free" and CC "clamped–clamped" end conditions:

For CF

$$\Big[x \mid h''''(\cdot)\in L_2[0,\,\ell];\; \theta''(\cdot)\in L_2[0,\ell], h(0)=0;\; h'(0)=0;$$
$$\theta(0)=0\; h'''(\ell)=0;\; h''(\ell)=0;\; \theta'(\ell)=0\Big].$$

For FF

$$\Big[x \mid h''''(\cdot)\in L_2[0,\,\ell];\; \theta''(\cdot)\in L_2[0,\ell], h'''(0)=0;\; h''(0)=0;$$
$$\theta'(0)=0\; h'''(\ell)=0;\; h''(\ell)=0;\; \theta'(\ell)=0\Big].$$

For CC

$$\Big[x \mid h''''(\cdot)\in L_2[0,\,\ell];\; \theta''(\cdot)\in L_2[0,\ell], h(0)=0;\; h'(0)=0;$$
$$\theta(0)=0\; h(\ell)=0;\; h'(\ell)=0;\; \theta(\ell)=0\Big].$$

Although strictly speaking we need to distinguish between these operators in principle, it is less messy to go ahead with one operator in the abstract theory, but making the necessary changes in the concrete calculations.

$$Ax = \begin{pmatrix} EI\,h'''' \\ -GJ\,\theta'' \end{pmatrix}.\tag{2.6}$$

Thus defined it is standard [16] to verify that the domain is dense in H and that A is closed on its domain. Note further that for x in $\mathcal{D}(A)$:

$$[Ax, y] = [Ay, x] \quad \text{for } y \text{ in } \mathcal{D}(A).$$

Or, A is self-adjoint. Furthermore we can calculate that
$[Ax, x] = 2$ potential energy ≥ 0.

Or, A is nonnegative definite.

Suppose for some x in the domain of A, we have

$$Ax = 0.$$

Then

$$[Ax, x] = 0.$$

Or the potential energy is zero, and hence

$$x = 0.$$

Or, zero is not in the spectrum of A. Any λ such that

$$Ax = \lambda x, \quad x \text{ nonzero}$$

is called an eigenvalue of A and x is the eigenvector corresponding to λ.

The set of eigenvalues is countable; the eigenvalues $\{\lambda_k\}$ of A are positive and the corresponding eigenfunction spaces are each of dimension 1. The eigenfunctions $\{\phi_k\}$ are orthogonal, "complete," span the space, and are orthonormalized for our purposes here. The resolvent, denoted $R(\lambda, A)$, is compact, because the beam is finite. Any complex number λ not equal to λ_k for any k is in the resolvent set. Also the resolvent is Hilbert–Schmidt [16]

$$\sum_{k=1}^{\infty} [[R(\lambda, A) \phi_k]]^2 = \sum_{k=1}^{\infty} \frac{1}{|\lambda - \lambda_k|^2} < \infty$$

and we have the eigenfunction expansion:

$$R(\lambda, A)x = \sum_{k=1}^{\infty} \frac{[x, \phi_k]}{\lambda - \lambda_k} \phi_k. \tag{2.7}$$

Moreover the eigenfunctions are "complete" in H; see [16]. Any element in H has the modal representation

$$x = \sum_{k=1}^{\infty} [x, \phi_k]\phi_k, \qquad \sum_{k=1}^{\infty} |[x, \phi_k]|^2 < \infty.$$

The square root operator

The potential energy can be defined for a larger class of elements in H than $\mathcal{D}(A)$. For this we need the square root of A denoted $\sqrt{A}$. It can be defined in terms of the eigenfunctions as

$$x \in D\left(\sqrt{A}\right) \quad \text{if} \quad \sum_{k=1}^{\infty} \lambda_k |[x, \phi_k]|^2 < \infty$$

and we define

$$\sqrt{A}x = \sum_{k=1}^{\infty} \sqrt{\lambda_k}[x, \phi_k]\phi_k \qquad (2.8)$$

from which it follows that zero is not an eigenvalue of $\sqrt{A}$.

The domain of $\sqrt{A}$ is thus larger than that of A. For x in $D(A)$

$$\left[\sqrt{A}x, \sqrt{A}x\right] = [Ax, x] = 2 \text{ Potential energy}.$$

We then define potential energy for all elements in $D(\sqrt{A})$ as:

$$\frac{1}{2}||\sqrt{A}x||^2.$$

This is the largest domain for which the potential energy can be defined. We rarely need to know the precise domain of $\sqrt{A}$ (see [55] for details on the domain) or the precise form; we should note, however, that it may not be a differential operator. See [63] and below.

The time domain solution: $S = 0$.

We begin by rewriting the dynamic equation (2.1) and (2.2) in abstract form as

$$\mathcal{M}\ddot{x} + Ax = \begin{pmatrix} L(t) \\ M(t) \end{pmatrix}, \qquad (2.9)$$

where $\mathcal{M}$ is the nonsingular nonnegative definite 2×2 matrix

$$\begin{pmatrix} m & s \\ s & I_\theta \end{pmatrix}$$

and is the mass/inertia operator; the operator A can be called the stiffness operator, extending the finite-dimensional definitions.

The Energy Space

Let us introduce next the "energy space" $\mathcal{H}$ consisting of elements of the form:

$$Y = \begin{pmatrix} x_l \\ x_2 \end{pmatrix} \quad x_1 \in D(\sqrt{A}), \; x_2 \in H$$

endowed with the energy inner product:

$$[Y, Z]_E = \left[\sqrt{A}x_1, \sqrt{A}z_1\right] + [\mathcal{M}x_2, z_2], \qquad (2.10)$$

where

$$Z = \begin{pmatrix} Z_1 \\ Z_2 \end{pmatrix} \quad Z_1 \in D\left(\sqrt{A}\right), \; z_2 \in H.$$

Denote this inner-product space by $\mathcal{H}$.

Theorem 2.1. $\mathcal{H}$ *is a Hilbert space.*

Proof. We need to show that every Cauchy sequence converges to an element in the space.

Let

$$Y_n = \begin{pmatrix} x_n \\ y_n \end{pmatrix}$$

be a Cauchy sequence in $\mathcal{H}$. Then

$$\sqrt{A}x_n$$

is a Cauchy sequence in H. Because 0 is not an eigenvalue of $\sqrt{A}$,

$$R\left(0, \sqrt{A}\right)\sqrt{A}x_n = x_n,$$

where $R\left(\lambda, \sqrt{A}\right)$ denotes the resolvent of $\sqrt{A}$, converges to an element x in H, actually in $D\left(\sqrt{A}\right)$, because $\sqrt{A}$ is closed.

$$[My, y] = \left[\sqrt{M}y, \sqrt{M}y\right],$$

thus it is seen that y_n is a Cauchy sequence in H, by a similar argument. Hence it follows that Y_n converges to an element in $\mathcal{H}$. Hence $\mathcal{H}$ is a Hilbert space. □

Note that what we have proved is that the domain of $\sqrt{A}$ is closed in H because $\sqrt{A}$ has a bounded inverse. This is the largest domain for which the potential energy can be defined.

We rarely need to know the precise domain of $\sqrt{A}$ (see [63] for details on the domain) or the precise form; we should note, however, that it may not be a differential operator. In our case we can see that

$$\sqrt{A}\begin{pmatrix} h \\ 0 \end{pmatrix} = \begin{pmatrix} -\sqrt{EI}\,h'' \\ 0 \end{pmatrix}.$$

But the square root for the torsion operator is no longer a differential operator.

For the clamped–clamped case, for example, it is given by

$$\sqrt{A_\theta}\,\theta = g; g(s) = \frac{1}{\ell}\int_0^\ell \frac{\operatorname{Sin}\frac{\pi s}{\ell}}{\operatorname{Cos}\frac{\pi s}{\ell} - \operatorname{Cos}\frac{\pi \sigma}{\ell}}\,\theta'(\sigma)d\sigma$$

$$0 < s < \ell \qquad \theta(0) = \theta(\ell) = 0.$$

See [63] for more on this.

Note that the norm in $\mathcal{H}$ has a physical significance. $||Y||^2 = 2$ total energy $= 2$ (potential energy + kinetic energy).

Hence we refer to it as energy space.

As in finite dimensions, we go on to the state space formulation of the problem. Let

$$Y(t) = \begin{pmatrix} x(t) \\ \dot{x}(t) \end{pmatrix}.$$

Then the second-order equation (2.5) goes over into the first-order in time equation:

$$\dot{Y}(t) = \mathcal{A}Y(t) + \mathcal{V}(t), \tag{2.11}$$

where

$$\mathcal{V}(t) = \begin{pmatrix} v_l(t) \\ v_2(t) \end{pmatrix},$$

where

$$v_1(t) = \begin{pmatrix} 0 \\ 0 \end{pmatrix},$$

$$v_2(t) = \begin{pmatrix} L\,(t) \\ M\,(t) \end{pmatrix},$$

where $L(t)$ denotes the element $L(t, y)$ and $M(t)$ the element $M(t, y)$ in $\mathcal{H}$, and the operator A is defined by:

$$A\begin{pmatrix} y_l \\ y_2 \end{pmatrix} = \begin{pmatrix} 0 & I \\ -\mathcal{M}^{-1}A & 0 \end{pmatrix}\begin{pmatrix} y_l \\ y_2 \end{pmatrix} = \begin{pmatrix} y_2 \\ -\mathcal{M}^{-1}Ay_1 \end{pmatrix}$$

with domain:

$$y_1 \in \mathcal{D}(A),$$

$$y_2 \in \mathcal{D}\left(\sqrt{A}\right).$$

Thus defined it is closed with domain dense in $\mathcal{H}$.

Moreover we can readily verify that the adjoint (in $\mathcal{H}$) $\mathcal{A}^*$ is given by $-\mathcal{A}$:

$$\mathcal{A} + \mathcal{A}^* = 0$$

and hence [16] $\mathcal{A}$ generates a C_0—semigroup $\mathcal{S}(t), t \geq 0$. This is not true incidentally if the space is $H \times H$. See [39]. The solution of the homogeneous equation

$$\dot{Y}(t) = \mathcal{A}Y(t)$$

is given by

$$Y(t) = S(t)\, Y(0) \quad \text{for } Y(0) \text{ in } D(\mathcal{A}).$$

Lemma 2.2. *The semigroup $S(\cdot)$ is isometric:*

$$||Y(t)|| = ||S(t)Y|| = ||Y||. \tag{2.12}$$

The energy remains a constant, in other words.

Proof. For Y in $\mathcal{H}$,

$$m(t) = [S(t)Y, S(t)Y]_E = ||Y(t)||^2.$$

Then for Y in $D(\mathcal{A})$

$$\dot{m}(t) = [(\mathcal{A} + \mathcal{A}^*)S(t)Y, B(t)Y] = 0$$

and hence for Y in $D(\mathcal{A})$

$$||S(t)Y|| = ||Y||.$$

But the domain of $\mathcal{A}$ is dense in $\mathcal{H}$, and hence (2.8) follows.   □

An equivalent statement is that

$$S(t)^{-1} = S(t)^*.$$

It should be noted that the differential equation:

$$\dot{Y}(t) = \mathcal{A}Y(t)$$

is satisfied only for $Y(0)$ in the domain of $\mathcal{A}$.

The nonhomogeneous equation we started with:

$$\dot{Y}(t) = \mathcal{A}Y(t) + V(t)$$

has the solution

$$Y(t) = S(t)Y(0) + \int_0^t S(t - \sigma)v(\sigma)d\sigma \qquad t > 0 \tag{2.13}$$

for $Y(0)$ in the domain of $\mathcal{A}$, and may need to be interpreted in the weak sense depending on the function $\nu(\cdot)$ (as a function of time) [3] but we do not need to go into this until later.

2.4 Structure Modes and Mode Shapes

Our main concern is the stability of the solution or of the semigroup $\mathcal{S}(\cdot)$. This means we need to study the spectrum of $\mathcal{A}$. Because $\mathcal{R}(\lambda, \mathcal{A})$ the resolvent of $\mathcal{A}$ is compact, we need only consider the eigenvalues of $\mathcal{A}$.

Unraveling

$$(\lambda I - \mathcal{A})Y = 0, \qquad Y = \begin{pmatrix} y_1 \\ y_2 \end{pmatrix},$$

we have:

$$\lambda y_1 - y_2 = 0,$$
$$\lambda y_2 + \mathcal{M}^{-1} A y_1 = 0.$$

Hence

$$y_2 = \lambda y_1,$$
$$\lambda^2 \mathcal{M} y_1 + A y_1 = 0; \; A y_1 = -\lambda^2 \mathcal{M} y_1,$$
$$[A y_1, y_1] = -\lambda^2 [\mathcal{M} y_1, y_1].$$

Hence it follows that $\lambda^2 \leq 0$; and hence $\lambda^2 = -\omega^2$, ω real. But λ cannot be zero, because if it is, so is y_1, y_2.

The eigenvalues then are defined by

$$y_2 = i\omega y_1,$$
$$A y_1 = \omega^2 \mathcal{M} y_1,$$

y_1 is an eigenvector of A but with respect to $\mathcal{M}$ rather than the identity. Again the eigenvectors are countable and complete in the Hilbert space H with an equivalent inner product:

$$[x, y]_{\mathcal{M}} = [x, \mathcal{M} y].$$

See [2].

Hence the eigenvalues are purely imaginary:

$$\lambda_k = \pm i\omega_k, \quad \omega_k > 0.$$

The ω_k are thus defined as the modes of the structure (in radians) and because $(1/\omega_k)$ goes to zero, can be arranged in increasing order of magnitude so that we can talk about the kth mode without ambiguity. We may note here that the asymptotic behavior of the modes is of little interest to us in practice, where only the "first few" modes play a role. Let Φ_k denote an eigenvector of $\mathcal{A}$ corresponding to the eigenvalue λ_k.

Note that Φ_k has the form

$$\Phi_k = \begin{pmatrix} \phi_k \\ \pm i\,\omega_k\,\phi_k \end{pmatrix}.$$

In our particular case we can calculate the modes ω_k and the "mode shapes" ϕ_k by solving the differential equations

$$\lambda^2 mh + EIh'''' = 0, \tag{2.14}$$

$$\lambda^2 I_\theta \theta - GJ\theta'' = 0. \tag{2.15}$$

Note that this is "no more" than taking Laplace transforms of the time domain equations, familiar in engineering.

Thus ϕ_k is of the form:

$$\begin{pmatrix} h_k \\ 0 \end{pmatrix} \text{ or } \begin{pmatrix} 0 \\ \theta_k \end{pmatrix}$$

and we can distinguish between the "bending" modes and the "torsion" modes.

We have corresponding to the bending motion

$$\omega_k^2 mh_k(y) = EIh_k''''(y) \tag{2.16}$$

plus the CF end conditions:

$$h(0) = 0 = h'(0),$$
$$h''(\ell) = 0 = h'''(\ell).$$

The ω_k then are pure bending modes.

As may be expected, this is a classical result already found in Timoshenko, 1928 [32] and in textbooks [34]. The modes ω_k are determined by the equations:

$$1 + \mathrm{Cosh}\,\gamma_k \mathrm{Cos}\,\gamma_k = 0 \tag{2.17}$$

and

$$\omega_k = \frac{1}{\ell}\left(\frac{EI}{m}\right)^{1/4} \gamma_k \qquad 0 < \gamma_k \uparrow.$$

Mode shape:

$$C_1(\sinh \omega_k\, y - \sin \omega_k\, y) + C_2(\sinh \omega_k\, y + \sin \omega_k\, y), \qquad 0 < y < \ell.$$

For more see [32, 34].

Torsion modes: These are the zeros of

$$\cos h\mu\omega_k \ell = \cos\mu\omega_k \ell = 0,$$

$$\omega_k = (2k-1)\frac{\pi}{2\ell}\sqrt{\frac{GJ}{I_\theta}}, \qquad k = 1, 2. \tag{2.18}$$

Mode shape

$$\text{const} \times \sin\frac{(2k-1)\pi}{2\ell}y \qquad 0 < y < \ell. \tag{2.19}$$

An obvious comment here is that the modes decrease linearly in magnitude as the span length increases. As a rule the structure also increases in flexibility so that the density of modes also increases. There are more modes to be considered as length increases.

We note here that for the clamped–clamped case the modes are determined as the roots of:

$$EI\sqrt{GJ}\left(-1 + \cos\left[\frac{\sqrt{2}\,\ell m^{1/4}\sqrt{|\lambda|}}{EI^{1/4}}\right]\cosh\left[\frac{\sqrt{2}\,\ell m^{1/4}\sqrt{|\lambda|}}{EI^{1/4}}\right]\right)$$

$$\times \sinh\left[\frac{\ell\lambda\sqrt{I_\theta}}{\sqrt{GJ}}\right] = 0$$

yielding an interesting variation of (2.4).

The modes are undamped. The energy remains constant and does not increase or decrease. The structure model is neutrally stable.

Modal Expansion: Green's Function

Theorem 2.3. *Any element Y in $\mathcal{H}$ has the "modal" representation expansion:*

$$Y = \sum_{k=1}^{\infty}\frac{1}{2\omega_k^2}\left([Y, \Phi_k]\Phi_k + [Y, \overline{\Phi}_k]\overline{\Phi}_k\right), \tag{2.20}$$

where

$$\Phi_k = \begin{pmatrix} \phi_k \\ i\omega_k\phi_k \end{pmatrix}; \qquad A\phi_k = \omega_k^2\phi_k; \qquad \mathcal{A}\Phi_k = i\omega_k\Phi_k;$$

$$\mathcal{A}^*\Phi_k = -i\omega_k\Phi_k, \qquad \omega_k > 0, \tag{2.21}$$

$$\sum_{k=1}^{\infty} \frac{1}{\omega_k^2} < \infty; \; [\Phi_k, \Phi_k] = \omega_k^2; \qquad [\Phi_k, \overline{\Phi}_k] = 0 \tag{2.22}$$

$$[\mathcal{M}\phi_k, \phi_j] = \delta_j^k \qquad \begin{pmatrix} \phi_k \\ -i\omega_k\phi_k \end{pmatrix} = \overline{\Phi_k}$$

$\{\{\Phi_k\}, \{\overline{\Phi_k}\}\}$ are orthogonal eigenvectors of both $\mathcal{A}$ and $\mathcal{A}^*$.

Then

$$\mathcal{S}(t)Y = \sum_{k=1}^{\infty} \frac{1}{2\omega_k^2}\left([Y, \Phi_k] e^{i\omega t} \Phi_k + [Y, \overline{\Phi}_k] e^{-i\omega t}\Psi_k\right), \tag{2.23}$$

$$\mathcal{R}(\lambda, \mathcal{A})Y = \sum_{k=1}^{\infty} \frac{1}{2\omega_k^2}\left([Y, \Phi_k]\frac{\Phi_k}{\lambda - i\omega_k} + [Y, \overline{\Phi}_k]\frac{\overline{\Phi}_k}{\lambda + i\omega_k}\right). \tag{2.24}$$

Proof. Follows [16, 54]. The modal expansion uses completeness of the $\{\phi_k\}$. Otherwise, because, the eigenfunctions are orthogonal, the proof of the expansions is straightforward. □

Green's Function Representation

A useful alternate representation of the semigroup and the resolvent is through the Green's function; see [44, vol. 1]. Here we develop it using the modal representation. Thus $\mathcal{S}(t)Y$ is the function

$$\sum_{k=1}^{\infty} \frac{1}{2\omega_k^2}\left([Y, \Phi_k]e^{i\omega_k t} \Phi_k(s) + [Y, \overline{\Phi}_k]e^{-i\omega_k t}\Psi_k(s)\right) = [Y, G(t, s, .)], \tag{2.25}$$

where

$$G(t, s, \zeta) = \sum_{k=1}^{\infty} \frac{1}{(2\omega_k^2)}\left(e^{i\omega_k t}\Phi_k(\zeta)\Phi_k(s) + e^{-i\omega t}\overline{\Phi}_k(\zeta)\Psi_k(s)\right) \tag{2.26}$$

is then the Green's function for $S(t)$. Similarly the resolvent is simply the Laplace transform which is analytic in λ excepting the imaginary axis.

$$\mathcal{R}(\lambda, \mathcal{A})Y = [Y, \hat{G}(\lambda, s, .)] \tag{2.27}$$

$$\hat{G}(\lambda, s, \zeta) = \sum_{k=1}^{\infty} \frac{1}{2\omega_k^2} \left(\frac{1}{\lambda - i\omega_k} \Phi_k(\zeta)\Phi_k(s) + \frac{1}{\lambda + i\omega_k} \overline{\Phi}_k(\zeta)\overline{\Phi}_k(s) \right), \lambda \neq i\omega_k$$

$$\tag{2.28}$$

and now the function is square integrable:

$$\int_0^\ell \int_0^\ell |\hat{G}(\lambda, s, \zeta)|^2 ds d\zeta < \infty. \tag{2.29}$$

General Case: Nonzero Coupling

Next we consider the general case allowing for nonzero S. The main question is how much the coupling changes the modes. We do expect the mode shapes to be coupled. We limit consideration to the CF case here.

The equations corresponding to (2.1, 2.2) are:

$$\lambda^2 mh(y) + \lambda^2 S\theta(y) + EI \, h''''(y) = 0, \tag{2.30}$$

$$\lambda^2 Sh(y) + \lambda^2 I_\theta \theta(y) - GJ \, \theta''(y) = 0 \tag{2.31}$$

with the same CF end conditions as before.

To solve this set of equations we proceed to consider the state-space version. Here we follow [56].

Thus let

$$Y(s) = \text{Col}.h(s), h'(s), h''(s), h'''(s), \theta(s), \theta'(s).$$

Then we have

$$Y'(s) = A(\lambda)Y(s) \qquad 0 < s < 1,$$

$$Y(0) = \text{Col } 0, 0, h''(0), h'''(0), 0, \theta'(0)$$

and $A(\lambda)$ is the 6×6 matrix

$$A(\lambda) = \begin{pmatrix} 0 & 1 & 0 & 0 & 0 & 0 \\ 0 & 0 & 1 & 0 & 0 & 0 \\ 0 & 0 & 0 & 1 & 0 & 0 \\ w_1 & 0 & 0 & 0 & w_2 & 0 \\ 0 & 0 & 0 & 0 & 0 & 1 \\ w_3 & 0 & 0 & 0 & w_4 & 0 \end{pmatrix} \tag{2.32}$$

$$w_1 = -\lambda^2 \frac{m}{EI},$$

$$w_2 = -\lambda^2 \frac{S}{EI},$$

$$w_3 = \lambda^2 \frac{S}{GJ},$$

$$w_4 = \lambda^2 \frac{I_\theta}{GJ},$$

where w_2 and w_3 are the coupling terms.

The eigenvalues in the CF case are the zeros of $d_c(\lambda) = \mathrm{Det}\, D_c(\lambda)$. We define the 3×3 matrix:

$$D_c(\lambda) = P e^{\ell A(\lambda)} Q_c,$$

$$P = \begin{pmatrix} 0\,0\,1\,0\,0\,0 \\ 0\,0\,0\,1\,0\,0 \\ 0\,0\,0\,0\,0\,1 \end{pmatrix} \qquad Q_c = \begin{pmatrix} 0\,0\,0 \\ 0\,0\,0 \\ 1\,0\,0 \\ 0\,1\,0 \\ 0\,0\,0 \\ 0\,0\,1 \end{pmatrix}.$$

The eigenvalues in the FF case are the zeros of

$$d_f(\lambda) = \mathrm{Det} D_f(\lambda),$$

$$D_f(\lambda) = P e^{\ell A(\lambda)} Q_f,$$

where

$$Q_f = \begin{pmatrix} 1\,0\,0 \\ 0\,1\,0 \\ 0\,0\,0 \\ 0\,0\,0 \\ 0\,0\,1 \\ 0\,0\,0 \end{pmatrix}$$

The eigenvalues in the CC case are the zeros of

$$d_c(\lambda) = \mathrm{Det}\, P_c e^{A(\lambda)\ell} Q_c$$

where

$$P_c = \begin{pmatrix} 1\,0\,0\,0\,0\,0 \\ 0\,1\,0\,0\,0\,0 \\ 0\,0\,0\,0\,1\,0 \end{pmatrix}.$$

Thus defined $d(\lambda)$ (generic for both d_c and d_f) is an entire function of finite order (see [10]).

For $S = 0$,

$$D_c(\lambda) = P e^{A(\lambda)} Q_c$$

$$= \begin{pmatrix} \frac{1}{2}(\cos[w1^{1/4}\ell] + \cosh[w1^{1/4}\ell]) & \frac{\sin[w1^{1/4}\ell] + \sinh[w1^{1/4}\ell]}{2w1^{1/4}} \\ \frac{1}{2}w1^{1/4}(-\sin[w1^{1/4}\ell] + \sinh[w1^{1/4}\ell]) & \frac{1}{2}(\cos[w1^{1/4}\ell] + \cosh[w1^{1/4}\ell]) \\ 0 & 0 \end{pmatrix}$$

this yields:

$$d_c(\lambda) = \frac{1}{2}(1 + \cos[\gamma\ell])\cosh[\gamma\ell]\cosh[\mu\ell],   \tag{2.33}$$

where

$$\gamma = w_1^{1/4};   \mu = w_4^{1/2},$$

which agrees with our previous calculation (2.4), (2.5).

To make explicit our interest in the dependence on S, let us modify the notation to:

$A(\lambda, S)$ in place of $A(\lambda)$
$d(\lambda, S)$ for $d(\lambda)$

so that $d(\lambda, 0)$ is given by $d(\lambda)$.

The function is clearly analytic in $\mathcal{S}$.
Let

$$A_p(\lambda, S) = (A(\lambda, S) - A(\lambda, 0))/S.$$

Note that $A_p(\lambda, S)$ is given by

$$\begin{pmatrix} 0 & 0 & 0 & 0 & 0 & 0 \\ 0 & 0 & 0 & 0 & 0 & 0 \\ 0 & 0 & 0 & 0 & 0 & 0 \\ 0 & 0 & 0 & 0 & -\dfrac{\lambda^2}{EI} & 0 \\ 0 & 0 & 0 & 0 & 0 & 0 \\ \dfrac{\lambda^2}{GJ} & 0 & 0 & 0 & 0 & 0 \end{pmatrix}.$$

Now

$$\frac{d}{dt}e^{tA(\lambda,S)} = A(\lambda, S)e^{tA(\lambda,S)}$$

$$= A(\lambda,0)e^{tA(\lambda,S)} + SA_p(\lambda,0)e^{tA(\lambda,S)}.$$

The familiar method of variation of parameters formula [16] yields:

$$e^{tA(\lambda,S)} = e^{tA(\lambda,0)} + S\int_0^t e^{(t-\sigma)A(\lambda,0)}A_p(\lambda,0)e^{\sigma A(\lambda,S)}d\sigma, \quad t > 0.$$

We can treat this as a matrix Volterra equation for $e^{tA(\lambda,S)}$ and the solution has the expansion:

$$e^{1A(\lambda,S)} = e^{1A(\lambda,0)} + \sum_{n=1}^{\infty} S^n F_n(1),$$

where

$$F_1(1) = \int_0^1 e^{A(\lambda,0)(1-\sigma)} \quad A_p(\lambda,0)e^{-\sigma A(\lambda,0)}d\sigma.$$

For terms of higher order, see [70]. As shown there, $d(\lambda, S)$ is a function of S^2 and using the first term in the series yields

$$d(\lambda, S) = d(\lambda,0) + S^2 d_2(\lambda,0), \tag{2.34}$$

where

$$d(\lambda,0) = \frac{1}{2}(1 + \cos[\gamma\ell]\cosh[\gamma\ell])\cosh[\mu\ell]$$

$$d_2(\lambda,0) = \frac{1}{2}(\gamma^2\lambda\mu^2(4(\gamma^2 - \lambda\mu^2)\cosh[l\lambda\ \mu]\sin[l\gamma\sqrt{\lambda}]$$

$$\times ((\gamma^2 + \lambda\mu^2)\sinh[l\gamma\sqrt{\lambda}] - \gamma\sqrt{\lambda}\mu\sinh[l\ \lambda\ \mu])$$

$$+ \cosh[l\gamma\sqrt{\lambda}](-(\gamma^2 + \lambda\mu^2)^2 + (\gamma^2 - \lambda\mu^2)^2\cos[2\ l\gamma\sqrt{\lambda}]$$

$$+ 4\gamma^2\lambda\mu^2(2\cos[l\gamma\sqrt{\lambda}]\cosh[l\ \lambda\ \mu] - \cosh[2\ l\lambda\mu])$$

$$- 4\gamma\sqrt{\lambda}\mu(\gamma^2 - \lambda\mu^2)\sin[l\gamma\sqrt{\lambda}]\sinh[l\ \lambda\ \mu]) + 4\gamma\sqrt{\lambda}\mu$$

$$\times \cosh[l\ \lambda\ \mu](2\gamma\sqrt{\lambda}\mu + (\gamma^2 + \lambda\mu^2)\sinh[l\gamma\sqrt{\lambda}]\sinh[l\ \lambda\ \mu])$$

$$+ \cos[l\gamma\sqrt{\lambda}]((\gamma^2 - \lambda\mu^2)^2 - (\gamma^2 + \lambda\mu^2)^2 \cosh[2\,l\gamma\sqrt{\lambda}]$$
$$- 4\gamma^2\lambda\mu^2 \cosh[2\,l\lambda\mu] + 4\gamma\sqrt{\lambda}\mu(\gamma^2 + \lambda\mu^2) \sinh l\gamma\sqrt{\lambda}]$$
$$\times \sinh[l\ \lambda\ \mu])))/(8m(\gamma^4 - \lambda^2\mu^4)^2 I_\theta). \tag{2.35}$$

Let λ_k denote a zero corresponding to $S = 0$. By applying one step of the Newton root finding algorithm, we can see that for small S it can be approximated by

$$\lambda_k + \Delta_k, \tag{2.36}$$

where

$$\Delta_k = -S^2 d_2(\lambda_k, 0)/d'(\lambda_k, 0)$$

and the main thing to note is that

$$\mathrm{Re}(\lambda_k + \Delta_k) < 0; \to 0 \quad \text{as } k \to \infty. \tag{2.37}$$

The modes are damped, and the mode shapes are now coupled. The bending mode-shape vector now has a nonzero torsion component proportional to S^2, and similarly for the torsion-mode shape there is a bending component. But the extra components being small, we continue to identify them as bending or torsion modes. Calculating these would take us too far from our main interest.

2.5 Robust Feedback Control Theory: Stability Enhancement

We have seen in Sect. 2.4 that the structure model with zero coupling is neutrally stable. The system energy neither decreases or increases whatever the initial condition. Experience shows that all modes decay to zero and the higher the damping the higher the mode. However, the lower modes may not decay as fast as we would like. Hence we need to instrument controls to increase the damping without, however, destabilizing the structure and without increasing the damping in any mode. Unfortunately, the inherent damping is difficult to model and we need to design a controller without knowing the damping and yet without degrading the inherent damping. Fortunately we can indeed design such a control, a triumph of control theory of structures. In any real system we can only provide for a finite number of control inputs. Hence we may assume that the controls denoted $u(\cdot)$,

$$u(\cdot) \in L_2[R^m, \ [0, \ T]] \quad \text{for every } 0 < T < \infty.$$

Let B denote the control operator so that (2.5) now includes controls and we have

$$\mathcal{M}\ddot{x}(t) + Bu(t) + Ax(t) = v(t).$$

To this we add a damping operator D even though we cannot specify it except to note that D is linear bounded self-adjoint and nonnegative definite. Thus we have finally:

$$\mathcal{M}\ddot{x} + D\dot{x}(t) + Ax(t) + Bu(t) = 0. \tag{2.38}$$

An example of a damping operator D is

$$2\zeta\sqrt{\mathcal{M}}\sqrt{T}\sqrt{\mathcal{M}}, \; T = \sqrt{\mathcal{M}^{-1}}\sqrt{A}\sqrt{\mathcal{M}^{-1}}, \qquad |\zeta| < 1,$$

see [61], corresponding to proportional damping, damping proportional to the velocity, going back to Timoshenko [32].

Here we simply assume it is compact and nonnegative definite, zero not in the spectrum.

With $\mathcal{H}$ as before we go to the state-space form:

$$\dot{Y}(t) = \mathcal{A}_d Y(t) + \mathcal{B}u(t),$$

where

$$\mathcal{A}_d = \begin{pmatrix} 0 & I \\ -\mathcal{M}^{-1}A & -\mathcal{M}^{-1}D \end{pmatrix},$$

$$\mathcal{B} = \begin{pmatrix} 0 \\ -\mathcal{M}^{-l}B \end{pmatrix},$$

$$[(\mathcal{A}_d + \mathcal{A}_d^*)Y, Y]_E = -[Dx_2, x_2],$$

where

$$Y = \begin{pmatrix} x_l \\ x_2 \end{pmatrix}.$$

Hence $\mathcal{A}_d$ now generates a contraction semigroup $\mathcal{S}_d(t)t \geq 0$:

$$[S_d Y, Y] \leq [Y, Y]$$

with compact resolvent. Our main result is to show that we can design a feedback controller that is robust in that it is not required to know either D or $\mathcal{A}$ and is such that the structure is strongly stable.

Controllability

For this purpose we introduce the important notion of "controllability"—see [16]—the original notion in finite dimensions extended to infinite dimensions. Let $\mathcal{R}$

denote the infinitesimal generator of a C_0 semigroup $\mathcal{S}(t), t \geq 0$, over the Hilbert space $\mathcal{H}$. We say that the semigroup is "strongly" stable if for each Y in $\mathcal{H}$,

$$||\mathcal{S}(t)Y|| \to 0 \quad \text{as } t \to \infty.$$

Let $\mathcal{B}$ denote a finite-dimensional operator on R^m into $\mathcal{H}$.

We say that the pair $\mathcal{R} \sim \mathcal{B}$ is controllable if $\bigcup_{t \geq 0} \mathcal{S}(t)\mathcal{B}u, u \epsilon R^m$ is dense in $\mathcal{H}$. Here we can state a basic result of control theory.

Theorem 2.4. *Suppose the semigroup $\mathcal{S}(\cdot)$ is a contraction, the generator has a compact resolvent, and $(\mathcal{A} \sim \mathcal{B})$ is controllable. Then the semigroup generated by*

$$\mathcal{A} - \mathcal{B}\mathcal{B}^*$$

is strongly stable.

Proof. See [16].   □

Remark. It is known (see [16]) that a finite-dimensional control cannot guarantee a uniform decay rate for all modes (exponential stability). The rate of decay will eventually go to zero as the mode number increases indefinitely. "Strong" stability means that the energy in any initial state will eventually decay to zero.

Let us apply this to our case, to the semigroup $\mathcal{S}(\cdot)$ with generator $\mathcal{A}$ and control operator $\mathcal{B}$.

Theorem 2.5. *The pair $\mathcal{A}, \mathcal{B}$ is controllable if and only if for any nonzero eigenvector Φ of $\mathcal{A}$,*
$$\mathcal{B}^*\Phi \neq 0.$$

Proof. Suppose contrarywise $\mathcal{B}^*\Phi = 0$ for some nonzero eigenvector Φ with eigenvalue $i\omega$.
Then

$$[\mathcal{B}^*\Phi,\, u] = [\Phi,\, \mathcal{B}\, u] = 0 \quad \text{for every } u \text{ in } R^m.$$

Now

$$[\Phi,\, \mathcal{S}(t)\mathcal{B}\, u] = [\mathcal{S}(t) \times \Phi,\, \mathcal{B}\, u] = e^{-it\omega}\,[\Phi,\, \mathcal{B}\, u] = 0.$$

Hence the set

$$\bigcup_{t \geq 0} \mathcal{S}(t)\mathcal{B}u, u \epsilon R^m \quad \text{is not dense in } \mathcal{H},$$

which is a contradiction.   □

And the argument can be retraced for the "only if" part readily.

We assume now that $\mathcal{A}, \mathcal{B}$ is controllable. The resolvent of $\mathcal{A}$ is compact and $\mathcal{B}$ is finite-dimensional and hence the semigroup generated by

$$\mathcal{A} - \mathcal{B}\mathcal{B}^*$$

is strongly stable.

Let us see what this implies. By modal stability we mean that each mode decays to zero, actually exponentially, with the rate determined by the real part of the eigenvalue. In the case where the dimension of the space is not finite this does not mean strong stability. It only implies that the system is damped but the rate of decay depends on the element. Exponential or uniform stability is defined in terms of the semigroup $S(\cdot)$ by requiring that the stability index:

$$\text{Inf} \left(\frac{1}{t} \right) \text{Log} ||S(t)|| = \omega_o < 0,$$

which implies that for C_0 semigroups that

$$\frac{\log ||S(t)||}{t} \to \omega_o \quad \text{as } t \to \infty.$$

And given any $\epsilon > 0$, we can find M such that

$$||S(t)|| \leq \text{Me}^{-(\omega_o + \epsilon)t}$$

and hence the name exponential stability.

In our present structure dynamics context we may interpret this as guaranteeing a uniform exponential decay rate for all initial states. Here we have a negative result [16] that this cannot be achieved by finite-dimensional controllers, hence not by means of active controllers.

Let us now return to show that the feedback control

$$u(t) = -\mathcal{B}^* Y(t)$$

does not destabilize the structure model whatever the damping operator D is, where

$$M \ddot{x}(t) + D \dot{x}(t) + Ax(t) + BB^* \dot{x}(t) = 0$$

so that we have a new damping operator

$$D + BB^*,$$

which is

$$\geq D.$$

If Φ is any mode of this system we see that the damping

$$[(D + BB^*)\Phi, \Phi] \geq [D\Phi, \Phi].$$

On the other hand, B being finite-dimensional

$$TrBB^* = \sum_{k=1}^{\infty} ||B^* \Phi_k||^2 < \infty,$$

where Φ_k denote the modes, so that the active damping introduced goes to zero as the mode number increases. It is usually much smaller than any natural damping. Another point is that one cannot guarantee a specific value of damping for any mode. Also, few reliable models of damping are known, so designing controls based on any damping model can be hazardous. For a successful use of this control law see [103].

Self-Straining Actuators

We have seen that finite-dimensional controls cannot yield exponential stability, so we would need an infinite-dimensional control—distributed control as opposed to point controllers on the boundary—to achieve this, which is of course not physically realizable.

A class of actuators using piezzo strips that are self-sensing and self-straining is described in (see [79] and the references therein) with the potential to yield exponential stability.

We can extend our theory to investigate this class. We begin with a beam model. The displacement of the piezzo-electric strip charges a condenser which is then bled as a current for actuation. The differentiator circuit is an integral component, and we have really "rate" feedback. We begin with the torsion dynamics:

$$I_\theta \ddot{\theta}(t, s) - GJ\theta''(t, s) = 0 \qquad 0 < s < \ell,$$

$$\theta(t, 1) = 0,$$

$$-GJ\theta'(t, 0) + g\dot{\theta}(t, 0) = 0,$$

where g is the actuator gain, and we need to study how the system stability depends on the gain. We may consider this as a singular mass matrix version of the boundary-control problem:

$$m\ddot{\theta}(t, \ell) - GJ\theta'(t, 0) + g\dot{\theta}(t, 0) = 0,$$

where now m is zero.

The novelty here is the inclusion of the values at the ends as separate from the functions.

Thus let H_1 denote the Hilbert Space:

$$L_2(0, \ell) \times R^1,$$

with elements

$$x = \begin{pmatrix} f(\cdot) \\ b \end{pmatrix}.$$

Let A denote the operator with domain in H_1:

$$\mathcal{D}(A) = \begin{pmatrix} f\ (\cdot)|f''(\cdot) \in L_2(0,\ell), \\ f(\ell) = 0 \\ f(0) \end{pmatrix}$$

and

$$A\ f = \begin{pmatrix} -GJ\ f''(\cdot) \\ -GJ\ f'(0) \end{pmatrix}.$$

Then A is self-adjoint and nonnegative definite with

$$[Af, f] = GJ \int_0^\ell |f'(s)|^2 ds,$$

which is twice the elastic energy. Denoting by $\sqrt{A}$ the positive square root of A, it can be seen following an argument similar to the one given in [11] that the domain:

$$\mathcal{D}(\sqrt{A}) = \left[\begin{pmatrix} f(\cdot) \\ f(0) \end{pmatrix}, \text{ where } f(\ell) = 0 \quad f'(\cdot) \in L_2(0,\ell) \right],$$

but $\sqrt{A}$ is not a differential operator. We note that zero is not in the spectrum of A and hence that of $\sqrt{A}$. Next, as before we define the energy space:

$$\mathcal{H}_E = \mathcal{D}(\sqrt{A}) \times L_2(0,\ell),$$

which is a Hilbert space with inner product:

$$\left[\begin{pmatrix} x \\ f \end{pmatrix}, \begin{pmatrix} y \\ g \end{pmatrix} \right]_E = \left[\sqrt{A}x, \sqrt{A}y \right] + [I_\theta f, g]$$

and

$$\left[\begin{pmatrix} x \\ f \end{pmatrix}, \begin{pmatrix} x \\ f \end{pmatrix} \right]_E = 2 \text{ (total energy: kinetic + potential)}.$$

We define next the operator $\mathcal{A}$ with domain in $\mathcal{H}_E$ with domain that is a little complicated:

$$\mathcal{D}(\mathcal{A}) = \left[Y = \begin{pmatrix} y \\ f_2(\cdot) \end{pmatrix} \text{ with } y \in \mathcal{D}(A), = \begin{pmatrix} f_1(\cdot) \\ f_1(0) \end{pmatrix}, \right.$$

$$\left. \text{and } \begin{pmatrix} f_2(\cdot) \\ \frac{GJ}{g} f_1'(0) \end{pmatrix} \in \mathcal{D}(\sqrt{A}) \right].$$

The condition on $f_2(\cdot)$ is equivalent to:

$$f_2(\ell) = 0,$$
$$f_2'(\cdot) \in L_2(0, \ell).$$

This domain is dense in $\mathcal{H}_E$. We now define $\mathcal{A}$ by:

$$\mathcal{A}Y = \begin{pmatrix} f_2(\cdot) \\ \frac{GJ}{g} f_1'(0) \\ \frac{GJ}{I_\theta} f_1''(\cdot) \end{pmatrix}.$$

Then $\mathcal{A}$ is closed linear with dense domain and compact resolvent. Also

$$[\mathcal{A}Y, Y]_E = -GJ[f_2(\cdot), f_1''(\cdot)] - \frac{(GJ)}{g} |f_1'(0)|^2 + GJ[f_1''(\cdot), f_2(\cdot)].$$

Hence

$$\text{Re} \cdot [\mathcal{A}Y, Y] = -(GJ)^2 \frac{|f_1'(0)|^2}{g} \le 0.$$

This is enough to prove that A generates a C_0 semigroup $\mathcal{S}(t)$, $t \ge 0$, a contraction semigroup, but actually exponentially stable, as we show presently. Of greater interest to us are the eigenvalues and how they depend on the gain g.

Spectrum of $\mathcal{A}$

Let $\mathcal{A}\Phi = \lambda\Phi$ which with

$$\Phi = \begin{pmatrix} f_1(\cdot) \\ f_1(0) \\ f_2(\cdot) \end{pmatrix}$$

yields

$$f_1(s) = c \sinh(\lambda v(\ell - s)), \quad 0 < s < \ell,$$

where

$$v^2 = \frac{I_\theta}{GJ},$$

where λ is the root of

$$GJ \cdot I_\theta \cosh \lambda \nu \ell + g \sinh \lambda \nu \ell = 0.$$

Now the function on the left is an entire function of order one and of completely regular growth [6] with a sequence of zeros given by

$$\tanh (\lambda \nu \ell) + \frac{\sqrt{GJI_\theta}}{g} = 0.$$

Or

$$e^{2\lambda \nu \ell} = \frac{g + \sqrt{GJI_\theta}}{g - \sqrt{GJI_\theta}},$$

$$\lambda_k = -|\sigma_k| \pm i\omega_k,$$

where the real part is a negative constant given by

$$-|\sigma_k| = -\frac{1}{2\ell} \sqrt{\frac{GJ}{I_\theta}} \log \left| \frac{g + \sqrt{GJI_\theta}}{g - \sqrt{GJI_\theta}} \right|,$$

$$\omega_k = \frac{(2k+1)\pi}{2\ell} \sqrt{\frac{GJ}{I_\theta}}, \quad g < \sqrt{GJI_\theta}$$

$$= k \frac{\pi}{\ell} \sqrt{\frac{GJ}{I_\theta}}, \quad g > \sqrt{GJI_\theta},$$

$$k = \text{nonnegative integers.}$$

Thus all modes are damped at the same rate. And

$$g_c = \sqrt{GJI_\theta}$$

may be called the critical gain.

A plot of the relative damping constant $\sigma \ell \nu$ versus the relative gain g/g_c is given in Fig. 2.2.

$$\frac{1}{2} \log \left[abs \left[\frac{1+x}{1-x} \right] \right].$$

As can be seen from Fig. 2.2, the limiting damping becomes infinite at the critical gain.

Let us explore further what happens at the critical gain.

We have

$$d(\lambda) = \cosh \lambda \nu \ell + \sinh \lambda \nu \ell = e^{\nu \lambda \ell}$$

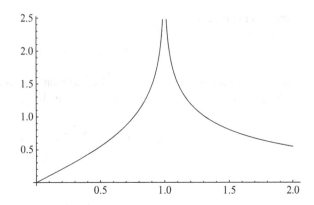

Fig. 2.2 Damping versus (relative) gain

and has no zeros at all; no modes at all! The stability index

$$= -\infty,$$

the system is superstable. It is interesting that the implication of superstability in this case is that the response vanishes in finite time T determined by g_c. All states are evanescent. This has been given the rather naive if picturesque name: a "disappearing solution", in another context, scattering theory. See [80].

But this is of course impossible in practice for structure response, for many reasons; see [62] for details. Primarily it is the degradation in the operational amplifier differentiator as the gain increases. The gain g_c is not physically attainable. See [79].

Although our primary interest is in the modal response, we complete the analysis for any noncritical gain by showing that we have exponential stability, that any initial state will decay at an exponential rate. The modes are given by

$$\Phi_k = \begin{pmatrix} \theta_k(\cdot) \\ \theta_k(0) \\ \lambda_k \theta_k(\cdot) \end{pmatrix},$$

where

$$\theta_k(s) = a_k \sinh[\lambda_k v(1-s)] \qquad 0 < s < \ell$$

are not orthogonal. But we can create a biorthogonal system; see [1] for more. Define

$$\Psi_k = \begin{pmatrix} \overline{\theta_k(\cdot)} \\ \overline{\theta_k(0)} \\ -\lambda_k \overline{\theta_k(\cdot)} \end{pmatrix},$$

the superbar denoting complex conjugate.

Thus defined it is readily verified that

$$[\Phi_k, \Psi_j] = 0 \quad \text{for } k \neq j.$$

Hence if

$$Y = \sum b_k \Phi_k, \quad \text{then } b_k = \frac{[Y, \Psi_k]}{[\Phi_k, \Psi_k]}$$

and

$$\begin{aligned}
\mathcal{S}(t)Y &= \sum e^{-\sigma t} e^{i\omega_k t} b_k \Phi_k \\
&= e^{-\sigma t} \sum e^{i\omega_k t} b_k \Phi_k,
\end{aligned}$$

which is enough to prove the "exponential" stability.

The corresponding theory for bending—Euler beams with self-straining actuators—is treated in [70], but there is no analagous superstability at the gain corresponding to maximum damping. This would indicate that we can attain more damping by torsion actuators, a fact corroborated by experiment.

Our interest is of course in the stabilization of the structure subject to aerodynamic loading. We would expect that if we have a controller that does well in still air, it should be a candidate for use in airflow as well. This is studied in Chap. 8.

2.6 Nonfixed Wing Models: Flying Wings

Free–Free Articulated Beam Model

We again consider the basic uniform Goland model, but it is no longer attached at one end point to the fuselage; it is simply a flying wing. Both ends are free, FF.

Hence we use the matrix Q_f in what follows. But it is articulated with discrete masses placed along points (nodes) on the beam. This is typical of the recent Helios UAV Flying Wing. See [81]. Such a model for a fixed-wing aircraft was also considered earlier by Goland and Luke [77]. There is only a single wing span, $0 < s < \ell$. The masses m_i, are at $s = s_i$, $i = 0, 1, \ldots, m + 1$, with $s_0 = 0, s_{m+1} = 1$. There are thus $(m + 2)$ masses maximum. We have only to set the mass to be zero if there is none at s_i for any i! See Fig. 2.3.

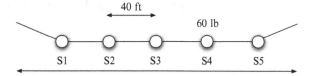

Fig. 2.3 Example of articulated beam

Between nodes: $s_i < s < s_{i+1}$, we have the Goland equations (cf. (2.1) and (2.2)):

$$m\ddot{h}(t,s) + S\ddot{\theta}(t,s) + EI\,h''''(t,s) = L(t,s), \quad t > 0; s_i < s < s_{i+1}, \quad (2.39)$$

$$I_\theta\ddot{\theta}(t,s) + S\ddot{h}(t,s) - GJ\theta''(t,s) = M(t,s), \quad t > 0, s_i < s < s_{i+1}. \quad (2.40)$$

To allow for the discontinuities at the nodes we follow the technique in [2]. We essentially incorporate the nodes into the function space: let

$$\mathcal{H}_\theta = L_2(0,\ell) \times \mathcal{R}^{m+2}.$$

Thus the elements in this space can be denoted:

$$f = \begin{pmatrix} f(\cdot) \\ f(0) \\ f(s_1) \\ f(s_i) \\ \vdots \\ f(s_m) \\ f(\ell) \end{pmatrix}.$$

Define the linear operator A_θ with domain in $\mathcal{H}_\theta$:

$$\mathcal{D}(A_\theta) = |f/f''(\cdot) \in L_2(0,\ell)|, \qquad A_\theta\,f = g;$$

$$g = -GJ \begin{pmatrix} f''(\cdot), s_i < s < s_{i+1} \\ f'(s_i+) - f'(s_i-) \\ i = 0, 1, \ldots, m, m+1 \end{pmatrix}, \quad (2.41)$$

where

$$s_i+ = \text{ limit } s_i + \delta, \quad 0 < \delta \to 0,$$

$$s_i- = \text{ limit } s_i - \delta, \quad 0 < \delta \to 0$$

$$0- = 0; \quad \ell+ = \ell.$$

Then it is readily verified by integration by parts that A_θ is self-adjoint with dense domain, nonnegative definite, and

$$[A_\theta\,f,\,f] = GJ \int_0^1 |f'(s)|^2 ds \geq 0, \quad (2.42)$$

which is recognized as the elastic energy. Indeed, the definition of g is designed to achieve this. The moment-balance end conditions we need at each node are:

$$r_i^2 m_i + m_i \ell_i \ddot{h}(t, s_i) - GJ(\theta'(t, s_i+) - \theta'(t.s_i-)) = 0, \qquad (2.43)$$

where r_i is the radius of gyration of the mass m_i about the elastic axis at normal distance ℓ_i. We note that if there are no masses at the ends 0 or ℓ, then

$$\theta'(t, 0+) = 0 = \theta'(t, \ell-). \qquad (2.44)$$

We proceed similarly for the bending. We define:

$$\mathcal{H}_h = L_2(0, \ell) \times \mathcal{R}^{m+4}$$

and the closed linear operator A_h with domain in $\mathcal{H}_h$:

$$D(A_h) = \begin{pmatrix} f(\cdot) \\ f'(0) \\ f(0) \\ f(s_i) \\ \vdots \\ f(s_m) \\ f(\ell) \\ f'(\ell) \end{pmatrix} \quad \text{with } f''''(\cdot) \in L_2(0, \ell)$$

and

$$A_h f = g; \quad g = EI \begin{pmatrix} f'''', (s)_i < s < s_{i+1} \\ f''(0) \\ f'''(s_i+) - f'''(s_i-), \\ i = 0, \ldots, m, m+1 \\ f''(\ell) \end{pmatrix} \qquad (2.45)$$

and it is readily verified by integration by parts that A_h is self-adjoint and nonnegative definite and

$$[A_h f, f] = EI \int_0^1 |f''(s)|^2 ds \qquad (2.46)$$

the elastic energy in the bending mode. In addition we have to impose the nodal end conditions

$$m_i (\ddot{h}(t, s_i) + \ell_i \ddot{\theta}(t, s_i)) + EI (h'''(t, s_i+) - h(t, s_i-)) = 0, \qquad (2.47)$$

at the nodes $s = s_i$, and if there are no masses at the ends, the free–free conditions:

$$h'''(t, \ell) = 0 = h'''(t, 0),$$
$$h''(t, \ell) = 0 = h''(t, 0).$$

Next let $\mathcal{H}$ denote the cross-product Hilbert space:

$$\mathcal{H} = \mathcal{H}_h \times \mathcal{H}_\theta$$

and define the operator:

$$\mathcal{A}_s = \begin{pmatrix} A_h & 0 \\ 0 & A_\theta \end{pmatrix}$$

on $\mathcal{H}$ with domain:

$$D(A_s) = D(A_h \times A_\theta)$$

and

$$A_s\, x = \begin{pmatrix} A_h h \\ A_\theta \theta \end{pmatrix} \text{ for } x = \begin{pmatrix} h \\ \theta \end{pmatrix} \text{ in } D(A_s).$$

Then $\mathcal{A}_s$ is self-adjoint and nonnegative definite with

$$[A_s\, x,\ x] = EI \int_0^\ell |h''(s)|^2 ds + GJ \int_0^\ell |\theta'(s)|^2 ds,$$

which is twice the potential energy stored in the structure. As before we note that $\mathcal{A}_s$ has a positive square root denoted $\sqrt{\mathcal{A}_s}$, and the elastic energy is defined on $D\left(\sqrt{\mathcal{A}_s}\right)$ as

$$E(x) = \left[\sqrt{\mathcal{A}_s}\, x,\ \sqrt{\mathcal{A}_s}\, x \right],$$

where we use the same inner-product notation for all three spaces we have introduced.

Finally the structure dynamic equations can now be expressed:

$$M\ddot{x}(t) + \mathcal{A}_s x(t) + Bu(t) = 0 \tag{2.48}$$

allowing for an actuator on the structure and control input $u(\cdot)$ as in the fixed-wing model, and M as before.

Structure Modes

We now calculate the structure modes for the uncoupled case $S = 0$. We use the same notation as before in Sect. 2.4, including $A(\lambda)$ but we need to introduce more to account for the fact that now we have several interconnected sections. Thus let E_i denote the 6×6 matrix

$$\begin{pmatrix} 1 & 0\,0\,0 & 0 & 0 \\ 0 & 1\,0\,0 & 0 & 0 \\ 0 & 0\,1\,0 & 0 & 0 \\ \dfrac{-\lambda^2 m_i}{EI} & 0\,0\,1 & \dfrac{-\lambda^2 m_i \ell_i}{EI} & 0 \\ 0 & 0\,0\,0 & 1 & 0 \\ \dfrac{-\lambda^2 m_i \ell_i}{GJ} & 0\,0\,0 & \dfrac{-\lambda^2 m_i r_i^2}{GJ} & 1 \end{pmatrix}. \tag{2.49}$$

Then the modes are the zeros of

$$d(\lambda) = Det.(PE_{m+1} e^{A(\lambda)(\ell - s_m)} E_m e^{A(\lambda)(s_m - s_{m-1})} E_1 e^{A(\lambda)s_1} E_0 Q_f). \tag{2.50}$$

If m_i, r_i, and ℓ_i are zero, then we revert to $d(\lambda)$ in Sect. 2.4. An important case is the symmetric case where all the masses are on the elastic axis so that r_i, ℓ_i are all zero.

The Symmetric Case

1. One point mass with $s_1 = \ell/2$. Here

$$d(\lambda) = P\, e^{(A(\lambda)\ell)/2} E_1\, e^{(A(\lambda)\ell)/2} Q_f, \tag{2.51}$$

which can be expressed:
$$F_1(\gamma, \mu)\sin \mu\ell,$$

where

$$F_1(\gamma, \mu) = \left(-\frac{1}{2}\gamma^4 \mu(-1 + \cosh \gamma\ell \cos \gamma\ell) - \frac{\lambda^2 m_1}{EI}(1 + \cosh \gamma^\ell \cos \gamma\ell) \right.$$

$$\left. \times \frac{1+i}{4}\gamma\mu \left(\sin\frac{(1+i)\gamma^\ell}{2} - \sinh\frac{(1+i)\gamma^\ell}{2} \right) \right).$$

As we expect, the mass at the center does not affect the torsion modes. Hence we call the roots of $F_1(\gamma, \mu) = 0$ the bending modes which now depend on μ as well. Obviously we get back the beam modes for large mass m_1.

2. Several Point Masses.

This result readily generalizes to the case of several point masses, but none at the end points. Following the Helios model [81], we consider $(N - 1)$ masses m_i at s_i, placed symmetrically:

$$s_i = i \, \frac{\ell}{N}, \qquad i = 1, \ldots, N - 1,$$

where N is even so that we have a mass at the center at $\ell/2$. Here
$d(\lambda) = \text{Det}\left(P e^{A(\lambda)(\ell/N)} E_{N-1} \ldots e^{A(\lambda)(\ell/N)} E_{\frac{N}{2}} \ldots e^{A(\lambda)(\ell/N)} Q \right)$ which again factors as

$$F_N(\gamma, \mu) \sin \mu \ell,$$

where the beam modes are the roots of

$$F_N(\gamma, \mu) = 0.$$

2.7   Nonlinear Structure Models

So far we have only considered linear models, where the dynamics are characterized by a linear equation. One may argue that "real life" structures are nonlinear. One can of course come up with nonlinear models but the question of whether they are "well formulated" in terms of existence and uniqueness of solutions is often ignored. See for example [75, 81, 82] where in fact the elastic part is eventually truncated—lumped—to yield ordinary equations in place of partial differential equations. One immediate difficulty with the nonlinear model is that there is no notion of modes; we have no concept of spectrum of the nonlinear operator. And even if one could define the notion, it is not clear what role it plays in the stability of the system, which is our primary concern; see below.

Also we can no longer define an inner product based on the energy. We consider two models that are essentially extensions of the generic Goland beam model.

The first one is due to Beran et al. [75]. We call it the Beran–Straganac model. This has only two degrees of freedom but is highly nonlinear.

The Beran–Straganac Model

We state this in our notation where we replace their $w(\cdot)$ by $h(\cdot)$ and $\alpha(\cdot)$ by $\theta(\cdot)$. We have, including the discrete masses, in their delta-function notation:

$$m\ddot{h} + S\ddot{\theta} + EIh'''' + M_s\left(\ddot{h} - x_s\ddot{\theta}(t,\, y)\right)\delta(y - y_s) = -D_x(h'(h'h'')')' + L(t) \tag{2.52}$$

plus cross-product terms involving time derivatives which we omit.

$$I_\theta\ddot{\theta} + S\ddot{h} - GJ\theta'' + (\cdot)\delta(y - y_s) = +D_x\theta(h'')^2$$

$$+ \text{ cross-product terms involving time derivatives which we omit}$$

$$+ M(t). \tag{2.53}$$

For the derivation and details we have omitted see [75]. Our main point here is that the equation is nonlinear so that there is no notion of eigenvalues, of the spectrum of the differential operator.

Of course the authors do not consider the question of whether these continuum equations have a unique solution, and all calculations are based on discretized models. Indeed, it would be a major task to establish this. We return to this model in the succeeding chapters.

The second example we consider is the Dowell–Hodges model [76].

The Dowell–Hodges Model

This is a nonlinear model with three degrees of freedom as in [61]: the torsion angle $\theta(\cdot)$ about the elastic axis and two bending variables, the plunge $h(\cdot)$ in the xz plane and an additional in-plane bending in the structure xy plane. We show existence and uniqueness by a constructive method of solution; our emphasis is on the differences from the linear Goland model.

The continuum equations (consistent with our notation) are:

$$m\ddot{h}(t,\, y) + EI_1h''''(t,\, y) + (EI_2 - EI_1)(\theta(t,\, y)v(t,\, y)'')''$$

$$= mg\sin\varphi + L(t, y), \tag{2.54}$$

$$m\ddot{v}(t,\, y) + EI_2v''''(t,\, y) + (EI_2 - EI_1)(\theta(t, y)h(t, y)'')'' = mg\cos\varphi, \tag{2.55}$$

$$I_{\theta_\theta}\ddot{\theta}(t,y)-GJ\theta''(t,y)+(EI_2EI_1)h(t,y)''v(t,y)'' = M(t,y), 0 < t; \ 0 < y < \ell,$$
$$(2.56)$$

where we have omitted the tip masses in [76] but retained the gravity terms as part of the forcing terms on the right. The angle φ is the angle between the x-axis and the gravity vector. For the derivation and other details reference should be made to the original paper [76].

The in-plane (in the xy-plane) displacement is denoted $v(\cdot)$ and does not impact the aerodynamics directly. This is a nonlinear system of equations (including the airflow, based on the nonlinear aerodynamics described in Chap. 3).

We shall first consider the case where the airspeed is zero so that there is no aerodynamic loading, as in the linear case. There is no notion of modes that determines stability any more. We include the gravity terms in the aerodynamic loading and hence omit them for the pure structure case. Then we have:

$$m\ddot{h}(t,y) + EI_1 h''''(t,y) + (EI_2 - EI_1)(\theta(t,y)v(t,y)'')'' = 0,$$
$$m\ddot{v}(t,y) + EI_2 v''''(t,y) + (EI_2 - EI_1)(\theta(t,y)h(t,y)'')'' = 0,$$
$$I_\theta\ddot{\theta}(t,y) - GJ\theta''(t,y) + (EI_2 - EI_1)v(t,y)''h(t,y)'' = 0,$$
$$0 < t, 0 < y < \ell, \qquad (2.57)$$

plus CF end conditions, with the notation:

$$x(t,y) = \begin{pmatrix} h\ (t,y) \\ v\ t,y) \\ \theta\ (t,y) \end{pmatrix} \quad t > 0, 0 < y < \ell.$$

One surprising feature of the model although still nonlinear, and not typical, is that we do have modes; in fact those of the linear system.

$$h(t,y) = 0,$$
$$v(t,y) = 0,$$
$$I_\theta\ddot{\theta}(t,y) - GJ\theta''(t,y) = 0$$

plus CF or FF end conditions, yields all the pitching modes (cf. (2.5)). In similar fashion, we also get all the bending modes. Thus we have:

$$\theta(t,y) = 0,$$
$$v(t,y) = 0,$$
$$m\ddot{h}(t,y) + EI_1 h''''(t,y) = 0$$

yields all the plunge modes and

$$\theta(t, y) = 0,$$
$$h(t, y) = 0,$$
$$m\ddot{v}(t, y) + EI_2 v''''(t, y) = 0$$

yields all the in-plane bending modes.

But of course we do not have the modal superposition property, except for the class of functions for which only one co-ordinate is nonzero for all functions. Hence it does not tell us much about the stability of the system. We return to this model in succeeding chapters.

Wing Camber Model

So far we have not taken in-thickness into account. Ideally of course one would want a plate model, however thin. For aircraft wings we have to model the tear-drop shape—the camber; see [18].

This brings considerable complication in the continuum model, explaining why all current works immediately discretize to finite dimensions. An approach to include camber is presented in [107] for flutter analysis.

2.8 Beams of Infinite Length

It is certainly of mathematical interest to examine the case of a uniform Goland beam, where the beam length is no longer finite. The main thing here is that we need to continue with the same notion of energy:

$$E = \frac{1}{2} EI_\theta \int_0^\infty |h''(s)|^2 ds + \frac{1}{2} GJ \int_0^\infty |\theta'(s)|^2 ds.$$

An interesting consequence of this is that if the beam is clamped at one end $s = 0$, then it would need to satisfy the same conditions at the end $s = \infty$. Thus let $\mathcal{H} = L_2[0, \infty]$ and let $h(\cdot)$ and $h'(\cdot), h''(\cdot), h'''$ be in $\mathcal{H}$ with

$$h(0) = 0,$$
$$h'(0) = 0.$$

Then

$$\frac{d}{ds} |h(s)|^2 = h'(s)\overline{h}(s) + h(s)(\overline{h}(s))'$$

and hence

$$\int_0^L \frac{d}{ds}|h(s)|^2 ds = |h(L)|^2 = \int_0^L (h'(s)\bar{h}(s) + h(s)(\bar{h}(s))')ds,$$

which converges as $L \to \infty$ to

$$2\,\text{Re}[h,\,h'].$$

Hence $h(L)$ converges as

$$L \to \infty$$

but the limit must be zero because $h(0)$ is in $L_2[0,\ \infty]$. In as much as h'' is also in $L_2[0,\ \infty]$, we have that h and h' are continuous with

$$h(\infty) = 0 = h'(\infty) = 0.$$

In a similar way for the torsion,

$$\theta(0) = 0 = \theta(\infty).$$

Thus C at one end implies C at the other end also. Thus we have the case CC.

The modes are inversely proportional to the length, therefore we see that the modes are not defined. Or, the point spectrum of $\mathcal{A}$ is empty.

Our interest is in the stability of the structure. Thus we can proceed as we did in Sect. 2.3 with CC end conditions, and ℓ replaced by ∞. The first problem then is the definition of the square root of A. Here we need to invoke Fourier transforms. Thus we denote $\tilde{h}(\cdot)$ as the Fourier transform of h, $\tilde{\theta}(\cdot)$ as the Fourier transform of $\theta(\cdot)$.

$$x(\cdot) = \begin{pmatrix} h \\ \theta \end{pmatrix}; \quad \tilde{x}(\cdot) = \begin{pmatrix} \tilde{h} \\ \tilde{\theta} \end{pmatrix},$$

$$\tilde{h}(\nu) = \int_0^\infty e^{-2\pi i \nu s} h(s)ds; \tilde{\theta}(\nu) = \int_0^\infty e^{-2\pi i \nu(s)} \theta(s)ds, \quad -\infty < \nu < \infty.$$

Thus defined, we have that $\tilde{h}(\cdot)$ and $\tilde{\theta}(\cdot)$ are in $L_2(-\infty,\ \infty)$ with the properties

$$\int_0^\infty |h(s)|^2 ds = \int_{-\infty}^\infty |\tilde{h}(\nu)|^2 d\nu.$$

Let $\mathcal{F}$ denote the operator on $L_2[0,\ \infty]$ into $L_2[-\infty,\ \infty]$

$$\tilde{h} = \mathcal{F}h.$$

Then by the Plancheral theorem (see, e.g., [10, 51])

$$[h, g] = [\tilde{h}, \tilde{g}]. \tag{2.58}$$

$\mathcal{F}$ maps $L_2[0, \infty]$ into a proper subspace of $L_2(-\infty, \infty)$ characterized by the Paley–Wiener theorem [51]:

$$\int_{-\infty}^{\infty} \frac{||\log||\tilde{h}(\omega)||||}{1 + \omega^2} d\omega.$$

The operator A becomes a "multiplier" [22, 28]:

$$(Ax)^\sim = m\tilde{x},$$

where m is the matrix:

$$\begin{pmatrix} EI(2\pi i v)^4 & 0 \\ 0 & -GJ(2\pi i v)^2 \end{pmatrix} = m, \tag{2.59}$$

which makes the definition of the square root as a multiplier

$$\sqrt{A} \sim\sim \sqrt{m} = \begin{pmatrix} \sqrt{EI(2\pi v)^4} & 0 \\ 0 & \sqrt{GJ(2\pi v)^2} \end{pmatrix}. \tag{2.60}$$

With $\mathcal{A}$ defined as before, it can be identified with the 4×4 matrix multiplier

$$\begin{pmatrix} 0 & I_2 \\ -M^{-l}m & 0 \end{pmatrix} \tag{2.61}$$

and the semigroup $S(t)$ $t \geq 0$ by the multiplier

$$\text{MatrixExp}\left[\begin{pmatrix} 0 & I_2 \\ -M^{-1}m & 0 \end{pmatrix} t\right], \tag{2.62}$$

which then yields an isometric group for $-\infty < t < \infty$. This should help answer the question of what happens as the beam length is infinite.

Notes and Comments

The wing model we use is perhaps the simplest. It represents an unswept wing so that it is rectangular; we neglect wing camber (see [18] for a detailed wing-shape description) and assume zero thickness.

The use of functional analysis is now common in applied mathematics but has yet to reach engineering and there is a disconnect here at present. Indeed, because of software packages the level of analytical skill is actually decreasing! It is interesting that the need for infinite-dimensional spaces is underscored by a recent publication, "Dynamics of Very High Dimensional Systems" [76].

For the articulated FF case the derivation of the nodal conditions is novel in that it is derived simply on the basis that the differential (stiffness) operator be nonnegative definite, and not from physical principles. It also requires the inclusion of the nodal values in the definition of the structure state. This technique was employed for the first time in [54]. This may be considered as a generalization of the Goland–Luke model [77].

It should be noted that feedback controllers are all rate controllers well known in classical control for stabilization.

For a detailed description of self-straining controllers, reference should be made to [79] where the performance limitation due to operational amplifiers is included.

There are, of course, a great many nonlinear structures too numerous for inclusion here even restricted to aircraft wings, neglecting thickness. Here we have chosen two that relate closely to the Goland model. There is no general theory for nonlinear models such as that for the linear case. The notion of elastic energy does not seem adequate. We are unable to provide a time-domain solution at the level of the Goland model. As we have noted, a generic feature here is that the need to show that they have unique solutions seems not to bother the originators who go on immediately to discretization of the model after the elaborate effort for constructing the model; see, for example, [76]. Of course the problem here is much less complicated than the viscous flow case (Chap. 7) and we only need the linearized model (linearized about the steady state) for flutter analysis, as we show in Chap. 6. Hence we resort to a perturbation technique leading to a Volterra integral equation bootstrapping on the linear equation; see Chaps. 4–6.

All the beam models we consider have zero thickness. However, we can include wing camber and we do so briefly.

Finally we consider the case where the beam length is allowed to be infinite and this case is of mathematical interest in that there are no discrete modes any more, and the Fourier transform theory and the notion of multipliers provide the appropriate techniques to this case.

Chapter 3
The Air Flow Model/Boundary Fluid Structure Interaction/The Aeroelastic Problem

3.1 Introduction

In this chapter we make a precise mathematical statement of the aeroelastic problem that we wish to solve. Having described the structure model, we turn to the air flow model simplifying it to the most used case where we neglect viscosity and consider "nonviscous flow" but more importantly assume that the entropy is constant. This makes the flow vortex free so that the flow can be described in terms of the potential. Our concern is again more the structure response in air flow—"aeroelasticity"—and hence the fluid–structure boundary conditions play the dominant role in determining the aerodynamic loading on the wing structure.

Starting with the three basic conservation laws, we derive the fundamental field equation describing the air flow, The Eulerfull potential equation with the Kutta–Joukowsky boundary conditions. We present a complete statement of the aeroelastic problem at the end of the chapter for nonviscous flow and nonlinear structure models, including the simplification to Strip theory, the typical section theory.

3.2 Notation/Physical Constants

We begin with the notation we use for the basic parameters necessary to describe the flow generally throughout the book from now on. We also list the relevant physical constants we need in the process:

ρ Density
q Fluid velocity vector
q_∞ Far field velocity
U_∞ $= |q_\infty|$ This is a free parameter, the far field air speed
a_∞ $=$ Speed of sound
p Pressure
μ Viscosity

A V Balakrishnan, *Aeroelasticity: The Continuum Theory*,
DOI 10.1007/978-1-4614-3609-6_3, © Springer Science+Business Media, LLC 2012

S Entropy
T Temperature
e Energy per unit volume
Subscript ∞ denotes far field values

Perfect Gas Law

$p = \rho R T$
$R = c_p - c_v$
$\gamma = c_p/c_v$ Ratio of specific heats

Thermodynamic Relation:

$p\rho^\gamma e^{-s/c_v} = Const.$
h enthalpy per unit mass $= c_p T$
E internal energy per unit mass $= c_v T$
All are functions of time t and the spatial coordinates x, y, z:
$f = f(t, x, y, z)$
The scalar functions are all positive.

Physical Constants:

All at standard air 59 F

ρ 0.00238 slug/ft^3
     1.23 kg/m^3
μ 0.372×10^{-6}slug/ftsec
     17.8×10^{-6} kg/msec
p 2, 116, lb/ft^2
     1.0312×10^5N/m^2
a_∞ 336 m/sec
R_e Reynolds Number $U1\rho/\mu \sim\sim 10^6$
R Gas Constant   287 kg/msecunits
c_v 717
c_p 1,004
γ 1.4
k Diffusivity $= \mu c_p$
              17.8×10^{-3}
     Prandtl no. $\sim\sim$ 0.7 (here taken as 1)
     $$\frac{k}{R\rho_\infty} = \mu \quad \frac{c_p}{R\rho_\infty} = \frac{\gamma}{\gamma - 1}\frac{\mu}{\rho_\infty}$$

3.3 Nonviscous Flow: The Euler Full Potential Equation

Throughout this chapter we set $\mu = 0$.

The field equation of fluid flow in 3D space (R^3, orthogonal coordinates x, y, z) is derived from three basic laws of conservation which we state here in differential form (as opposed to the integral form) with t representing the time co-ordinate.

We begin with the premordial.

Conservation Laws

1. Conservation of mass: Continuity equation

$$\frac{\partial \rho}{\partial t} + \nabla \cdot (\rho q) = 0. \tag{3.1}$$

2. Conservation of momentum. The Euler momentum equation (nonviscous flow, no heat conduction)

$$\rho \frac{\partial q}{\partial t} + \rho(q \cdot \nabla)q + \nabla p = 0 \tag{3.2}$$

in the usual notation [4, 12], where $(q \cdot \nabla)q$ is the vector:

$$i q \cdot \nabla(q \cdot i) + j q \cdot \nabla(q \cdot j) + k q \cdot \nabla(q \cdot k) \quad q = i q_1 + j q_2 + k q_3,$$

where i, i, k are the orthogonal unit vectors.

Note that second spatial derivatives of q are not involved, unlike the viscous case (see Chap. 7).

3. Conservation of energy: First law of thermodynamics in Eulerian form for perfect gas [14]

$$E = \frac{1}{\gamma - 1} \frac{p}{\rho},$$

$$\rho \frac{DE}{Dt} + p \nabla \cdot q = 0. \tag{3.3}$$

Or

$$\rho \left(\frac{\partial E}{\partial t} + q \cdot \nabla E \right) + p \cdot \nabla q = 0. \tag{3.4}$$

Note that we need three equations: conservation of mass, conservation of momentum, and conservation of energy. All are expressed in differential form rather than in integral form. This makes it possible to state the dynamic boundary conditions crucial for aeroelasticity.

This equation (3.4) implies [12, 14] that the total derivative of the entropy S is zero:

$$\frac{DS}{Dt} = 0,\tag{3.5}$$

or the flow is homentropic [14].

Note that:

$$p\ \rho\ T\ S$$

are thermodynamic state variables any one of which is determined by the other three. Thus, under our assumption that the specific heats are constants, we have [14]:

$$p\rho^{-\gamma} = \text{constant} \times e^{s/c_v}.\tag{3.6}$$

Isentropic Flow

A flow is said to be isentropic if

$$S = \text{constant in } t, x, y, z.$$

Homentropy does not imply isentropy. But we now assume that the flow is isentropic. In particular then the energy equation is satisfied. And from (3.6) we have the important conclusion that the pressure is a function of density. More specifically:

$$p = A\rho^{\gamma},\tag{3.7}$$

where A is a constant determined from

$$p_{\infty}\rho_{\infty}^{-\gamma} = A.$$

Next

$$\begin{aligned}
\frac{\nabla p}{\rho} &= \frac{A}{\rho}\nabla\rho^{\gamma}\\
&= A\rho^{\gamma-1}\nabla\log\rho^{\gamma}\\
&= A\gamma\rho^{\gamma-1}\nabla\log\rho\\
&= A\frac{\gamma}{\gamma-1}\rho^{\gamma-1}\nabla\log\rho^{\gamma-1}\\
&= A\frac{\gamma}{\gamma-1}\nabla\rho^{\gamma-1}.
\end{aligned}\tag{3.8}$$

Hence the Euler equation can be purged of the pressure variable p to yield:

$$\frac{\partial q}{\partial t} + (q \cdot \nabla)q + A \frac{\gamma}{\gamma - 1} \nabla \rho^{\gamma-1} = 0. \qquad (3.9)$$

This is what enables us to deduce that the flow is curl or vortex free. Thus let

$$\Omega = \nabla \times q.$$

Then taking the curl on both sides of (3.9), we obtain:

$$\nabla \times \frac{\partial q}{\partial t} + \nabla \times (q \cdot \nabla)q = 0$$

because $\nabla \times \nabla \rho^{\gamma-1} =$ the curl of a gradient is zero.

    Hence

$$\frac{\partial \Omega}{\partial t} + \nabla \times (q \cdot \nabla)q = 0.$$

But

$$(q \cdot \nabla)q = \frac{1}{2} \nabla(q \cdot q) - q \times \Omega. \qquad (3.10)$$

Hence finally:

$$\frac{\partial \Omega}{\partial t} + \nabla \times (q \times \Omega) = 0. \qquad (3.11)$$

Consistent with our assumption that the far field is q_∞ is that

$$q(0, x, y, z) = q_\infty = \nabla \phi_\infty$$

and hence

$$\Omega(0,\ x,\ y,\ z) = 0.$$

Now for any (smooth enough) solution $q(t, x, y, z)$, we may consider $\Omega(\cdot)$ as a solution of (3.9) which is a linear equation with zero initial condition and has the identically zero solution

$$\Omega(t, x, y, z) = 0 = \nabla \times q(t, x, y, z).$$

Or the velocity q $(\cdot)$ is curl-free. Hence it is expressible as

$$q = \nabla \phi.$$

And $\phi(t, x, y, z)$ is the velocity potential such that the far field potential

$$\phi(t, x, y, z) \text{ where } r = \sqrt{z^2 + y^2 + x^2} \to \infty,$$
$$= x(q_\infty \cdot i) + y(q_\infty \cdot j) + z(q_\infty \cdot k)$$
$$= \phi_\infty(0, x, y, z).$$

The Euler Full Potential Equation

Assuming entropic flow, we are now ready to derive the field equation for the velocity potential $\phi(\cdot)$ using the continuity equation. First though we rewrite the Euler equation using (3.7) and (3.8) as:

$$\frac{\partial \nabla \phi}{\partial t} + \frac{1}{2} \nabla [\nabla \phi, \nabla \phi] + \frac{A\gamma}{\gamma - 1} \nabla \rho^{\gamma - 1} = 0. \tag{3.12}$$

Hence

$$\frac{\partial \phi}{\partial t} + \frac{1}{2} [\nabla \phi, \nabla \phi] + A\frac{\gamma}{\gamma - 1} \rho^{\gamma - 1} = \text{constant} = \text{far field values}$$

$$= \frac{1}{2} [q_\infty, q_\infty] + \frac{\gamma}{\gamma - 1} A\rho_\infty^{\gamma - 1}.$$

Now because p is a function of ρ,

$$\frac{dp}{d\rho} = \gamma A\rho^{\gamma - 1} = a^2,$$

where a is the local speed of sound with the far field value a_∞, so that

$$a_\infty^2 = \gamma A \, \rho_\infty^{\gamma - 1},$$

or

$$\gamma A = \frac{a_\infty^2}{\rho_\infty^{\gamma - 1}}. \tag{3.13}$$

Hence we have

$$\frac{\partial \phi}{\partial t} + \frac{1}{2} \left(U^2 - U_\infty^2 \right) = \frac{a_\infty^2}{\gamma - 1} \left(1 - \frac{\rho^{\gamma - 1}}{\rho_\infty^{\gamma - 1}} \right),$$

where we use U for the flow speed. Hence we have for the density:

$$\rho^{\gamma - 1} = \rho_\infty^{\gamma - 1} \left(1 - \left(\frac{\gamma - 1}{a_\infty^2} \right) \left(\frac{1}{2} \left(U^2 - U_\infty^2 \right) + \partial_t \phi \right) \right). \tag{3.14}$$

Now using the continuity equation:

$$\frac{\partial \rho^{\gamma - 1}}{\partial t} = (\gamma - 1)\rho^{\gamma - 2}\frac{\partial \rho}{\partial t}$$

$$= -(\gamma - 1)\rho^{\gamma - 2}\nabla \cdot (\rho \nabla \phi)$$

$$= -(\gamma - 1)\rho^{\gamma - 2}(\rho \Delta \phi + \nabla \phi \cdot \nabla \rho)$$

$$= -(\gamma - 1)\rho^{\gamma - 1}\Delta \phi - \nabla \phi \cdot \nabla \rho^{\gamma - 1}$$

using

$$(\gamma - 1)\frac{\nabla \rho}{\rho} = \frac{\nabla \rho^{\gamma-1}}{\rho^{\gamma-1}}.$$

From (3.14) we have:

$$\frac{\partial \rho^{\gamma-1}}{\partial t} = -\rho_\infty^{\gamma-1}\left(\frac{\gamma-1}{a_\infty^2}\right)\frac{\partial^2 \phi}{\partial t^2} + \nabla\phi \cdot \partial t \nabla\phi)$$

$$-\nabla\rho^{\gamma-1} \cdot \nabla\phi = -\rho_\infty^{\gamma-1}\left(\frac{\gamma-1}{a_\infty^2}\right)\left(\frac{1}{2}\nabla\left|\nabla\phi\right|^2 + \partial t\nabla\phi\right)\cdot\nabla\phi.$$

Hence

$$\rho_\infty^{\gamma-1}\left(\frac{\gamma-1}{a_\infty^2}\right)\left(\frac{\partial^2 \phi}{\partial t^2} + \nabla\phi\cdot\frac{\partial\nabla\phi}{\partial t}\right)$$

$$= (\gamma-1)\rho_\infty^{\gamma-1}\left(1-\left(\frac{\gamma-1}{a_\infty^2}\right)\left(\frac{1}{2}\left(U_\infty^2 - U^2\right) + \frac{\partial\phi}{\partial t}\right)\right)\Delta\phi$$

$$-\rho_\infty^{\gamma-1}\left(\frac{\gamma-1}{a_\infty^2}\right)\left(\frac{1}{2}\nabla||\nabla\phi||^2 + \frac{\partial\nabla\phi}{\partial t}\right)\cdot\nabla\phi.$$

Hence finally:

$$\frac{\partial^2 \phi}{\partial t^2} + \nabla\phi\cdot\frac{\partial\nabla\phi}{\partial t} = a_\infty^2\Delta\phi\left(1 - \frac{\gamma-1}{a_\infty^2}\left(\frac{1}{2}\left(U^2 - U_\infty^2\right) + \frac{\partial\phi}{\partial t}\right)\right)$$

$$-\frac{1}{2}\nabla\phi\cdot\nabla(||\nabla\phi||^2) - \frac{\partial\nabla\phi}{\partial t}\cdot\nabla\phi,$$

which we can rewrite in the form:

$$\frac{\partial^2 \phi}{\partial t^2} + \frac{\partial}{\partial t}||\nabla\phi||^2 = a_\infty^2\left(1 + \frac{\gamma-1}{a_\infty^2}\left(\frac{U_\infty^2}{2} - \frac{||\nabla\phi||^2}{2} - \frac{\partial\phi}{\partial t}\right)\right)\Delta\phi$$

$$-\nabla\phi\cdot\nabla\left(\frac{||\nabla\phi||^2}{2}\right). \tag{3.15}$$

This is the 3D Euler full potential equation valid except on the boundary specified later.

Acceleration Potential and Pressure

We can now derive an explicit expression for the pressure in terms of the fluid velocity. The acceleration is defined by the total derivative of the velocity:

$$a(t) = \frac{Dq}{Dt} = \frac{\partial q}{\partial t} + (q \cdot \nabla)q$$

and for potential flow by (3.10)

$$(q \cdot \nabla)q = \frac{1}{2}\nabla(q \cdot q).$$

Hence

$$a(t) = \nabla\left(\partial t\phi + \frac{1}{2}|\nabla\phi|^2\right) = \nabla\psi(t),$$

where the acceleration potential

$$\psi(t) = \partial t\phi + \frac{|\nabla\phi|^2}{2} \tag{3.16}$$

and the far field value

$$\psi_\infty = \frac{U_\infty^2}{2}.$$

The flow being isentropic

$$p = A\rho^\gamma$$

and by (3.14)

$$\rho = \rho_\infty\left(1 - (\gamma - 1)\left(\frac{\psi - \psi_\infty}{a_\infty^2}\right)\right)^{1/(\gamma-1)} \tag{3.17}$$

and hence

$$p = A\rho_\infty^\gamma\left(1 - (\gamma - 1)\frac{(\psi - \psi_\infty)}{a_\infty^2}\right)^{\gamma/(\gamma-1)}. \tag{3.18}$$

And a very reasonable (and universal) approximation here is to take:

$$p = A\rho_\infty^\gamma\left(1 - \gamma\frac{(\psi - \psi_\infty)}{a_\infty^2}\right) \tag{3.19}$$

consistent with

$$0 < M = \frac{U_\infty}{a_\infty} < 1,$$

where M is the Mach number, and that the "perturbation" of the flow by the airplane is small compared to the far field speed. In any event we use (3.19) for p throughout for isentropic flow.

The main relations we have are thus (3.19) and (3.15).

What distinguishes aeroelasticity from aerodynamics is of course the interaction with the structure dynamics on the boundary, a singular boundary that complicates the problem further.

Boundary Conditions

The far field condition

$$\phi(t, \infty) = U_\infty(x \cos \alpha + y \sin \alpha),$$

where α is the specified "angle of attack."

As with any field equation, the conditions on the boundary determine the solution. Here it is further complicated by the fact we have taken into account the structure dynamics as well.

Flow-Structure Interaction:
Hence we first need to specify the structure model.

The Simplest Wing Structure Model
The simplest wing model is a slender "thin" plate whose thickness is then taken to be zero, rectangular in shape (unswept wing) uniform, with ℓ denoting the half wing span and $2b$ the width or chord length, so that b is the halfchord.

We only consider wings of high-aspect ratio:

$$\ell \gg b,$$

which would justify the flexible model. However, we do consider a finite rectangular plane (Finite plane) as the boundary for the aerodynamic equations. Thus we have 3D aerodynamics and a 2D wing boundary.

We choose the spatial co-ordinate system consistent with the aircraft rigid body dynamics. Thus the x-axis is along the airplane axis, the y-axis is the span axis and the negative z-axis is the "plunge" axis. Thus the boundary for the field equation is the rectangle in the xy-plane described by:

$$\Gamma = \{-b \le x \le b; \quad 0 \le y \le \ell\}.$$

For the structure dynamics, however, we specialize to a beam model, ignoring the dependence on the chord variable. Such a model was described by Goland, as we have seen in Chap. 2 where the structure is endowed with two degrees of freedom: plunge and pitch.

The plunge displacement denoted h (t, y) is then the displacement of the wing along the negative z-axis. It is uniform along the chord. The pitch $\vartheta(t, y)$ is the twist angle in radians about an axis parallel to the y-axis at distance ab from the x-axis and again does not depend on the chord variable x. Hence $h(t, y)$ and $\vartheta(t, y)$ are defined on:

$$0 \le t; \quad 0 \le y \le \ell.$$

The wing is fixed to the fusilage. Thus we have a clamped–free (CF) model with:

$$h(t, 0) = 0 = h'(t, 0); \quad \vartheta(t, 0) = 0.$$

The wing is free at the other end, and hence

$$h'''(t, \ell) = 0 = h''(t, \ell) = 0; \quad \vartheta'(t, \ell) = 0.$$

These are then the "end" conditions to be satisfied in addition to the force/momentum balance equations.

The resulting structure dynamic equations are described in Chap. 2. Here we shall describe the air–wing interaction dynamics.

Thus we have two sets of conditions.

I. The Attached Flow Condition

The normal velocity of the air along the wing is equal to the normal velocity of the wing. We may need to distinguish between the top and bottom of the wing if we allow for discontinuity in the fluid velocity across the wing, even though the thickness is zero.

The total displacement of the wing is given by:

$$\overrightarrow{k}\, z(t, x, y) \quad \text{where} \quad z(t, x, y) = -(h(t, \ y) + (x - ab)\, \vartheta \ (t, y))$$

and hence the normal velocity of the wing is given by:

$$\overrightarrow{k}\, \frac{Dz(t)}{D(t)} = \overrightarrow{k} \left(\frac{\partial z(t)}{\partial t} + q(t, x, y, 0) \cdot \nabla z(t) \right)$$

$$= -\overrightarrow{k} \left(\dot{h}(t, y) + (x - a)\dot{\vartheta}(t, y) + (q(t, x, y, 0) \cdot \overrightarrow{i})\vartheta(t, y) \right.$$

$$\left. + \left(q(t, x, y, 0) \cdot \overrightarrow{j} \right) \left(\frac{\partial}{\partial y}(h(t, y) + (x - a)\vartheta(t, y)) \right) \right).$$

Hence allowing for discontinuity in the flow, the flow tangency conditions become:

$$\frac{\partial \phi(t, x, y, 0+)}{\partial z} = \frac{\partial \phi_\infty}{\partial z} + (-1) \left[\dot{h}(t, y) + (x - a)\dot{\vartheta}(t, y) \right.$$

$$+ (q(t, x, y, 0+) \cdot \overrightarrow{i})\vartheta(t, y) + (q(t, x, y, 0+) \cdot \overrightarrow{j})$$

$$\left. \times \left(\frac{\partial}{\partial y}(h(t, y) + (x - a)\#(t, y)) \right) \right], x, y \epsilon \Gamma, \qquad (3.20)$$

$$\frac{\partial \phi(t,x,y,0-)}{\partial z} = \frac{\partial \phi_\infty}{\partial z} + (-1)\left[\dot{h}(t,y) + (x-a)\dot{\vartheta}(t,y)(q(t,x,y,0-)\cdot \vec{i}\,)\right.$$
$$\left. \times \vartheta(t,y) + (q(t,x,y,0-)\cdot \vec{j}\,)\left(\frac{\partial}{\partial y}(h(t,y)+(x-a)\vartheta(t,y))\right)\right],$$
$$\times x, y \epsilon \Gamma. \tag{3.21}$$

II. Kutta–Joukowski Conditions

These conditions are peculiar to aeroelasticity and are given in terms of the pressure which is discontinuous across the wing. We define the pressure Jump by δp:

$$\delta p(t,x,y) = p(t,x,y,0+) - p(t,x,y,0-). \tag{3.22}$$

Defining

$$\delta \psi(t,x,y) = \psi(t,x,y,0+) - \psi(t,x,y,0-),$$

we have from (3.20) that

$$\delta p = -A\rho_\gamma^\infty \frac{\gamma}{a_2^\infty}\delta\psi = -\rho_\infty \delta \psi. \tag{3.23}$$

II. The pressure jump is zero off the wing

$$\delta p(t,x,y) = 0 \quad \text{for } x, y \text{ not in } \Gamma. \tag{3.24}$$

Or

$$\delta \psi(t,x,y) = 0 \quad \text{for } x, y \text{ not in } \Gamma. \tag{3.25}$$

II. Kutta condition

$$\delta p(t,x,y) = 0 \text{ as } x \to b-, \ x, y \text{ in } \Gamma. \tag{3.26}$$

Or

$$\delta \psi(t,x,y) = 0 \text{ as } x \to b-, \ x, y \text{ in } \Gamma. \tag{3.27}$$

Structure Dynamics

We are now ready to complete the structure dynamics models begun in Chap. 2 by including the aerodynamic lift and moment which are expressed in terms of the structure state variables.

Goland Model

$$m\ddot{h}(t, y) + S\ddot{\vartheta}(t, y) + EI\,h''''(t, y) = L(t, y) = \int_{-b}^{b} \delta p(t, x, y)dx,$$

$$0 < y < \ell, \quad (3.28)$$

$$I_{\vartheta}\ddot{\vartheta}(t, y) + S\ddot{h}(t, y) - GJ\vartheta''(t, y) = M(t, y) = \int_{-b}^{b} (x - ab)\delta p(t, x, y)dx$$

$$0 < y < \ell. \quad (3.29)$$

Dowell–Hodges Model

$$m\ddot{h}(t, y) + EI_1 h''''(t, y) + (EI_2 - EI_1)(\vartheta(t, y)v(t, y)'')'' = mg\,\sin\varphi + L(t, y),$$
$$m\ddot{v}(t, y) + EI_2 v''''(t, y) + (EI_2 - EI_1)(\vartheta(t, y)h(t, y)'')'' = mg\cos\varphi,$$
$$I_{\vartheta}\ddot{\vartheta}(t, y) - GJ\vartheta''(t, y) + (EI_2 - EI_1)h(t, y)''v(t, y)'' = M(t, y),$$
$$0 < t;\ 0 < y < \ell.$$

3.4   Problem Statement

The Aeroelastic Problem: 3D/Nonzero Angle of Attack/2D Boundary

We can now present a complete statement of the combined 3D aeroelastic problem by the following equations.

Field equation

$$\frac{\partial^2 \phi}{\partial t^2} + \partial t\,||\nabla\phi||^2 = a_{\infty}^2 \left(1 + \frac{\gamma - 1}{a_{\infty}^2}\left(\frac{U_{\infty}^2}{2} - \frac{||\nabla\phi||^2}{2} - \partial t\phi\right)\right)\Delta\phi$$

$$- \nabla\phi \bullet \nabla\left(\frac{||\nabla\phi||^2}{2}\right). \quad (3.30)$$

Far field

$$\phi(t, \infty) = xU_\infty \cos\alpha + yU_\infty \sin\alpha.$$

Boundary conditions

$$\Gamma = \{\{x, y\}, -b < x < b; 0 < y < \ell\}$$

$$\times \frac{\partial \phi(t, x, y, 0+)}{\partial z} = \frac{\partial \phi_\infty}{\partial z} + (-1)\left[\dot{h}(t, y) + (x - a)\dot{\vartheta}(t, y)\right.$$

$$+ (q(t, x, y, 0+) \cdot \vec{i})\vartheta(t, y) + (q(t, x, y, 0+) \cdot \vec{j})$$

$$\times \left.\left(\frac{\partial}{\partial y}(h(t, y) + (x - a)\vartheta(t, y))\right)\right], \quad x, y \epsilon \Gamma. \tag{3.31}$$

And

$$\frac{\partial \phi(t, x, y, 0-)}{\partial z} = \frac{\partial \phi_\infty}{\partial z} + (-1)\left[\dot{h}(t, y) + (x - a)\dot{\vartheta}(t, y)\right.$$

$$\times (q(t, x, y, 0-) \cdot \vec{i})\vartheta(t, y) (q(t, x, y, 0-) \cdot \vec{j})$$

$$\times \left.\left(\frac{\partial}{\partial y}(h(t, y) + (x - a)\vartheta(t, y))\right)\right], x, y \in \Gamma. \tag{3.32}$$

Kutta–Joukowski conditions

$$\psi(t) = \frac{\partial \phi}{\partial t} + \frac{\|\nabla \phi\|^2}{2}, \tag{3.33}$$

$$\delta\psi(t, x, y) = 0 \quad \text{for } x, y \text{ not in } \Gamma, \tag{3.34}$$

$$\delta\psi(t, x, y) = 0 \quad \text{as } x \to b-, x, y \text{ in } \Gamma. \tag{3.35}$$

Structure Dynamics: Linear (Goland)

$$\ddot{h}(t, y) + S\ddot{\vartheta}(t, y) + EIh''''(t, y) = -\rho_\infty \int_{-b}^{b} \delta\psi(t, x, y) \, dx, \quad 0 < y < \ell,$$

$$\tag{3.36}$$

$$I_\vartheta \ddot{\vartheta}(t,y) + S\ddot{h}(t,y) - GJ\vartheta''(t,y) = -\rho_\infty \int_{-b}^{b} (x - ab)\delta\psi(t,x,y)dx \quad 0 < y < \ell,$$

(3.37)

plus CF end conditions:

$$h(t,0) = 0 = h'(t,0); \vartheta(t,0) = 0,$$
$$h'''(t,\ell) = 0 = h''(t,\ell) = 0; \vartheta'(t,\ell) = 0,$$

or FF end conditions:

$$h'''(t,0) = 0 = h''(t,0) = 0; \vartheta'(t,0) = 0,$$
$$h'''(t,\ell) = 0 = h''(t,1) = 0; \vartheta'(t,\ell) = 0.$$

Structure Dynamics Nonlinear

Beran Straganac

The state variables are the same as in the Goland. For the nonlinearities see Chap. 2, Sect. 2.7. The end conditions are also the same as in the Goland model.

Dowell–Hodges

Here there is an extra state variable $v(t,\cdot)$:

$$m\ddot{h}(t,y) + EI_1 h''''(t,y) + (EI_2 - EI_1)(\vartheta(t,y)v(t,y)'')''$$

$$= mg\sin\varphi - \rho_\infty \int_{-b}^{b} \delta\psi(t,x,y)dx, \quad 0 < y < \ell,$$

(3.38)

$$m\ddot{v}(t,y) + EI_2 v''''(t,y) + (EI_2 - EI_1)(\vartheta(t,y)h(t,y)'')'' = mg\cos\varphi, \quad (3.39)$$

$$I_\vartheta \ddot{\vartheta}(t,y) - GJ\vartheta''(t,y) + (EI_2 - EI_1)h(t,y)''v(t,y)''$$

$$= -\rho_\infty \int_{-b}^{b} (x - ab)\delta\psi(t,x,y)dx0 < t; \quad 0 < y < \ell.$$

(3.40)

The end conditions for $v(t,.)$ are the same as for $h(t,.)$.

Typical Section (Strip) Theory

A universally invoked simplification is to neglect the dependence on the y-co-ordinate in consequence of the assumed high-aspect ratio ℓ/b of the wing.

Problem Statement

Typical Section Theory

2D Aerodynamics 1D Structure

The field equations: with the y-co-ordinate omitted:

$$\frac{\partial^2 \phi}{\partial t^2} + \frac{\partial}{\partial t}\left(\left(\frac{\partial \phi}{\partial x}\right)^2 + \left(\frac{\partial \phi}{\partial z}\right)^2\right)$$

$$= a_\infty^2 \left(1 + \frac{\gamma - 1}{a_\infty^2}\left(\frac{U_\infty^2}{2} - \frac{\left(\frac{\partial \phi}{\partial x}\right)^2 + \left(\frac{\partial \phi}{\partial z}\right)^2}{2} - \frac{\partial \phi}{\partial t}\right)\right)\Delta\phi$$

$$- \nabla\phi \cdot \nabla\left(\frac{\left(\frac{\partial \phi}{\partial x}\right)^2 + \left(\frac{\partial \phi}{\partial z}\right)^2}{2}\right) \qquad -\infty < x, z < \infty. \qquad (3.41)$$

Far field potential

$$\phi(t, \infty) = xU_\infty\cos\alpha + zU_\infty\sin\alpha.$$

Boundary conditions

Γ is now just the Chord: $-b < x < b$ and the flow tangency condition becomes:

$$\text{Total Displacement} = \overrightarrow{k}\, z(t, x) \text{ where}$$

$$z(t, x) = -(h(t, y) + (x - ab)\vartheta(t, y))$$

and hence the normal velocity of the wing is given by

$$\vec{k}\frac{Dz(t)}{D(t)} = \vec{k}\left(\frac{\partial z(t)}{\partial t} + q(t, x, y, 0) \cdot \nabla z(t)\right),$$

$$= -\vec{k}\left(\dot{h}(t, y) + (x - ab)\dot{\vartheta}(t, y) + \left(q(t, x, y, 0) \cdot \vec{i}\right)\vartheta(t, y)\right).$$

Hence allowing for discontinuity in the flow, the flow tangency conditions become:

$$\frac{\partial\phi(t, x, 0+)}{\partial z} = \frac{\partial\phi_\infty}{\partial z} + (-1)\left[\dot{h}(t, y) + (x - a)\dot{\vartheta}(t, y)\right.$$

$$\left. + \left(q(t, x, 0+) \cdot \vec{i}\right)\vartheta(t, y)\right]. \tag{3.42}$$

And

$$\frac{\partial\phi(t, x, 0-)}{\partial z} = \frac{\partial\phi_\infty}{\partial z} + (-1)\left[\dot{h}(t, y) + (x - a)\dot{\vartheta}(t, y)\right.$$

$$\left. + \left(q(t, x, 0-) \cdot \vec{i}\right)\vartheta(t, y)\right] x\epsilon\Gamma. \tag{3.43}$$

Kutta–Joukowski Conditions

$$\delta p(t, x) = p(t, x, 0+) - p(t, x, 0-),$$

$$\delta\psi(t, x) = \psi(t, x, 0+) - \psi(t, x, 0-),$$

$$\delta p = -\rho_\infty\delta\psi.$$

The pressure jump is zero off the wing.

$$\delta p(t, x) = 0 \quad \text{for } |x| > b,$$

Or

$$\delta\psi(t, x) = 0 \quad \text{for } |x| > b.$$

Kutta condition

$$\delta p(t, x) \to 0 \text{ as } x \to b-, \text{ in } \Gamma.$$

Or,

$$\delta\psi(t, x)| \to 0 \text{ as } x \to b-, \ |x| < b.$$

The structure equations with the indicated aerodynamic loading remain the same.

This is then the precise statement of the aeroelastic problem continuum equations. The objective is to determine the stability of the structure state as a function of U_∞, the air speed.

To anticipate the theory that follows, the main conclusion is that for a given value of M (equivalently, speed of sound, equivalently altitude) there is a speed, called flutter speed, denoted U_F, for $U_\infty < U_F$, the structure is stable (see later for precise definition of stability), and for $U > U_F$ the structure is unstable.

Notes and Comments

It is interesting to note that none of the books on aeroelasticity, including Dowell et al. [17], Hodges and Pierce [5] or Bisplinghof, Ashley, and Halfman [6] care to make a precise statement of the aeroelastic problem as we do in this chapter. Indeed without such a statement it is not clear what it is that the computer codes used universally (see the many recent papers on aeroelasticity, for example [75, 81–83, 93]) are providing the (approximate) solution to, even omitting the cases where the solvability of the problem cannot be established.

Indeed without such a formulation it is not possible to define the Flutter Speed calculating which is a main objective of the theory.

We should note that most progress has been made for the typical section case (2D aerodynamics) and it is fortunate that flexibility is consistent with high-aspect ratio so that the typical section approximation is reasonable (without necessarily being very high).

Regarding the foundational conservation laws, following [4, 12] we have invoked three of them rather than the first two as in [4, 17], for example. The triad is essential for aeroelasticity. As Meyer [14] notes the third is the Euler version of the first law of thermodynamics.

Chapter 4
The Steady-State (Static) Solution of the Aeroelastic Equation

4.1 Introduction

In this chapter we specialize to the time-invariant version of the aeroelastic equations of Chap. 3 where we set all the time derivatives to zero and there is no input. It is called the *static solution* in that there is no change with time. It is of interest on its own—it is in fact central to the study of stability—but it also serves to illustrate the solution techniques used for the general case in Chap. 6.

Often this solution is also referred to as the steady-state solution, and so we need to clarify the terminology. The term *steady-state* response is used to indicate the response of a linear system to a steady input, such as a sinusoid input which is then not dependent on the initial condition. Here we have the case of zero input. So we take an arbitrary initial condition (and far field speed) and let time "march" (literally in computational programs) until there is no longer any change with time; in other words we consider the asymptotic response in time. Physically we are considering a pointwise limit of the potential as well as the structure state. Of course depending on the assumed speed parameter and the initial conditions there may not be such a solution. We largely follow [67]. To begin with, we can state the following.

Theorem 4.1. *Every steady-state solution is a static solution.*

Proof. If a steady-state limit exists, and there is no change in time, then it must satisfy the static equation. □

In our case in addition to the initial condition we have a parameter to specify the far field speed. And we show that for a sequence of speeds there is indeed more than one solution to the static aeroelastic equation. However, as we show below, there is one static solution for all speeds which is also a steady-state solution as well. This provides us a simple example where the continuum equations need not have a unique solution depending on the far field speed.

The steady-state solution depends obviously on the structure model used. We begin with the workhorse model, the linear Goland model.

A V Balakrishnan, *Aeroelasticity: The Continuum Theory*,
DOI 10.1007/978-1-4614-3609-6_4, © Springer Science+Business Media, LLC 2012

4.2   Goland Structure Model

We first consider the Goland linear structure model. We set all time derivatives in the field and boundary equations to zero. And all functions are independent of time so we drop the time variable. Thus we begin with the velocity potential $\phi(x, y, z)$ which satisfies the 3D time-invariant Euler full potential equation, purging the time derivatives in (3.15), yielding now:

$$\left(a_\infty^2 + \frac{\gamma - 1}{2}(|\nabla \phi_\infty|^2 - |\nabla \phi|^2)\right) \Delta \phi - \frac{1}{2} \nabla \phi \cdot \nabla(|\nabla \phi|^2) = 0,$$

which we may expand as

$$0 = a_\infty^2 \left[1 + \frac{(\gamma - 1)}{2a_\infty^2}\left(U_\infty^2 - \left(\frac{\partial \phi}{\partial x}\right)^2 - \left(\frac{\partial \phi}{\partial y}\right)^2 - \left(\frac{\partial \phi}{\partial z}\right)^2\right)\right](\Delta \phi)$$
$$- \frac{1}{2}\left[\frac{\partial \phi}{\partial x}\frac{\partial}{\partial x}\left|\nabla \phi\right|^2 + \frac{\partial \phi}{\partial y}\frac{\partial}{\partial y}\left|\nabla \phi\right|^2 + \frac{\partial \phi}{\partial z}\frac{\partial}{\partial z}\left|\nabla \phi\right|^2\right]. \tag{4.1}$$

Angle of Attack

We assume that the far field velocity vector is in the XZ plane making an angle α with the X-axis. This angle in flight control rigid body model [9] is called the *angle of attack*. Hence

$$\phi_\infty = U_\infty(x \cos \alpha + z \sin \alpha),$$
$$q_\infty = U_\infty(i \cos \alpha + k \sin \alpha).$$

We next specialize to the time-invariant structure dynamics.

Time Invariant Structure Dynamics

$$EI \frac{\partial^4 h(y)}{\partial y^4} = L(y), \tag{4.2}$$

$$-GJ \frac{\partial^2 \theta(y)}{\partial y^2} = M(y) \tag{4.3}$$

with the CF boundary conditions:

$$h(0) = 0; \quad h'(0) = 0; \quad \theta(0) = 0,$$
$$h''(1) = 0; \quad h'''(1) = 0; \quad \theta'(1) = 0.$$

Next the fluid-structure boundary conditions.

I. Flow Tangency Condition

$$\frac{\partial \phi}{\partial z}(x, y, 0+) = U_\infty \sin \alpha - (q(x, y, 0+) \cdot i)\theta(y) - (q(x, y, 0+) \cdot j)(h'(y)$$

$$+(x - ab) \, \theta'(y)), \tag{4.4}$$

$$\frac{\partial \phi}{\partial z}(x, y, 0-) = U_\infty \sin \alpha - (q(x, y, 0-) \cdot i)\theta(y) - (q(x, y, 0-) \cdot j)(h'(y)$$

$$+(x - ab) \, {}^{\backprime}(y)). \tag{4.5}$$

II. The Kutta–Joukowsky conditions

$$\delta p(x, y) = 0 \quad x, y \notin \Gamma \quad \delta p(b-, y) = 0. \tag{4.6}$$

Theorem 4.2. *The time-invariant equations are satisfied by the "rest" or equilibrium solution:*

$$\phi(x, y, z) = \phi_\infty(x, y, z).$$

Structure at rest:

$$h(y) = 0 \quad \text{and} \quad \theta(y) = 0 \qquad 0 \le y \le 1$$

for all far field speeds.

In this case the fluid velocity and pressure are continuous across the wing.

Proof. Verified by direct substitution into the equations. The solution holds for every value of the far field speed. □

We now show the existence of a nonzero static solution.

The Nonzero Static Solution

To find the nonzero static solution we need to solve the nonlinear static Euler equation. We show that because of the nature of the boundary conditions, we can obtain a series expansion, a solution technique that works also in the dynamic case, but is of course less complex here.

In the structure equations we need to calculate $L(y)$ and $M(y)$ for each $y, 0 < y < \ell$. If we fix y, however, the structure variables $h(y)$ and $\theta(y)$ may be looked upon as parameters which enter linearly in fact in (4.4) and (4.5) and we may postulate that the solution is analytic in these parameters, at least locally. The solution in other words can be expanded in a power series (for some nonzero radius of convergence) about

$$h(y) = 0; \quad \theta(y) = 0.$$

This is conveniently done by introducing the complex number parameter λ and considering the response to

$$\lambda h(y), \lambda \theta(y)$$

for given $h(y), \theta(y)$. More specifically, let the solution be denoted $\phi(\lambda, x, y, z)$ corresponding to $\lambda h(y), \lambda \theta(y)$, so that

$$\phi(0, x, y, z) = \phi_\infty(x, y, z).$$

And we have the expansion

$$\phi(\lambda, x, y, z) = \phi_\infty(x, y, z) + \sum_{k=1}^{\infty} \frac{\lambda^k}{k!} \phi_k(x, y, z), \tag{4.7}$$

where

$$\phi_k(x, y, z) = \frac{\partial^k}{\partial \lambda^k} \phi(0, x, y, z). \tag{4.8}$$

Before discussing the sense in which the series converges, let us first see how we can calculate the derivatives for each point (x, y, z). We go back to the field equation (4.1) and differentiating it once with respect to λ and setting $\lambda = 0$, we obtain for the first term therein:

$$a_\infty^2 \Delta \phi_1$$

and for the second term:

$$-U_\infty^2 \cos^2 \alpha \frac{\partial^2 \phi_1}{\partial x^2} - U_\infty^2 \sin^2 \alpha \frac{\partial^2 \phi_1}{\partial z^2} - 2U_\infty^2 \cos \alpha \sin \alpha \frac{\partial^2 \phi}{\partial z \partial x}.$$

Hence ϕ_1 satisfies the field equation:

$$\Xi(\phi_1) = 0,$$

where

$$\Xi(\phi) = (1 - M^2 \cos^2 \alpha) \frac{\partial^2 \phi}{\partial x^2} + \frac{\partial^2 \phi}{\partial y^2} + (1 - M^2 \sin^2 \alpha) \frac{\partial^2 \phi}{\partial z^2}$$

$$- 2M^2 \sin \alpha \cosff \frac{\partial^2 \phi}{\partial x \partial z}, \tag{4.9}$$

where the Mach number

$$M = \frac{U_\infty}{a_\infty} < 1.$$

Thus ϕ_1 satisfies a linear homogeneous equation, with boundary conditions described presently. Note that the thermodynamic constant γ does not appear.

For the higher-order potentials ϕ_k we need to work a little harder. Let us fix the spatial coordinates and let:

$$f(\lambda) = \left(a_\infty^2 + \frac{\gamma - 1}{2}(U_\infty^2 - |\nabla\phi(\lambda)|^2) \right), \tag{4.10}$$

$$g(\lambda) = \Delta\phi(\lambda). \tag{4.11}$$

Then we need to expand the first term in (4.1) – the product

$$f(\lambda)g(\lambda)$$

in a Taylor series. Now $g(\lambda)$ has the expansion

$$g(\lambda) = \sum_{k=1}^{\infty} \frac{\lambda^k}{k!}\Delta\phi_k.$$

Hence

$$\begin{aligned}
f(\lambda)g(\lambda) &= a_\infty^2 \sum_{k=1}^{\infty} \frac{\lambda^k}{k!}\Delta\phi_k + \left(\frac{\gamma-1}{2}\right)\left[U_\infty^2 - \left(q_\infty + \sum_{k=1}^{\infty}\frac{\lambda^k}{k!}\nabla\phi_k\right)\right.\\
&\qquad \left. \times \left(q_\infty + \sum_{k=1}^{\infty}\frac{\lambda^k}{k!}\nabla\phi_k\right)\right]\left(\sum_{k=1}^{\infty}\frac{\lambda^k}{k!}\Delta\phi_k\right)\\
&= a_\infty^2 \sum_{k=1}^{\infty} \frac{\lambda^k}{k!}\Delta\phi_k + \left(\sum_{k=1}^{\infty}\frac{\lambda^k}{k!}\Delta\phi_k\right)\left[(1-\gamma)\,q_\infty \cdot \sum_{k=1}^{\infty}\frac{\lambda^k}{k!}q_k\right.\\
&\qquad \left. + \frac{1-\gamma}{2}\left(\sum_{k=1}^{\infty}\frac{\lambda^k}{k!}q_k\right)\cdot\left(\sum_{k=1}^{\infty}\frac{\lambda^k}{k!}q_k\right)\right]\\
&= a_\infty^2 \sum_{k=1}^{\infty} \frac{\lambda^k}{k!}\Delta\phi_k + (1-\gamma)\left(\sum_{j=1}^{\infty}\sum_{k=1}^{\infty}\frac{\lambda^j}{j!}\frac{\lambda^k}{k!}\Delta\phi_j q_\infty \cdot q_k\right)\\
&\qquad + \frac{1-\gamma}{2}\sum_{i=1}^{\infty}\sum_{j=1}^{\infty}\sum_{k=1}^{\infty}\frac{\lambda^i}{i!}\frac{\lambda^j}{j!}\frac{\lambda^k}{k!}q_i \cdot q_j \Delta\phi_k, \tag{4.12}
\end{aligned}$$

where we have used the notation

$$q_k = \nabla \phi_k .$$

Next the second term in (4.1) is best handled using x_1, x_2, x_3, respectively, in place of x, y, z. Thus it can be expressed:

$$-\sum_{i=1}^{3} \frac{\partial \phi}{\partial x_i} \sum_{j=1}^{3} \frac{\partial \phi}{\partial x_j} \frac{\partial^2 \phi}{\partial x_i \partial x_j}$$

$$= -\sum_{i=1}^{3} \sum_{j=1}^{3} \sum_{n=0}^{\infty} \sum_{m=0}^{\infty} \sum_{p=1}^{\infty} \frac{1}{n!m!p!} \lambda^{(n+m+p)} \frac{\partial \phi_n}{\partial x_i} \frac{\partial \phi_m}{\partial x_j} \frac{\partial^2 \phi_p}{\partial x_i \partial x_j},$$

where

$$\phi_0 = \phi_\infty$$

so that

$$\frac{\partial^k \phi_0}{\partial x^k} = 0 \quad k > 1.$$

Hence we have the expansion for (4.1):

$$a_\infty^2 \sum_{k=1}^{\infty} \frac{\lambda^k}{k!} \Delta \phi_k + (1-\gamma) \left(\sum_{j=1}^{\infty} \sum_{k=1}^{\infty} \frac{\lambda^j}{j!} \frac{\lambda^k}{k!} \Delta \phi_j q_\infty \cdot q_k \right)$$

$$+ \frac{1-\gamma}{2} \sum_{i=1}^{\infty} \sum_{j=1}^{\infty} \sum_{k=1}^{\infty} \frac{\lambda^i}{i!} \frac{\lambda^j}{j!} \frac{\lambda^k}{k!} q_i \cdot q_j \Delta \phi_k$$

$$- \sum_{i=1}^{3} \sum_{j=1}^{3} \sum_{n=0}^{\infty} \sum_{m=0}^{\infty} \sum_{p=1}^{\infty} \frac{1}{n!m!p!} \lambda^{(n+m+p)} \frac{\partial \phi_n}{\partial x_i} \frac{\partial \phi_m}{\partial x_j} \frac{\partial^2 \phi_p}{\partial x_i \partial x_j} = 0, \qquad (4.13)$$

where the third term:

$$-\sum_{i=1}^{3} \sum_{j=1}^{3} \sum_{n=0}^{\infty} \sum_{m=0}^{\infty} \sum_{p=1}^{\infty} \frac{1}{n!m!p!} \lambda^{(n+m+p)} \frac{\partial \phi_n}{\partial x_i} \frac{\partial \phi_m}{\partial x_j} \frac{\partial^2 \phi_p}{\partial x_i \partial x_j}$$

$$= -\sum_{i=1}^{3} \sum_{j=1}^{3} \sum_{p=1}^{\infty} \frac{1}{p!} \lambda^p \frac{\partial \phi_0}{\partial x_i} \frac{\partial \phi_0}{\partial x_j} \frac{\partial^2 \phi_p}{\partial x_i \partial x_j}$$

$$- 2\sum_{i=1}^{3} \sum_{j=1}^{3} \sum_{m=1}^{\infty} \frac{1}{m!p!} \lambda^{m+p} \frac{\partial \phi_0}{\partial x_i} \frac{\partial \phi_m}{\partial x_j} \frac{\partial^2 \phi_p}{\partial x_i \partial x_j}$$

$$- \sum_{i=1}^{3} \sum_{j=1}^{3} \sum_{n=1}^{\infty} \sum_{m=1}^{\infty} \sum_{p=1}^{\infty} \frac{1}{n!m!p!} \lambda^{(n+m+p)} \frac{\partial \phi_n}{\partial x_i} \frac{\partial \phi_m}{\partial x_j} \frac{\partial^2 \phi_p}{\partial x_i \partial x_j} .$$

Next let us examine the boundary conditions:

$$\partial_z \phi(\lambda, x, y, 0+) = U_\infty \sin \alpha - \lambda \left(\partial_x \phi(\lambda, x, y, 0+) \theta(y) \right.$$
$$\left. + \partial_y \phi(\lambda, x, y, 0+) \left(\frac{\partial h}{\partial y} + (x - ab) \frac{\partial \theta}{\partial y} \right) \right).$$

Hence

$$\frac{\partial \phi_1(x, y, 0+)}{\partial z} = -U_\infty \cos \alpha \; \theta(y) = \frac{\partial \phi_1(x, y, 0-)}{\partial z} \qquad (4.14)$$

and for $k \geq 2$:

$$\frac{\partial \phi_k(x, y, 0+)}{\partial z} = -k \left[\frac{\partial \phi_{k-1}(x, y, 0+)}{\partial x} \theta(y) + (\partial \phi_{k-1}(x, y, 0+)) \right.$$
$$\left. \Big/ \partial y \left(\frac{\partial h}{\partial y} + (x - ab) \frac{\partial \theta}{\partial y} \right) \right]. \qquad (4.15)$$

Kutta–Joukowsky Condition

$$\psi = \frac{1}{2} \left(U_\infty^2 + \sum_{k=1}^\infty \sum_{j=1}^\infty \frac{\lambda^{k+j}}{k!j!} \nabla \phi_k \cdot \nabla \phi_j + 2 \sum_{k=1}^\infty \frac{\lambda^k}{k!} \nabla \phi_\infty \cdot \nabla \phi_k \right). \qquad (4.16)$$

From which we can derive the power series expansion:

$$\psi(\lambda, x, y, z) = \frac{U_\infty^2}{2} + \sum_{k=1}^\infty \frac{\lambda^k}{k!} \psi_k(x, y, z),$$
$$\psi_1(x, y, z) = \nabla \phi_\infty \cdot \nabla \phi_1. \qquad (4.17)$$

And hence

$$\delta \psi_1 = \psi_1(x, y, 0+) - \psi_1(x, y, 0-)$$
$$= U_\infty \cos \alpha ((\partial \phi_1(x, y, 0+))/\partial x - (\partial \phi_1(x, y, 0-))/\partial x) \qquad (4.18)$$

using (4.15) and (4.16), and as we have noted $= 0$ off the wing.

We pause here at this point to examine in detail the linear case:

$$k = 1.$$

4.3   Linear Aeroelasticity Theory: The Finite Plane Case

Linear theory has an importance all its own and in addition we actually bootstrap about it for the nonlinear case. And we can carry the solution some distance for the general 3D aerodynamics. We do not discretize it now (as in most work in aeroelasticity, as we have noted already). The field equation is, as we have seen:

$$\Xi(\phi_1) = 0.$$

We rewrite this as an equation in z because $z = 0$ contains the boundary.

$$(1 - M^2\sin^2\alpha)\frac{\partial^2\phi_1}{\partial z^2} - 2M^2\sin\alpha\cos\alpha\frac{\partial^2\phi_1}{\partial x\partial z}$$

$$= -(1 - M^2\cos^2\alpha)\frac{\partial^2\phi_1}{\partial x^2} - \frac{\partial^2\phi_1}{\partial y^2}. \tag{4.19}$$

The first step is to choose the function space for the solution.

For each z we seek a solution in

$$L_p[R^2], \quad 1 < p < 2.$$

We define the $L_p - L_q$ Fourier transform:

$$\hat{\phi}_1(z, \omega_1, \omega_2) = \int_{-\infty}^{\infty}\int_{-\infty}^{\infty}\phi_1(x, y, z)e^{-(i\omega_1 + i\omega_2)}dxdy \qquad \omega_1, \omega_2 \in R^2 \tag{4.20}$$

and note that the Fourier transform of the right side of (4.21) yields

$$(\omega_1^2(1 - M^2\cos^2\alpha) + \omega_2^2)\hat{\phi}_1(z, \omega_1, \omega_2).$$

Hence in terms of the Fourier transforms, (4.20) becomes an equation in $L_q[R^2]$:

$$(1 - M^2\sin^2\alpha)\frac{\partial^2\hat{\phi}_1}{\partial z^2} + 2M^2\sin\alpha\cos\alpha\frac{\partial^2\hat{\phi}_1}{\partial x\partial z}$$

$$= (\omega_1^2(1 - M^2\cos^2\alpha) + \omega_2^2)\hat{\phi}_1(z, \omega_1, \omega_2). \tag{4.21}$$

Let

$$\hat{v}(i\omega_1, i\omega_2) = \frac{\partial}{\partial z}\hat{\phi}_1(0\pm, i\omega_1, i\omega_2).$$

Then we can show that the only solution which goes to zero (in the L_p norm) as $|z| \to \infty$ is given by

$$\hat{\phi}_1(z, \omega_1, \omega_2) = \frac{1}{r_1}e^{zr_1}\hat{v}(i\omega_1, i\omega_2) \quad \text{for } z > 0 \tag{4.22}$$

$$= \frac{1}{r_2}e^{zr_2}\hat{v}(i\omega_1, i\omega_2) \quad \text{for } z < 0, \tag{4.23}$$

where

$$r_1 = (-M^2(i\omega_1)\sin\alpha\cos\alpha - \sqrt{(\omega_2^2(1 - M^2\sin^2\alpha)}$$
$$+ \omega_1^2(1 - M^2)))/(1 - M^2\mathrm{Sin}^2\alpha), \qquad (4.24)$$

$$r_2 = (-M^2(i\omega_1)\sin\alpha\cos\alpha + \sqrt{(\omega_2^2(1 - M^2\sin^2\alpha)}$$
$$+ \omega_1^2(1 - M^2)))/(1 - M^2\sin^2\alpha) = -\overline{r_1}. \qquad (4.25)$$

We only need to calculate the pressure jump across the wing:

$$\delta p = -\rho_\infty \delta\psi.$$

Let

$$A_1(x, y) = \frac{-\delta\psi_1}{U_\infty} \qquad (4.26)$$

called the Kushner doublet function; see [6] which by (4.15)

$$= -\cos\alpha\,\delta\phi_1.$$

Hence the L_1 Fourier transform

$$\hat{A}_1(i\omega_1, i\omega_2) = \int_{-b}^{b}\int_{0}^{\ell} e^{-ix\omega_1 - iy\omega_2} A(x, y)\mathrm{d}x\mathrm{d}y$$
$$= -i\omega_1\cos\alpha\left(\frac{1}{r_1} - \frac{1}{r_2}\right)\left(\hat{v}_1(i\omega_1, i\omega_2)\right),$$

or,

$$\hat{v}_1(i\omega_1, i\omega_2) = \frac{-1}{2}\frac{\hat{A}(i\omega_1, i\omega_2)}{i\omega_1\cos\alpha} \times [(\omega_1^2(1 - M^2\cos^2\alpha)$$
$$+ \omega_2^2)/(\sqrt{(\omega_1^2(1 - M^2) + \omega_2^2)})]$$

and by (4.15)

$$= -U_\infty\cos\alpha\,\hat{\theta}(i\omega_2), \qquad (4.27)$$

where

$$\hat{\theta}(i\omega) = \int_0^\ell \grave{}(y)e^{-iy\omega}dy.$$

Hence we have that:

$$\hat{P}(i\omega_1, i\omega_2)\hat{A}(i\omega_1, i\omega_2) = U_\infty \cos\alpha\hat{\theta}(i\omega_2),$$

where

$$\hat{P}(i\omega_1, i\omega_2) = \frac{1}{2}\frac{1}{i\omega_1\cos\alpha} \times \left[\left(\omega_1^2(1 - M^2\cos^2\alpha)\right.\right.$$
$$\left.\left. + \omega_2^2\right)\Big/\left(\sqrt{\omega_1^2(1 - M^2) + \omega_2^2}\right)\right] \qquad (4.28)$$

$$\hat{P}(i\omega_1, i\omega_2) = \int_{-\infty}^\infty \int_{-\infty}^\infty P(x, y)e^{-ix\omega_1 - iy\omega_2}dxdy.$$

This is the "static" version of the 2D integral equation of Possio which we describe later and which plays a crucial role in our theory.

$$\int_{-b}^b \int_0^\ell P(x - \xi, y - \zeta)A(\xi, \zeta)d\xi d\zeta = U_\infty\theta(y), \quad 0 < y < 1, |x| < b, \quad (4.29)$$

where the Kutta condition becomes

$$A(b-, y) = 0 \qquad 0 < y < 1. \qquad (4.30)$$

Let us assume that this equation has a solution in $L_p[-b, b]$. Then

$$\delta p = \rho_\infty U_\infty A(x, y). \qquad (4.31)$$

We expect the solution to be of the form:

$$A(x, y) = \int_0^\ell L(x, y, \zeta)\theta(\zeta)d\zeta, \qquad x, y \text{ on the wing} \qquad (4.32)$$

and hence, getting back to the structure equations (4.2) and (4.3), we have:

$$GJ\theta''(y) = \int_{-b}^b (x - ab) \int_0^\ell L(x, y, \zeta)\theta(\zeta)d\zeta dx$$
$$= \int_0^\ell M(y, \zeta)\theta(\zeta)d\zeta. \qquad (4.33)$$

And with the CF end conditions we have thus a linear eigenvalue problem that would have a solution at most for a sequence of values of the far field speeds which we may call divergence speeds. But not being able to calculate the kernel M $(., .)$, we cannot do much more. So we proceed to simplify the problem where we can say more.

The first and most important simplification is to consider the following.

Typical Section: "Airfoil Theory"

We assume that the aspect ratio ℓ/b is high enough so we may consider the air flow on the wing to be uniform so that there is no dependence on the span variable y, also called the strip theory. In this case the linear field equation (4.20) simplifies to:

$$(1 - M^2\sin^2\alpha)\frac{\partial^2\phi_1}{\partial z^2} - 2M^2\sin\alpha\,\cos\alpha\,\frac{\partial^2\phi_1}{\partial x\partial z} = -(1 - M^2\cos^2\alpha)\frac{\partial^2\phi_1}{\partial x^2} \quad (4.34)$$

and we define, modifying (4.21)

$$\hat{\phi}_1(z, i\omega) = \int_{-\infty}^{\infty} \phi_1(z, x)e^{-ix\omega}dx$$

$$\hat{v}(i\omega) = \phi_1(0, i\omega).$$

$\phi_1(z, x)$ is the solution of (4.35) subject to the boundary condition:

$$\frac{1}{\partial z}\partial\phi_1(0, x) = -U_\infty\cos\alpha\,\theta(y), |x| < b,$$

where y is fixed.

We want

$$\delta p(x) = -\rho_\infty\delta\psi(x)$$

and

$$\delta\psi_1(x) = -U_\infty\delta\phi_1 = 0 \qquad |x| > b.$$

We define:

$$A_1(x) = -\frac{\delta\psi_1}{U_\infty}, \qquad -b < x < b$$

so that

$$\delta p = \rho_\infty U_\infty A_1.$$

We now turn to Fourier transforms which are obtained by simply setting in the previous expressions:

$$\omega_1 = \omega; \qquad \omega_2 = 0.$$

Then

$$\hat{A}_1(i\omega) = -i\omega \left(\frac{1}{r_1} - \frac{1}{r_2}\right) \hat{v}(i\omega).$$

Hence

$$\hat{v}(i\omega) = -\frac{1}{2}\frac{\hat{A}(i\omega)}{i\omega\cos\alpha}\left[\left(\omega^2(1 - M^2\cos^2\alpha)\right)\Big/\left(\sqrt{\left(1 - M^2\right)\omega^2}\right)\right].$$

Let

$$\hat{P}(i\omega) = -\frac{1}{2}\frac{1}{i\omega\cos\alpha}\left[\omega^2\left(1 - M^2\cos^2\alpha\right)\Big/\sqrt{\left(1 - M^2\right)\omega^2}\right];$$

$$\hat{P}(i\omega) = \int_{-\infty}^{\infty} P(x)e^{-ix\omega}dx, \qquad -\infty < x < \infty$$

and we have:

$$\int_{-b}^{b} P(x - \xi)A(\xi)d\zeta = U_\infty\theta(y), \quad |x| < b$$

$$A(b-) = 0. \tag{4.35}$$

The Airfoil Equation

We should note that (4.36) is a special case of the Possio integral equation which we go into in far more detail in Chap. 5, where it is now the airfoil equation or the finite Hilbert transform equation. $\hat{P}(i\omega)$ can be expressed as

$$\hat{P}(i\omega) = \text{const} \cdot \frac{1}{2}\left(\frac{|\omega|}{i\omega}\right), \tag{4.36}$$

where

$$\text{const} = \frac{1 - M^2\cos^2\alpha}{\sqrt{1 - M^2}}\sec\alpha$$

and

$$\left(\frac{|\omega|}{i\omega}\right)$$

Fig. 4.1 Pressure
distribution along chord

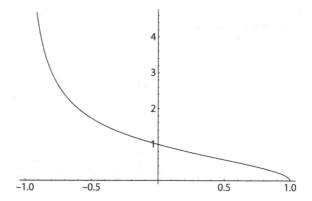

is the "multiplier" corresponding to the Hilbert transform. Thus (4.36) becomes

$$\text{const} \cdot \frac{1}{2\pi} \int_{-b}^{b} \frac{1}{x-s} A(s)\mathrm{d}s = U_\alpha \theta(y), \quad A(b-) = 0, \quad |x| < b. \tag{4.37}$$

And we have the explicit solution, due to Tricomi (see [11], and the more general
treatment in [31]).

$$A(x) = \frac{1}{\text{Const}} U_\infty \theta(y) \frac{1}{\pi} \sqrt{\frac{b-x}{b+x}} \int_{-b}^{b} \sqrt{\frac{b+\xi}{b-\xi}} \frac{1}{\xi - x} \mathrm{d}\xi, \quad |x| < b.$$

In as much as

$$\frac{1}{\pi} \int_{-b}^{b} \sqrt{\frac{b+\xi}{b-\xi}} \frac{1}{\xi - x} \mathrm{d}\xi = b,$$

we thus have the explicit solution:

$$A(x) = bU_\infty \theta(y) \sqrt{\frac{b-x}{b+x}} \frac{2\sqrt{1-M^2}}{1-M^2 \cos^2 \alpha} \cos \alpha, \quad |x| < b. \tag{4.38}$$

Note that $A(b-) = 0$. The function $A(.)$ is in $L_p[-b, b]$ for $1 < p < 2$ but Not for
$p = 2$. In other words this is Not an L_2 theory, as opposed to structure dynamics,
where it is and we have a notion of energy.

See Fig. 4.1 for the canonical static pressure distribution along the chord for
$b = 1$.

Note the discontinuity with respect to the angle of attack. For $\alpha = 0$, the pressure
distribution at midchord increases without bound as M increases to 1. But for small
nonzero α, the pressure peaks at some point and then decreases to zero. We have a

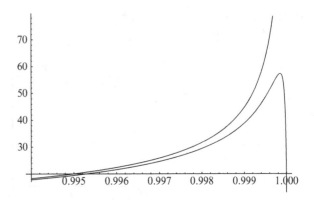

Fig. 4.2 Pressure distribution at midchord $\alpha = 0$ and $\alpha = 1$ degree

"transonic peak." Here the word transonic simply means that M is close to 1. See Fig. 4.2.

Structure Response: Divergence Speed

Because the pressure jump depends only on the pitch angle the structure response equations reduce to

$$GJ\theta''(y) = U_\infty^2 \theta(y) \left[\left(2\sqrt{1 - M^2} \right) / \left(1 - M^2\cos^2\alpha \right) \right]$$

$$\rho_\infty\cos\alpha \int_{-b}^{b} (x - ab)\sqrt{\frac{b - x}{b + x}}\,dx\theta(0) = 0\theta'(1) = 0, \qquad (4.39)$$

where the integral:

$$2\int_{-b}^{b} (x - ab)\sqrt{\frac{b - x}{b + x}}\,dx = 2b^2 \int_{-1}^{1} (x - a)\sqrt{\frac{1 - x}{1 + x}}\,dx = -b^2(1 + 2a)\pi,$$

where we are assuming that

$$|a| < \frac{1}{2}.$$

The structure equation for the torsion angle $\theta(y)$,

$$-GJ\theta''(y) = \int_{-b}^{b} (x - ab)\delta p(x)\,dx,$$

becomes $\theta''(y) = -\lambda^2\theta(y), 0 < y < 1$, with $\theta(0) = \theta'(0) = 0$ and has the solution $\theta(y) = B\sin[\mu y]$, where B is an arbitrary constant including zero and we must have

$$B\cos[\ell\mu] = 0.$$

Thus $B = 0$ is always a solution, whatever U_∞.

If we want a nonzero solution—B is not zero—we have an eigenvalue problem and thus we have a nonzero solution only for

$$\mu = (2n - 1)\frac{\pi}{2\ell}, \quad n \geq 1$$

and

$$\theta(y) = B \sin\left[(2n - 1)\frac{\pi y}{2\ell}\right] \quad 0 < y < \ell,$$

where:

$$\mu^2 = U_\infty^2 \rho_\infty \cos\alpha(1 - M^2)^{1/2}\Big/(1 - M^2\cos^2\alpha)b^2(1 + 2a),$$

which means that U_∞ cannot be arbitrary and must equal

$$U_\infty = (2n - 1)\left((1 - M^2\cos^2\alpha)\Big/\left(\sqrt{1 - M^2}\right)\right)^{1/2}$$
$$\times \frac{1}{2b1}(\sqrt{\pi GJ})\Big/\left(\sqrt{(\rho_\infty(1 + 2a))}\right) \sec\alpha,$$

n positive integer.

The smallest value is called the divergence Speed:

$$U_d = \left(\left(1 - M^2\cos^2\alpha\right)\Big/\left(\sqrt{1 - M^2}\right)\right)^{1/2}$$
$$\times \frac{1}{2b1}\left(\sqrt{\pi GJ}\right)\Big/\left(\sqrt{(\rho_\infty(1 + 2a))}\right) \sec\alpha. \qquad (4.40)$$

It is interesting to see the dependence on M, and the difference between

$$\alpha = 0 \quad \text{and} \quad \alpha \neq 0.$$

For $\alpha = 0$, we have

$$U_d = (1 - M^2)^{1/4}\frac{1}{2b\ell}\left(\sqrt{\pi GJ}\right)\Big/\left(\sqrt{(\rho_\infty(1 + 2a))}\right) \qquad (4.41)$$

and U_d decreases monotonically to zero.

On the other hand for nonzero α we have a nonzero minimum and U_d increases to infinity at $M = 1$. We have thus what is referred to as the transonic dip which occurs at

$$M = \sqrt{(1 - \tan^2\alpha)},$$

our main interest being small $\alpha \sim\sim \pi/180$. See Fig. 4.3.

$$U_d/U_d(0).$$

It is interesting to contrast this with the way in which it is done in CFD computation; see [99] where it has to be extrapolated from literally a few points on the graph.

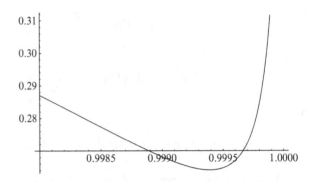

Fig. 4.3 Transonic dip $\alpha = 1^o$

The bending or plunge shape at the divergence speed is determined by

$$EIh''''(y) = L(y)$$

$$= \rho_\infty U_d^2 2\theta(y)\left(\sqrt{1-M^2}\right) \Big/ (1-M^2\cos^2\alpha) \int_{-b}^{b} \sqrt{\frac{b-x}{b+x}}$$

$$= 2\pi b\left(\sqrt{1-M^2}\right) \Big/ (1-M^2\cos^2\alpha)\rho_\infty U_d^2 \sin\left[\frac{\pi y}{2\ell}\right], \quad 0 < y < \ell,$$

a nonhomogeneous equation with the end conditions:

$$h(0) = h'(0) = 0 = h''(1) = h'''(1),$$

which is determined by the solution to the pitch equation.

4.4   The General Nonlinear Case

Getting back now to the general case, let D denote the derivative with respect to λ. Then with $f(\lambda)$ given by (4.11) and $g(\lambda)$ by (4.12), we have:

$$D^k(f(\lambda)g(\lambda)) = f(\lambda)D^k g(\lambda) + g(\lambda)D^k f(\lambda) + \sum_{j=1}^{k-1} C_j^k D^{k-j} g(\lambda)D^j f(\lambda),$$

where C_m^n are the binomial coefficients, which at $\lambda = 0$ yields

$$\Delta\phi_k + \frac{1-\gamma}{2}\sum_{j=1}^{k-1} c_j^k \Delta\phi_{k-j} \sum_{m=0}^{j} C_m^j \frac{\partial\phi_{j-m}}{\partial x_i}\frac{\partial\phi_m}{\partial x_i},$$

where the second term involves ϕ_j only for $j < k$.

Next in (4.14), the coefficient of λ^k is given by

$$\sum_{j=1}^{3}\sum_{i=1}^{3} -\frac{\partial^2\phi_k}{\partial x_i \partial x_j}\left(\frac{\partial\phi_0}{\partial x_i}\right)\left(\frac{\partial\phi_0}{\partial x_j}\right) - \sum_{l=1}^{k-2}\sum_{i=1}^{3}\sum_{j=1}^{3}\frac{\partial^2\phi_{k-1}}{\partial x_i \partial x_j}\sum_{m=0}^{1}c_m^1\frac{\partial\phi_{1-m}}{\partial x_i}\frac{\partial\phi_m}{\partial x_j},$$

(4.42)

where the second term again involves ϕ_j, $j \le k - 1$.

Hence we obtain the nonhomogeneous linear equation for ϕ_k:

$$\Xi(\phi_k) = g_{k-1},$$

(4.43)

where

$$g_{k-1} = \frac{1-\gamma}{2}\sum_{i=1}^{3}\sum_{j=1}^{k-1}C_j^k\,\Delta\phi_{k-j}\sum_{m=1}^{j}C_m^j\frac{\partial\phi_{j-m}}{\partial x_i}\frac{\partial\phi_m}{\partial x_i}$$

$$+\sum_{l=1}^{k-2}\sum_{i=1}^{3}\sum_{j=1}^{3}\frac{\partial^2\phi_{k-1}}{\partial x_i \partial x_j}\sum_{m=0}^{1}C_m^1\frac{\partial\phi_{1-m}}{\partial x_i}\frac{\partial\phi_m}{\partial x_j}$$

$$+2\sum_{i=1}^{3}\sum_{j=1}^{3}\sum_{m=1}^{k-1}C_m^k\frac{\partial\phi_0}{\partial x_i}\frac{\partial\phi_m}{\partial x_i}\frac{\partial^2\phi_{k-m}}{\partial x_i \partial x_j}$$

$$+(1-\gamma)\sum_{j=1}^{k-1}\Delta\phi_j q_\infty \cdot q_{k-j}$$

and we note that

$$g_0 = 0.$$

And in particular

$$g_1 = 2(1-\gamma)\Delta\phi_1 q_\infty \cdot q_1 + 4\sum_{i=1}^{3}\sum_{j=1}^{3}\frac{\partial\phi_0}{\partial x_i}\frac{\partial\phi_1}{\partial x_i}\frac{\partial^2\phi_1}{\partial x_i \partial x_j},$$

illustrating the nature of the functions g_k. Note that the derivatives are well defined for any k.

Next we need to go deeper into the boundary conditions. We begin with the typical section case where the flow is 2D and brings considerable simplification well justified for high-aspect ratio wings and is the best developed part of the theory. We specialize to typical section theory, but keep a nonzero angle of attack.

Typical Section Theory: Nonzero α

Flow Tangency Condition: Specializing (4.15), (4.16) we have:

$$\frac{\partial \phi_k(x, 0+)}{\partial z} = -C_1^k D^{k-1} \frac{\partial \phi(0, x, 0+)}{\partial x} \theta(y)$$

$$= -k \frac{\partial \phi_{k-1}(x, 0+)}{\partial x} \theta(y) \qquad (4.44)$$

$$\frac{\partial \phi_k(x, 0-)}{\partial z} = -C_1^k D^{k-1} \frac{\partial \phi(0, x, 0-)}{\partial x} \theta(y)$$

$$= -k \frac{\partial \phi_{k-1}(x, 0-)}{\partial x} \theta(y). \qquad (4.45)$$

A crucial next step is as follows.

Flow Decomposition

To solve (4.43), we decompose the flow field as the sum of two interdependent terms: for each k

$$\phi_k(x, z) = \phi_{k_L}(x, z) + \phi_{k0}(x, z), \qquad (4.46)$$

where ϕ_{kL} satisfies the homogeneous equation:

$$\Xi(\phi_{k_L}) = 0$$

subject to the boundary conditions (4.44) and (4.45) and ϕ_{k0} satisfies the nonhomogeneous equation

$$\Xi(\phi_{k0}) = g_{k-1}$$

with no boundary conditions imposed. In particular

$$\delta(\nabla \phi_k) = \delta(\nabla \phi_{kL}) \qquad \phi_1 = \phi_{1L}. \qquad (4.47)$$

We define

$$\Phi_0 = \phi_0 + \sum_{k=2}^{\infty} \frac{\lambda^k}{k!} \phi_{k0}, \tag{4.48}$$

$$\Phi_L = \sum_{k=1}^{\infty} \frac{\lambda^k}{k!} \phi_{kL} \tag{4.49}$$

so that

$$\phi = \Phi_0 + \Phi_L, \tag{4.50}$$

where Φ_0 has no discontinuities on $z = 0$ and hence produces no lift, and Φ_L produces all the lift.

Our main concern is of course the pressure jump across the wing, or equivalently the acceleration potential jump. From (4.7)

$$\delta\psi = \delta\left(\left(\frac{\partial\phi}{\partial x}\right)^2 + \left(\frac{\partial\phi}{\partial z}\right)^2\right) = 0 \quad z = 0, \quad |x| > b.$$

By analyticity therefore

$$\delta\psi_k = 0 \text{ for } k \geq 1 \quad \text{off the wing.}$$

Again our main concern is to evaluate this on the structure. We begin with

Theorem 4.3. *Let φ denote the disturbance potential*

$$\varphi = \phi - \phi_0 \tag{4.51}$$

Suppose $\delta|\nabla\varphi|^2 = 0, \quad |x| < b$.

Then

$$\delta\psi_k = U_\infty\cos\alpha\,\delta\left(\frac{\partial\phi_k}{\partial x}\right) + U_\infty\sin\alpha\,\delta\left(\frac{\partial\phi_k}{\partial z}\right), \quad -b < x < b \tag{4.52}$$

the same as for $k = 1$.

Proof. We calculate that:

$$|\nabla\phi|^2 = U_\infty^2 + 2\nabla\phi_0 \cdot \nabla\varphi + |\nabla\varphi|^2. \tag{4.53}$$

Hence

$$\delta|\nabla\phi|^2 = 2\nabla\phi_0 \cdot \delta\nabla\varphi + \delta|\nabla\varphi|^2 = 2\nabla\phi_0 \cdot \delta\nabla\varphi$$

$$\delta\psi = \frac{1}{2}\delta|\nabla\phi|^2 = \nabla\phi_0 \cdot \delta\nabla\varphi.$$

Hence

$$\delta \psi_k = \nabla \phi_0 \cdot \delta \nabla \phi_k = U_\infty \left(\cos \alpha \, \delta \frac{\partial \phi_k}{\partial x} + \sin \alpha \, \delta \frac{\partial \phi_k}{\partial z} \right). \qquad \square$$

Note that φ is the "disturbance" potential and we expect that $|\nabla \varphi|^2$ is small compared to U_∞^2 so that we may neglect it in (4.53) anyway. The angle of attack is usually less than a degree and can be neglected in the first look, simplifying the analysis. And sensitivity to the angle of attack is analyzed subsequently.

Zero Angle of Attack

Let us now specialize to the case where the angle of attack is zero.
   Then we have the following remarkable result.

Theorem 4.4. *Suppose* $\alpha = 0$. *Then*

$$\delta |\nabla \varphi|^2 = 0, \qquad |x| < b. \tag{4.54}$$

The speed is continuous across the wing, in other words.

Proof. First we note that

$$\nabla \varphi = \sum_{k=1}^{\infty} \frac{\lambda^k}{k!} \nabla \phi_k.$$

From

$$\phi_1(x, z) = -\phi_1(x, -z)$$

we have:

$$\phi_1(x, 0+) + \phi_1(x, 0-) = 0.$$

Hence

$$\frac{\partial \phi_1(x, 0+)}{\partial x} + \frac{\partial \phi_1(x, 0-)}{\partial x} = 0.$$

Hence

$$\delta \left(\frac{\partial \phi_1}{\partial x} \right)^2 = 0.$$

Also from

$$\delta \left(\frac{\partial \phi_1}{\partial z} \right) = 0$$

we have

$$\delta \left(\frac{\partial \phi_1}{\partial z} \right)^2 = 0.$$

Hence:

$$\delta |\nabla \phi_1|^2 = 0. \qquad \square$$

To show that this holds for every k, we use an induction argument 5.

Lemma 4.5. *Suppose*

$$\delta |\nabla \phi_j|^2 = 0, \qquad j \le k - 1.$$

Then

$$\delta |\nabla \phi_k|^2 = 0.$$

Proof. First from the flow tangency condition:

$$\frac{\partial \phi_k}{\partial z} = -k\theta(y) \frac{\partial \phi_{k-1}}{\partial x} \qquad |x| < b$$

and hence

$$\delta \left(\frac{\partial \phi_k}{\partial z} \right)^2 = k^2 \theta(y)^2 \delta \left(\frac{\partial \phi_{k-1}}{\partial x} \right)^2 \qquad |x| < b.$$

Hence

$$\delta \left(\frac{\partial \phi_k}{\partial z} \right)^2 = 0 \qquad |x| < b.$$

Let us use the notation

$$\gamma_j = \frac{\partial \phi_j}{\partial x}$$

along with

$$\nu_j \text{ for } \frac{\partial \phi_j}{\partial z}$$

and the Fourier transforms by $\hat{\gamma}_j$, $\hat{\nu}_j$, respectively. We are only concerned with ϕ_{jL}.
$\square$

Following the proof for $j = 1$, we take Fourier transforms and obtain

$$\hat{\gamma}_j(i\omega, 0+) = \frac{i\omega}{r_1} \hat{\nu}_j(i\omega, \ 0+), \qquad (4.55)$$

$$\hat{\gamma}_j(i\omega, 0-) = -\frac{i\omega}{r_1} \hat{\nu}_j(i\omega, \ 0-), \qquad (4.56)$$

where

$$r_1 = -|\omega|\sqrt{1 - M^2}.$$

Hence, subtracting, we get:

$$\hat{v}_j(i\omega, 0+) + \hat{v}_j(i\omega, 0-) = \frac{r_1}{i\omega}\delta\hat{v}_j \qquad (4.57)$$

and adding, we get

$$\frac{r_1}{i\omega}\overline{\gamma}_j(i\omega) = (\hat{v}_j(i\omega, 0+) - \hat{v}_j(i\omega, 0-)), \qquad (4.58)$$

where we use the notation $\overline{\gamma}_j = \gamma_j(x, \ 0+) + \gamma_j(x, \ 0-)$; the super bar does Not denote complex conjugate.

Next we use the boundary condition (4.44) and (4.45)

$$v_j(x, 0+) + v_j(x, 0-) = -j`(y)\overline{\gamma}_{j-1}(x) \qquad |x| < b.$$

By Theorem 4.2,

$$\delta\psi_k(x) = U\delta_{\gamma_k}(x), \qquad |x| < b$$

and hence

$$A_k(x) = -\frac{\delta\psi_k(x)}{U} = -\delta_{\gamma_k}(x), \quad |x| < b.$$

Now

$$v_j(x, \ 0+) + v_j(x, 0-) = -j\theta(y)\overline{\gamma}_{j-1}(x), \quad |x| < b.$$

But

$$\hat{v}_j(i\omega, 0+) + \hat{v}_j(i\omega, 0-) = \frac{r_1}{i\omega}\delta\hat{\gamma}_j(i\omega).$$

Hence

$$\frac{r_1}{i\omega}\hat{A}_j(\omega) = j\theta(y)\hat{\overline{\gamma}}_{j-1}(i\omega). \qquad (4.59)$$

Next

$$v_j(x, 0+) - v_j(x, 0-) = -j\theta(y)\delta\gamma_{j-1}(x), \quad |x| < b$$
$$= j\theta(y)A_{j-1}(x), \qquad |x| < b.$$

Hence taking Fourier transforms, because

$$\delta \gamma_{j-1}(x) = 0 \; |x| > b$$

and

$$\overline{\gamma}_j(i\omega) = \frac{i\omega}{r_1}(\hat{v}_j(i\omega, 0+) - \hat{v}_j(i\omega, 0-)),$$

hence

$$\frac{i\omega}{r_1}\hat{A}_{j-1}(i\omega) j\theta(y) = \overline{\gamma}_j(i\omega),$$

or

$$\frac{|\omega|}{i\omega\sqrt{1 - M^2}}\hat{A}_{j-1}(i\omega) j\theta(y) = \overline{\gamma}_j(i\omega). \tag{4.60}$$

In (4.60) and (4.59) we have a pair of Possio equations for the time-invariant case, similar to (4.38). This time it is convenient to define the Tricomi transform:

$$\mathcal{T}f = h, h(x) = \frac{1}{\pi}\sqrt{\frac{b-x}{b+x}} \int_{-b}^{b} \sqrt{\frac{b+s}{b-s}} \frac{f(s)}{s-x} ds, \quad |x| < b,$$

which we discuss in more detail in Chap. 5. Then we have similar to (4.40):

$$\sqrt{1 - M^2}A_j = -j\theta(y)\,\mathcal{T}P\overline{\gamma}_{j-1}$$

$$\frac{|\omega|}{i\omega\sqrt{1 - M^2}}\hat{A}_{j-2}(i\omega)(j-1)\theta(y) = \hat{\overline{\gamma}}_{j-1}(i\omega).$$

Hence

$$A_{j-2}(j-1)\theta(y) = \sqrt{1 - M^2}\,\mathcal{T}P\overline{\gamma}_{j-1},$$

where we are taking advantage of the fact that

$$\frac{1}{\sqrt{1 - M^2}}\frac{r}{i\omega} = -\frac{i\omega}{r}\sqrt{1 - M^2}.$$

Hence

$$j\theta(y)A_{j-2}(j-1)\theta(y) = j\theta(y)\sqrt{1 - M^2}\mathcal{T}P\overline{\gamma}_{j-1} = -(1 - M^2)A_j.$$

Thus we have our recursive relation

$$(1 - M^2)A_j = -j(j - 1)\theta(y)^2 A_{j-2}. \tag{4.61}$$

An immediate conclusion is that

$$A_j = 0 \quad \text{for } j \text{ even}$$

and hence

$$\overline{\gamma}_j = 0 \quad \text{for } j \text{ odd}.$$

And for k odd:

$$\frac{A_k}{k!} = \theta(y)^{k-1}(1 - M^2)^{((k-1)/2)} A_1$$

and

$$A_{2k} = 0.$$

Hence

$$A = \sum_{k=1}^{\infty} \frac{A_k}{k!} = \sum_{k=1}^{\infty} \theta(y)^{2k-2}(1 - M^2)^{((2k-1)/2)} A_1 = \frac{1}{1 + \frac{\theta(y)^2}{1-M^2}} A_1. \tag{4.62}$$

It is important to note that the series converges for small enough $\theta(y)$, but the closed-form solution holds without any smallness assumption on $\theta(y)$. Also the convergence is uniform in x and z.

We can now consider the convergence of the power series expansion:

$$\sum_{k=0}^{\infty} \frac{\phi_{2kL}(x, z)}{(2k)!} + \sum_{k=0}^{\infty} (\phi_{(2k+1)L}(x, z)) \Big/ (2k + 1)! \tag{4.63}$$

For k even, combining (4.56) and (4.57), we have

$$\frac{1}{k!}\hat{\phi}_{kL}(i\omega, z) = e^{|z|r_1} \frac{(k - 1)}{k!}\theta(y)^{k-1}(1 - M^2)^{-((k-2)/2)} \hat{A}_1(i\omega).$$

Hence for k even:

$$\frac{\phi_{kL}(x, z)}{k!} = \phi_1(x, z)\frac{1}{k!}(k - 1)\theta(y)^{k-1}(1 - M^2)^{-((k-2)/2)} \tag{4.64}$$

and we see that the convergence of the first series in (4.57) is uniform in x and z.

For k odd similarly we obtain using (4.61):

$$\phi_{kL}(x, z) = \phi_1(x, z)\theta(y)^{k-1}(1 - M^2)^{-((k-1)/2)}.$$

Again we see that the convergence of the second series in (4.57) is uniform in x and z. And we have an explicit solution that we can now describe. First of all we see that for each y it is of the form:

$$\phi_L(x, z) = N(M, U_\infty, \theta(y))\phi_1(x, z), \qquad (4.65)$$

where

$$N(M, U_\infty, \theta(y)) \sum_{k=0}^{\infty} \frac{(2k)\theta(y)^{k-1}}{(2k)!}(1 - M^2)^{-(k-1)}$$

$$+ \sum_{k=0}^{\infty} \frac{1}{(2k + 1)!}(y)^{2k}(1 - M^2)^{-k},$$

where $\theta(y)$ is determined by the structure response.

Thus in particular the differentiability properties of ϕ_L are those of $\phi_1(x, z)$ and there are no discontinuities in the velocity off the wing.

Steady-State Structure Response

Let us next calculate the steady-state structure response. We have for the aerodynamic moment:

$$M(y) = \frac{1}{1 + \frac{\theta(y)^2}{1-M^2}} \int_{-b}^{b} \rho U_\infty (x - ab) A_1(x) dx,$$

where the integral

$$= -\theta(y)\pi\rho b^2 U_\infty^2 (1 + 2a)\frac{1}{\sqrt{1 - M^2}}.$$

Hence

$$M(y) = \frac{-\theta(y)}{1 + \frac{\theta(y)^2}{1-M^2}}\pi\rho b^2 U_\infty^2 (1 + 2a)\frac{1}{\sqrt{1 - M^2}}.$$

Let

$$\mu^2 = \pi\rho b^2 \frac{(1 + 2a)}{GJ}\frac{1}{\sqrt{1 - M^2}}.$$

Then the pitch equation (4.3) becomes:

$$\theta''(y) = -\mu^2 U_\infty^2 \frac{\theta(y)}{1 + \frac{\theta(y)^2}{1-M^2}} \qquad 0 < y < 1, \qquad (4.66)$$

with the end conditions:

$$\theta(0) = 0; \quad \theta'(1) = 0.$$

We have the trivial solution

$$\theta(y) = 0; \quad \text{and} \quad h(y) = 0,$$

which holds for all values of U_∞.

To find the nonzero solution we see that we have to solve a nonlinear eigenvalue problem. Making a change of variable

$$\alpha(y) = \frac{\theta(y)}{\sqrt{1 - M^2}},$$

this becomes

$$\alpha''(y) = -\mu^2 U_\infty^2 \frac{\alpha(y)}{1 + \alpha(y)^2} \qquad 0 < y < 1 \qquad (4.67)$$

with the end conditions

$$\alpha(0) = 0 \qquad \alpha'(1) = 0.$$

We now establish that (4.67) has a solution only for a countable values of U_∞.

Multiplying (4.67) by $\alpha'(y)$ on both sides we obtain:

$$\frac{d}{d\alpha}\alpha'(y) = -\mu^2 U_\infty^2 \frac{d}{d\alpha}\frac{\log(1 + \alpha(y)^2)}{2}.$$

Or

$$\alpha'(y) + \mu^2 U_\infty^2 \frac{\log(1 + \alpha(y)^2)}{2} = \text{const} = \alpha'(0). \qquad (4.68)$$

Putting $y = 1$ we get:

$$\mu^2 U_\infty^2 \frac{\log(1 + \alpha(1)^2)}{2} = \alpha'(0), \qquad (4.69)$$

which is an homogeneous boundary condition. Now the solution of the first-order differential equation (4.68) is analytic in the coefficient

$$\mu^2 U_\infty^2$$

and in (4.69) $\alpha(1)$ and $\alpha'(0)$ are entire functions of $\mu^2 U_\infty^2$. Hence the zeros of

$$\mu^2 U_\infty^2 \frac{\log(1 + \alpha(1)^2)}{2} - \alpha'(0) = 0$$

are at most countable and cannot have a finite accumulation point. Thus we have at most a sequence of values of U_∞ or which (4.67) holds. An approximation is provided by the divergence speeds.

We could also solve the nonlinear Volterra equation:

$$\alpha(y) + \mu^2 U_\infty^2 \int_0^y \frac{\log(1 + \alpha(s)^2)}{2} ds = \alpha'(0)y, \quad 0 < y < 1,$$

which follows from (4.68). But the sequence of speeds does not play a role in stability analysis, so we don't pursue this any more.

Next the lift

$$L(y) = \rho U_\infty \int_{-b}^{b} A(x)dx = -2\pi\rho b U_\infty^2 \frac{\theta(y)}{\sqrt{1 - M^2}}$$

and is determined solely by the beam torsion.

Next the plunge equation is Eulerian:

$$EI \frac{\partial^4 h(y)}{\partial y^4} = -2\pi\rho b U_\infty^2 \theta(y) \quad 0 < y < 1 \tag{4.70}$$

and is thus determined completely by the pitch, and we do not continue the calculation because the solution is of little interest.

The Velocity Potential

To complete the determination of the velocity potential we go on to calculate the part Φ_0 in (4.50), even though it plays no role in the stability of the structure.

We first need to solve the nonhomogeneous equation (4.43) for $\alpha = 0$:

$$(1 - M^2) \frac{\partial^2 \phi_k}{\partial x^2} + \frac{\partial^2 \phi_k}{\partial z^2} = g_{k-1}.$$

Let us begin with $k = 2$. We have specializing (4.44)

$$g_1(x, z) = 2(\gamma - 1)\Delta\phi_1 \frac{\partial \phi_1}{\partial x},$$

where

$$\frac{\partial \phi_1}{\partial x} = -\int_{-b}^{b} |z| \left(\sqrt{1 - M^2}\right) \Big/ ((x - \xi)^2 + z^2(1 - M^2)) A(\xi) d\xi$$

$$A(x) = U_\infty`(y) \frac{2}{\sqrt{1 - M^2}} \sqrt{\frac{b - x}{b + x}},$$

which is bounded continuous in R^2, and

$$\phi_{20}(x, z) = \int_{-\infty}^{\infty} \int_{-\infty}^{\infty} L(x - \xi, z - \eta) g_1(\xi, \eta) d\xi d\eta,$$

where

$$L(x, z) = \frac{1}{\pi} \frac{|z| \sqrt{1 - M^2}}{x^2 + (1 - M^2)z^2}.$$

It follows that

$$\Phi_0(x, z) = \phi_0(x) + \sum_{k=2}^{\infty} \int_{-\infty}^{\infty} \int_{-\infty}^{\infty} L(x - \xi, z - \eta) g_{k-1} d\xi d\eta. \qquad (4.71)$$

We stop here because as before our interest is again only in the structure dynamics which does not depend on this part of the flow.

Note, however, we have deduced a constructive solution to the Euler potential equation in the form of a convergent power series.

It is expressible as the sum of two terms neither of which satisfies the equation. But only one is required for the structure response. Hence there is no point in calculating the second part as in CFD codes.

A Variation on the Problem

An interesting (and actually useful, as we show in Chap. 7) variation on this problem is to consider a virtual structure with no dynamics, but we prescribe a normal velocity on the boundary and seek the solution to the steady state 2D solution for zero angle of attack aerodynamic field equations. Thus we specify, with the same notation as before:

$$\frac{\partial \phi(x, 0)}{\partial z} = w(x), \quad |x| < b \qquad (4.72)$$

and the Kutta–Joukowsky conditions in addition. Proceeding as before we denote the solution corresponding to $\lambda w(\cdot)$ by $\phi(\lambda, x, z)$, assumed analytic in λ with a nonzero radius of convergence uniformly in x, z.

$$\phi(0, x, z) = x\, U_\infty,$$

$$\frac{\partial \phi(\lambda, x, 0)}{\partial z} = \lambda w(x), \quad |x| < b,$$

and we define $\phi_k(x, z)$, $\psi_k(x, z)$ as before.

For $k = 1$, we obtain as before the Fourier transform

$$\hat{\phi}_1(i\omega,\, z) = \frac{e^{|z|r_1}}{r_1}\hat{v}_1(i\omega,\, 0+), \quad z > 0$$

$$= -\frac{e^{|z|r_1}}{r_1}\hat{v}_1(i\omega,\, 0-), \quad z < 0. \tag{4.73}$$

In particular then

$$\delta\hat{v}_1 = 2\hat{v}_1(i\omega,\, 0).$$

We have also that the linear velocity potential

$$\psi_1(x, z) = U_\infty \frac{\partial \phi_1}{\partial x}(x, z).$$

Hence with

$$A(x) = -\frac{\delta\psi_1}{U_\infty},$$

to satisfy Kutta–Joukowski, we need:

$$A(x) = 0 \qquad |x| > b,$$

$$A(b-) = 0.$$

Hence the Fourier Transform:

$$\hat{A}(i\omega) = -2\frac{1}{r_1}i\omega\hat{v}_1(i\omega),$$

or

$$\hat{v}_1(i\omega) = \frac{1}{2}A(\hat{i}\omega)\frac{|\omega|}{i\omega}\sqrt{1 - M^2}.$$

Hence using the Tricomi operator we have:

$$A(x) = \frac{2}{\sqrt{1 - M^2}}\frac{1}{\pi}\sqrt{\frac{b - x}{b + x}}\int_{-b}^{b}\sqrt{\frac{b + \xi}{b - \xi}}w(\xi)d\xi, \quad |x| < b, \tag{4.74}$$

where to satisfy the Kutta condition we assume that $w(\cdot)$ satisfies a Lipschitz condition of order α, $\frac{1}{2} < \alpha \leq 1$.

The corresponding solution for the velocity potential is then given by:

$$\hat{\phi}_1(i\omega, z) = -\frac{e^{|z|r_1}}{2i\omega}\hat{A}(i\omega) \qquad z > 0$$

$$= +\frac{e^{|z|r_1}}{2i\omega}\hat{A}(i\omega) \qquad z < 0, \qquad (4.75)$$

where

$$e^{|z|r_1} = e^{-|\omega||z|\sqrt{1-M^2}} = \int_{-\infty}^{\infty} e^{-i\omega x}\left(2|z|\sqrt{1-M^2}\right)\Big/(z^2(1-M^2) + x^2)dx,$$

$$z \neq 0.$$

Hence

$$\frac{\partial\phi_1(x, z)}{\partial x} = z\sqrt{1 - M^2}\int_{-b}^{b}(A(\xi)d\xi)/z^2(1-M^2) + (x-\xi)^2, \qquad |x| < \infty.$$

$$(4.76)$$

Next

$$\frac{\partial\hat{\phi}_1(i\omega, z)}{\partial z} = -r_1\frac{e^{|z|r_1}}{2i\omega}\hat{A}(i\omega) = e^{|z|r_1}\frac{1}{2}\frac{|\omega|}{i\omega}\sqrt{1-M^2}\,\hat{A}(i\omega).$$

Hence taking the inverse Fourier transform,

$$\frac{\partial\phi_1(x, z)}{\partial z} = z\frac{\sqrt{1-M^2}}{2}\int_{-\infty}^{\infty}(B(\xi))/(z^2(1-M^2) + (x-\xi)^2)d\xi, \qquad (4.77)$$

where

$$B = \mathcal{H}A(\cdot); \qquad B(x) = \frac{1}{\pi}\int_{-b}^{b}\frac{A(\xi)}{x-\xi}d\xi, \qquad |x| < \infty,$$

$\mathcal{H}$ being the Hilbert transform. This is an $L_p - L_q$ transform, $1 < p < 2$, $A(\cdot)$ being in L_p, $1 \leq p < 2$.

A canonical example is:

$$w(x) = \frac{1}{\sqrt{b+x}}, \qquad |x| < b, \text{ in which case,} \qquad (4.78)$$

$$A(x) = \frac{2}{\sqrt{1-M^2}}\sqrt{\frac{b-x}{b+x}}\int_{-b}^{b}\sqrt{\frac{1}{b-\xi}}\frac{1}{\xi-x}d\xi, \qquad |x| < b. \qquad (4.79)$$

The integral does not have a closed form and has to be evaluated numerically.

Fig. 4.4 Pressure
distribution along chord

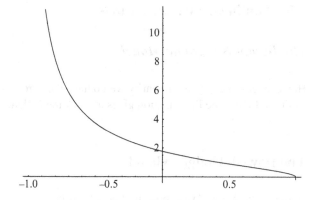

Figure 4.4 shows the plot for $b = 1$ (within a gain factor). For $k \geq 2$, abbreviating
the notation for this section $\phi_k = \phi_{k0}$ satisfies the nonhomogeneous equation

$$(1 - M^2)\frac{\partial^2 \phi_k}{\partial x^2} + \frac{\partial^2 \phi_k}{\partial z^2} = g_{k-1}. \tag{4.80}$$

For $k = 2$,

$$g_1 = U_\infty(\gamma - 1)\Delta\phi_1 \frac{\partial\phi_1}{\partial x} \tag{4.81}$$

and is a bounded continuous function in R^2. So in fact is g_k, for every k.

The nonhomogeneous equation (4.80) is readily solved by taking Fourier transforms. We have

$$\phi_k(x, z) = \int_{-\infty}^{\infty} \int_{-\infty}^{\infty} L(x - \xi, z - \eta)g_{k-1}(\xi, \eta)d\xi d\eta, \qquad x, z \text{ in } R^2, \tag{4.82}$$

where

$$L(\xi, \eta) = \frac{1}{2\pi}\left(2|\eta|\sqrt{1 - M^2}\right)\Big/(\eta^2(1 - M^2) + \xi^2). \tag{4.83}$$

Hence it follows that $\phi_k(\cdot)$ is a bounded continuous function in R^2, with continuous
derivative as well, no discontinuities in flow velocity. Hence the series:

$$\Phi_0(x, z) = \phi_0(x, z) + \sum_{k=2}^{\infty} \frac{1}{k!}\phi_{k,0}(x, z) \tag{4.84}$$

converges pointwise and has a continuous derivative. Hence so does

$$\Phi = \Phi_0 + \Phi_L.$$

The flow thus has no discontinuities. And hence there are no shocks in the flow.

4.5   Nonlinear Structure Models

The Beran–Straganac Model

However complex the nonlinearity we do have the zero or rest solution as the static solution. In turn the linearization gives us back the Goland equation.

The Dowell–Hodges Model

Here the static aeroelastic system of equations is:

$$EI_1 h''''(y) + (EI_2 - EI_1)(\theta(y)v(y)'')'' = mg \cos\varphi - \rho \int_{-b}^{b} \delta\psi(x, y)dx, \quad (4.85)$$

$$EI_2 v''''(y) + (EI_2 - EI_1)(\theta(y)h(y)'')'' = mg \sin\varphi \quad (4.86)$$

$$GJ\theta''(y) + (EI_2 - EI_1)v(y)''h''(y) = -\rho \int_{-b}^{b} (x - ab)\delta\psi(x, y)dx, \quad (4.87)$$

plus CF end conditions.
    Field Equation

$$0 = a_\infty^2 \left[1 + \frac{(\gamma - 1)}{2a_\infty^2} \left(U_\infty^2 - \left(\frac{\partial\phi}{\partial x}\right)^2 - \left(\frac{\partial\phi}{\partial z}\right)^2 \right) \right] (\Delta\phi)$$

$$- \frac{1}{2} \left[\frac{\partial\phi}{\partial x} \frac{\partial}{\partial x} |\nabla\phi|^2 + \frac{\partial\phi}{\partial z} \frac{\partial}{\partial z} |\nabla\phi|^2 \right] \quad (4.88)$$

$$\psi = \frac{|\nabla\phi|^2}{2}.$$

Fluid–Structure Boundary Conditions

$$\frac{\partial\phi(x, 0\pm)}{\partial z} = -\frac{\partial\phi(x, 0\pm)}{\partial x}\theta(y).$$

Plus Kutta–Joukowsky Conditions

The gravity terms on the right—$mg \cos\varphi, mg \sin\varphi$—contribute to the complexity.

If we set g in the gravity term to be zero, we see that we have the zero solution for the structure and constant flow for the aerodynamic equation is a static solution for all U_∞. In particular we need to set g to be zero for determining the "ground modes" corresponding to $U = 0$, and we see that we go back to the linear Goland model.

If the difference term

$$EI_2 - EI_1$$

is set to zero, then of course, the equations become linear, and we go back to the solution for the Goland model we started with, except for the addition of the $v(\cdot)$ equation:

$$EI_2 v''''(y) = 0,$$

which stands alone, and has only the zero solution.

Let us examine then the dependence on

$$(EI_2 - EI_1).$$

Let us consider the solution for

$$\lambda(EI_2 - EI_1),$$

where λ is a small parameter so that we have the solution:

$$h(\lambda, y) \quad v(\lambda, y) \quad \text{and} \quad \theta(\lambda, y)$$

satisfying:

$$EI_1 h''''(\lambda, \ y) + \lambda(EI_2 - EI_1)(\theta(\lambda, \ y)v(\lambda, \ y)'')''$$

$$= \mathrm{mg}\cos\varphi - 2\pi\rho b U_\infty^2 \frac{\theta(\lambda, y)}{\sqrt{1 - M^2}}$$

$$EI_2 v''''(\lambda, \ y) + \lambda(EI_2 - EI_1)(\theta(\lambda, y)h(\lambda, y)'')'' = \ \mathrm{mg}\sin\varphi$$

$$GJ\theta''(\lambda, y) + \lambda(EI_2 - EI_1)v(\lambda, y)''h''(\lambda, y) = -\mu\frac{\theta(\lambda, y)}{1 + \dfrac{\theta(\lambda, y)^2}{1 - M^2}},$$

where

$$\mu = \pi\rho b^2 U_\infty^2(1 + 2a)\frac{1}{\sqrt{1 - M^2}}.$$

Let

$$h_k(y) = \frac{\mathrm{d}^k}{\mathrm{d}\lambda^k}h(\lambda, y) \ \Big|\lambda = 0$$

and the power series expansion:

$$h(\lambda, y) = \sum_{k=0}^{\infty} \frac{h_k(y)\lambda^k}{k!}$$

with a similar notation for the others.

We can calculate that

$$EI_1 h_1''''(y) + (EI_2 - EI_1)(\theta(0, y)v(0, y)'')'' = -2\pi\rho b U_\infty^2 \frac{\theta(0, y)}{\sqrt{1 - M^2}}$$

$$EI_2 v_1''''(y) + (EI_2 - EI_1)(\theta(0, y)h(0, y)'')'' = 0$$

$$GJ\theta_1''(y) + (EI_2 - EI_1)v(0, y)''h''(0, y) = -\mu\theta_1(y),$$

where we have used the fact that $\theta(0) = 0$.

More generally we have for $k \geq 2$:

$$GJ\theta_k''(y) + \mu\theta_k(y) = r_{\theta,\ k-1}(y)$$

$$EI_1 h_k''''(y) = r_{h,\ k-1}(y)$$

$$EI_2 v_k''''(\lambda,\ y) = r_{v,\ k-1}(y)$$

with zero end conditions, because the derivatives with respect to λ are zero. The functions

$$\gamma_{\theta,k-1}(y) \quad \gamma_{h,k-1}(y) \quad \gamma_{v,k-1}(y)$$

depend only on the functions with index $\leq (k - 1)$.

Theorem 4.6. *The pitch equation*

$$GJ\theta_k''(y) + \mu`_k(y) = \gamma_{\theta,k-1}(y) \qquad 0 < y < \ell$$

with end conditions

$$CF : \theta_k(0) = 0; \qquad \theta_k'(\ell) = 0$$

$$FF : \theta_k'(0) = 0; \qquad \theta_k'(\ell) = 0$$

has the solution given by:

$$\theta_k(y) = \int_0^\ell G_\theta(y, s)\gamma_{\theta,k-1}(s)ds, \qquad 0 < y < \ell,$$

except for a sequence of divergence speeds U_n for which

$$GJ\theta_k''(y) + \mu_n\theta_k(y) = 0, \qquad 0 < y < \ell$$

has a nonzero solution.

Proof.

$$\text{Let } \omega = \sqrt{\frac{\mu}{GJ}}.$$

Then

$$\theta_k'(s) = \left(\frac{-1}{\cos \omega \ell} \int_0^\ell \cos \omega(\ell - \sigma)\gamma_{\theta,k-1}(\sigma)d\sigma\right) + \int_0^s \frac{\sin \omega(s - \sigma)}{\omega}\gamma_{,k-1}(\sigma)d\sigma$$

excepting ω such that $\cos \omega \ell = 0$, which is the sequence U_n.   □

The solution procedure is similar for the other two equations.

Finally then we have the series expansion for the solution which we can express in the form

$$x(y) = \sum_{k=0}^{\infty} \frac{\tilde{x}_k(y)}{k!}(EI_2 - EI_1)^k, \quad 0 < y < \ell, \tag{4.89}$$

where

$$\tilde{x}_k(y) = \frac{x_k(y)}{(EI_2 - EI_1)^k},$$

$$x(y) = \begin{pmatrix} h\ (0, y) \\ v\ (0, y) \\ \theta\ (0, y) \end{pmatrix}$$

is then the static solution to the problem (CF end conditions)

$$x_k(y) = \int_0^\ell G(y,s)x_{k-1}(s)ds, 0 < y < \ell,$$

$$G(y, s) = \begin{pmatrix} G_h\ (y, s) \\ G_v\ (y, s) \\ G_\theta\ (y, s) \end{pmatrix}.$$

Let us calculate $x_1(.)$. We have

$$EI_1 h_1(y)'''' = 0,$$

$$EI_2 v_1''''(y) = 0,$$

$$GJ\theta_1''(y) + \mu\theta_1(y) = (EI_2 - EI_1)v(0, y)''h''(0, y)$$

$$= (EI_2 - EI_1)\left(m^2 g^2 \cos \varphi \, \sin \varphi \left(\frac{y2}{4} - 2\ell y + (6\ell - 3\ell^2)\right)^2\right)$$

$$= \gamma(y)$$

and

$$\theta_1'(y) = \theta_k'(s) = \left(\frac{-1}{\cos \omega \ell} \int_0^\ell \cos \omega(\ell - \sigma)\gamma(\sigma)d\sigma\right) + \int_0^s \frac{\sin \omega(s - \sigma)}{\omega}\gamma(\sigma)d\sigma.$$

And

$$h_1(y) = h(0, y); \quad v_1(y) = v(0, y),$$

which gives some idea about the nature of the solution.

Linearization

Having evaluated the static solution, we now go on to linearize the solution to the aeroelastic equations about this solution for U nonzero so that g has to be nonzero. We anticipate that it is more complex than for the Goland model. Thus let

$$h(\lambda, t, y) = h(0, y) + \lambda h_1(t, y),$$

$$v(\lambda, t, y) = v(0, y) + \lambda v_1(t, y),$$

$$\theta(\lambda, t, y) = \theta(0, y) + \lambda \theta_1(t, y),$$

$$\phi(\lambda, t, x, z) = \phi(0, x, z) + \lambda \phi_1(t, x, z),$$

$$\frac{\partial \phi(\lambda, t, x, 0)}{\partial z} = \frac{\partial \phi(0, x, 0+)}{\partial z} + \lambda \frac{\partial \phi_1(t, x, 0+)}{\partial z},$$

where

$$\begin{pmatrix} h\ (0, y) \\ v\ (0, y) \\ \theta\ (0, y) \end{pmatrix} = x(y),$$

$$L(\lambda, t, y) = L(y) + \lambda L_1(t, y),$$

$$M(\lambda, t, y) = M(y) + \lambda M_1(t, y),$$

$$m\ddot{h}(\lambda, t, y) + EI_1 h'''(\lambda, t, y) + (EI_2 - EI_1)(\theta(\lambda, t, y)v(\lambda, t, y)'')''$$

$$= m\, g \sin \varphi + L(\lambda, t, y), \tag{4.90}$$

$$m\ddot{v}(\lambda, t, y) + EI_2 v''''(\lambda, t, y) + (EI_2 - EI_1)(\theta(\lambda, t, y)h(\lambda, t, y)'')'' = mg \cos \varphi \tag{4.91}$$

$$I\ddot{\theta}(\lambda, t, y) - GJ^{\cdot\prime\prime}(\lambda, t, y) + (EI_2 - EI_1)h(\lambda, t, y)''v(\lambda, t, y)''$$
$$= M(\lambda, t, y), \quad 0 < t; \quad 0 < y < \ell. \tag{4.92}$$

Taking derivative with respect to λ at zero, we have:

$$m\ddot{h}_1(t, y) + EI_1 h_1''''(t, y) + (EI_2 - EI_1)(\theta_1(t, y)v(0, y)''$$
$$+ \theta(0, y)v_1(t, y)'')'' = L_1(t, y), \tag{4.93}$$

$$m\ddot{v}_1(t, y) + EI_1 v_1''''(t, y) + (EI_2 - EI_1)(\theta_1(t, y)h(0, y)''$$
$$+ \theta(0, y)h_1(t, y)'')'' = 0, \tag{4.94}$$

$$I\ddot{\theta}_1(t, y) - GJ\theta_1''(t, y) + (EI_2 - EI_1)(h_1(t, y)''v(0, y)''$$
$$+ h(0, y)''v_1(t, y)) = M_1(t, y), \tag{4.95}$$

where the linearized lift and moment

$$L_1(t, y) = \frac{\mathrm{d}}{\mathrm{d}\lambda} L(0, t, y),$$

$$M_1(t, y) = \frac{\mathrm{d}}{\mathrm{d}\lambda} M(0, t, y).$$

This is the linearized equation that will determine system stability and is now more complex than the Goland case because the coefficients in the equation are no longer constants. It also then requires an existence and uniqueness proof. We consider this as part of the convolution/evolution equation in Chap. 5.

Notes and Comments

The crucial result used here is the explicit solution to the finite Hilbert transform given by Tricomi [11] who notes also the work of Sohngen, the German pioneer in aeroelasticity theory, and shows the Kutta condition as necessary for uniqueness of solution. There is also a large body of related Russian work (mainly that of Gohberg; see [31]) which seems to ignore the Tricomi work.

Note that the aerodynamics is replaced by the Hilbert transform, which we look upon as the static case of the Possio integral equation. A feature of our entire theory is the systematic use of the Possio equation. As here and more generally, the best-developed results are for the typical section case where the air flow is 2D and the Possio equation is 1D. The angle of attack becomes important in that the functionals turn out to be discontinuous at zero for large M. The point here is that for large enough M there may be regions in the interior for which the local flow is supersonic, in which case the flow is termed transonic. The precise value of M for which this happens is not clear. Meyer [14, p 146] gives an inequality for the interior Mach number. One author [48] suggests $M \geq 0.8$ as transonic.

We have in this chapter a simple instance of nonunique solutions for the static solution for a sequence of values of the far field speed determined by a nonlinear eigenvalue problem. We could argue that we won't see this nonuniqueness in time marching because we will not have the exact values for the sequence. In any case it does raise a question for CFD solutions. Our main result concerning the air flow would appear to be new: that the potential can be expressed as the sum of two, one which produces lift and the other that does not, consistent with our theme that we are concerned only with the structure dynamics whereas the second part holds little interest for us.

Chapter 5
Linear Aeroelasticity Theory/
The Possio Integral Equation

5.1 Introduction

In this chapter we present the linear theory which plays an essential role in the nonlinear Hopf bifurcation stability theory to come in Chap. 6. This is the most studied part of the theory beginning with the pioneers: Kussner, Sohngen, Wagner, Garrick, and Theodorsen; see the classic treatise [6]. This is also the longest chapter in the book and covers the theory buttressing a wide variety of topics loosely referred to as "Flutter Analysis": the every day "bread-and-butter" part of the subject.

5.2 Power Series Expansion of the 3D Potential

We begin with a power series expansion of the solution of (3.15) with the associated boundary dynamics about the rest or equilibrium aeroelastic state solution (Chap. 4):

$$\phi_0(t, x, y, z) = U_\infty(x \cos \alpha + z \sin \alpha),$$

$$h(t, y) = 0; \ \theta(t, y) = 0.$$

We call this an "Attractor" and note that it is unique in that it holds for every value of U_∞ unlike the Divergence speed (Chap. 4).

To determine the Linear potential field equation, we consider the structure variables as parameters and expand the solution in a power series about zero. Thus with λ denoting a complex variable we consider the solution $\phi(\lambda, t, x, y, z)$ corresponding to $\lambda h(t, y), \lambda \theta(t, y)$:

$$\phi(\lambda, t, x, y, z) = \phi_0(t, x, y, z) + \sum_{k=1}^{\infty} \frac{\lambda^k}{k!} \phi_k(t, x, y, z), \tag{5.1}$$

which satisfies the field equation (3.15).

A V Balakrishnan, *Aeroelasticity: The Continuum Theory*,
DOI 10.1007/978-1-4614-3609-6_5, © Springer Science+Business Media, LLC 2012

Leaving for later the sense in which the convergence is taken, as in the steady state case in Chap. 4, we first calculate the kth order potentials ϕ_k given by

$$\phi_k(t, x, y, z) = \left(\partial^k \phi(0, t, x, y, z)\right) / \partial \lambda^k.$$

Differentiating (3.15) with respect to λ we get the field equation for each k:

$$\Xi(\phi_k) = g_{k-1}, \tag{5.2}$$

a linear nonhomogeneous equation, where

$$\Xi(\phi) = \frac{\partial^2 \phi}{\partial t^2} + 2U_\infty \left(\mathrm{Cos}\,\alpha \frac{\partial^2 \phi}{\partial t \partial x} + \mathrm{Sin}\,\mathrm{ff}\frac{\partial^2 \phi}{\partial t \partial z}\right)$$

$$- a_\infty^2 \left[\left(1 - M^2 \mathrm{Cos}^2 \alpha\right) \frac{\partial^2 \phi}{\partial x^2} + \left(1 - M^2 \mathrm{Sin}^2 \alpha\right) \frac{\partial^2 \phi}{\partial z^2}\right.$$

$$\left. + \frac{\partial^2 \phi}{\partial y^2} - 2M^2 \mathrm{Cos}\,\alpha\,\mathrm{Sin}\,\alpha \frac{\partial^2 \phi}{\partial x \partial z}\right] \tag{5.3}$$

and g_k involves the potentials up to the kth order, and $g_0 = 0$. We begin with ϕ_1 the Linear potential. The boundary dynamics are determined from (3.19): Flow Tangency:

$$(\partial \phi(\lambda, t, x, y, 0\pm))/\partial z = \frac{\partial \phi_0}{\partial z} + \lambda(-1)\left[\dot{h}(t, y) + (x - a)\dot{\theta}(t, y)\right.$$

$$+ \left(((\partial \phi(\lambda, t, x, y, 0\pm))/\partial x)\theta(t, y)\right)$$

$$+ \left((\partial \phi(\lambda, t, x, y, 0\pm))/\partial y\right)$$

$$\left. \times \left(\frac{\partial}{\partial y}(h(t, y) + (x - a)\theta(t, y))\right)\right], \quad x, y \in \Gamma. \tag{5.4}$$

Hence taking the derivative with respect to λ at $\lambda = 0$, we obtain:

$$(\partial \phi_1(t, x, y, 0\pm))/\partial z = (-1)\left[\dot{h}(t, y) + (x - a)\dot{\theta}(t, y)\right.$$

$$\left. + U_\infty \mathrm{Cos}\,\alpha\,\theta(t, y)\right], \quad x, y \in \Gamma. \tag{5.5}$$

Let us denote the right-hand side by $w_a(t, x, y)$—this is referred to as the "downwash" in the aeroelasticity parlance.

In particular we see that the normal derivative of the linear potential is continuous across the wing. Also no spanwise derivatives are involved.

The pressure jump

$$\delta p(\lambda, t, x, y) = -\rho_\infty \delta \psi(\lambda, t, x, y),$$

where

$$\psi(\lambda, t, x, y, z) = (\partial \phi(\lambda, t, x, y, z))/\partial t + \frac{1}{2} \left| \nabla \phi(\lambda, t, x, y, z) \right|^2. \tag{5.6}$$

Hence in the power series expansion for $\psi(\lambda, t, x, y, z)$:

$$\psi(\lambda, t, x, y, z) = \frac{1}{2} U_\infty^2 + \sum_{k=1}^{\infty} \frac{\lambda^k}{k!} \psi_k(t, x, y, z), \tag{5.7}$$

we need to calculate

$$\left(\partial^k \psi(0, t, x, y, z) \right)/\partial \lambda^k = \psi_k(t, x, y, z).$$

For $k = 1$, this yields

$$\psi_1(x, y, z) = (\partial \phi_1(t, x, y, z))/\partial t + U_\infty \mathrm{Cos}\, \alpha (\partial \phi_1(t, x, y, z)/\partial x)$$
$$+ U_\infty \mathrm{Sin}\, \alpha (\partial \phi_1(t, x, y, z)/\partial z).$$

Hence defining (Kussner Doublet Function)

$$A_1(t, x, y) = -\delta \psi_1(t, x, y)/U_\infty,$$

we have:

$$A_1(t, x, y) = \frac{\partial}{\partial t} \delta \phi_1(t, x, y) + U_\infty \mathrm{Cos}\, \alpha \frac{\partial}{\partial x} \delta \phi_1(t, x, y). \tag{5.8}$$

And the Kutta–Joukowsky conditions are:

$$A_1(t, x, y) = 0 \quad \text{for } x, y \text{ NOT in } \Gamma, \tag{5.9}$$

$$A_1(t, b-, y) = 0, \quad 0 < y < \ell, \tag{5.10}$$

which are then the boundary conditions for the linear equation (5.3), along with (5.5). Thus we can state the

Linear Aeroelasticity Problem: 3D Flow/Finite Plane

Field Equation

$$
\frac{\partial^2 \phi}{\partial t^2} + 2U_\infty \left(\text{Cos}\,\alpha \frac{\partial^2 \phi}{\partial t\,\partial x} + \text{Sin}\,\alpha \frac{\partial^2 \phi}{\partial t\,\partial z} \right) - a_\infty^2 \left[\left(1 - M^2\text{Cos}^2\alpha \right) \frac{\partial^2 \phi}{\partial x^2} \right.
$$

$$
\left. + \left(1 - M^2\text{Sin}^2\alpha \right) \frac{\partial^2 \phi}{\partial z^2} + \frac{\partial^2 \phi}{\partial y^2} - 2M^2\text{Cos}\,\alpha\,\text{Sin}\,\alpha \frac{\partial^2 \phi}{\partial x\,\partial z} \right] = 0.
$$

Boundary Conditions: Far field

$$
\phi(t, \infty) = U_\infty(x\,\text{Cos}\,\alpha + z\,\text{Sin}\,\alpha).
$$

Flow Tangency:

$$
(\partial \phi_1(t, x, y, 0\pm))/\partial z
$$

$$
= (-1)\left[\dot{h}(t, y) + (x - a)\dot{\theta}(t, y) + U_\infty \text{Cosff}\,\theta(t, y) \right], \quad x, y \in \Gamma.
$$

Kutta–Joukowsky:

$$
\psi(x, y, z) = (\partial \phi_1(t, x, y, z))/\partial t + U_\infty \text{Cos}\,\alpha(\partial \phi_1(t, x, y, z))/\partial x,
$$

$$
+ U_\infty \text{Sin}\,\alpha(\partial \phi_1(t, x, y, z))/\partial z,
$$

$$
A(t, x, y) = -\delta\psi(t, x, y)/U_\infty,
$$

$$
A(t, x, y) = 0 \text{ for } x, y \text{ NOT in } \Gamma,
$$

$$
A(t, b-, y) = 0, \quad 0 < y < \ell. \tag{5.11}
$$

Our technique of solution is to convert it to an integral equation: The Possio integral equation. This is not too surprising because basically we have a Neumann-type boundary value problem which is usually treated this way (cf [1]).

5.3   The Linear Possio Integral Equation

See [55, 56, 72, 96]. The Possio Integral equation plays a central role in our theory and we shall revisit it in many places in many forms, in both Laplace Domain and Time Domain.

The Input Output Relation

We may look at what we need from the air flow model as an "input–output" problem where the downwash is the "Input" and the pressure jump is the "Output." The "Input–Output Relation" is then the Possio Integral Equation.

The Connecting Glue: Lagrangian to Eulerian

We may also consider it as providing the link between Lagrangian Dynamics and the Fluid Eulerian Dynamics. This gluing is usually the most "mysterious" part in computational aeroelasticity.

We take the Laplace Transform of (5.3) in the time domain and the $L_p - L_q$ Fourier Transforms, $1 < p < 2$, in the spatial x, y coordinates. We use super^ to denote Laplace Transform and super $\wedge$ to denote the Fourier Transform:

$$\tilde{\hat{\phi}}(\lambda, \omega_1, \omega_2, z) = \int_{-\infty}^{\infty} \int_{-\infty}^{\infty} \int_{0}^{\infty} e^{-\lambda t} e^{-ix\omega_1} e^{-iy\omega_2} \phi(t, x, y, z) dx\, dy\, dt$$

$$\text{Re}\,\lambda > \sigma_a \geq 0; \quad \omega_1,\, \omega_2\, \varepsilon\, R^2$$

with a similar definition for other functions, such as $w_a(.), \psi(.)$.

It is convenient from now on to use the "reduced" frequency

$$k = \frac{\lambda b}{U_\infty}$$

$$= \frac{\lambda}{U_\infty} \text{ for } b = 1.$$

Then taking the Laplace Transform of the field equation (5.3), we obtain:

$$\hat{\phi}(\lambda, x, y, z) = \int_{0}^{\infty} e^{-\lambda t} \phi(t, x, y, z) dt; \quad \phi(0, x, y, z) = 0; \quad \frac{\partial \phi}{\partial t}(0, x, y, z) = 0$$

($\phi(t, x, y, z)$ may be called the "disturbance potential") we have:

$$0 = \lambda^2 \hat{\phi} + 2\lambda U_\infty \left(\text{Cos}\,\alpha \frac{\partial \hat{\phi}}{\partial x} + \text{Sin}\,\alpha \frac{\partial \hat{\phi}}{\partial z} \right) - a_\infty^2 \left[(1 - M^2\text{Cos}^2\alpha) \frac{\partial^2 \hat{\phi}}{\partial x^2} \right.$$

$$\left. + (1 - M^2\text{Sin}^2\alpha) \frac{\partial^2 \hat{\phi}}{\partial z^2} + \frac{\partial^2 \hat{\phi}}{\partial y^2} - 2M^2 \text{Cos}\,\alpha\,\text{Sin}\,\alpha \frac{\partial^2 \hat{\phi}}{\partial x \partial z} \right]. \tag{5.12}$$

Using $\varphi(\lambda, \omega_1, \omega_2, z)$ in place of $\tilde{\hat{\phi}}(\lambda, \omega_1, \omega_2, z)$, we have, taking Fourier Transforms in (5.12) and dividing by a_∞^2:

$$\left(1 - M^2 \mathrm{Sin}^2 \alpha\right) \frac{\partial^2 \varphi}{\partial z^2} - 2\left(M^2 i \omega_1 \mathrm{Cos}\,\alpha\, \mathrm{Sin}\,\alpha + M^2 k\, \mathrm{Sin}\,\alpha\right) \frac{\partial \varphi}{\partial z}$$

$$- \left(M^2 k^2 + 2M^2 k\, \mathrm{Cos}\,\alpha\, i \omega_1 + \omega_1^2 (1 - M^2 \mathrm{Cos}^2 \alpha) + \omega_2^2\right) \varphi = 0. \qquad (5.13)$$

We note that

$$\varphi = c\, e^{rz},$$

where c is not a function of z, is a solution of this equation provided r satisfies the algebraic equation

$$ar^2 - 2br - c = 0,$$

where

$$a = (1 - M^2 \mathrm{Sin}^2 \alpha),$$

$$b = M^2(k + i \omega_1 \mathrm{Cos}\,\alpha) \mathrm{Sin}\,\alpha,$$

$$c = k^2 M^2 + 2M^2 i \omega_1 \mathrm{Cos}\,\alpha + \omega_1^2 (1 - M^2 \mathrm{Cos}^2 \alpha) + \omega_2^2.$$

The equation has exactly two roots:

$$r_1 = \left(\frac{1}{(1 - M^2 \mathrm{Sin}^2 \alpha)}\right) \left(M^2(k + i \omega_1 \mathrm{Cos}\,\alpha) \mathrm{Sin}\,\alpha\right.$$

$$\left. - \sqrt{\left(k^2 M^2 + 2k M^2 i \omega_1 \mathrm{Cos}\,\alpha + \omega_1^2 (1 - M^2) + \omega_2^2 (1 - M^2 \mathrm{Sin}^2 \alpha)\right)}\right),$$

$$(5.14)$$

$$r_2 = \left(\frac{1}{1 - M^2 \mathrm{Sin}^2 \alpha}\right) \left(M^2(k + i \omega_1 \mathrm{Cos}\,\mathrm{ff}) \mathrm{Sin}\,\alpha\right.$$

$$\left. + \sqrt{\left(k^2 M^2 + 2k M^2 i \omega_1 \mathrm{Cos}\,\alpha + \omega_1^2 (1 - M^2) + \omega_2^2 (1 - M^2 \mathrm{Sin}^2 \alpha)\right)}\right).$$

$$(5.15)$$

Before we proceed further we need to clarify the definition of the square root in (5.14) and (5.15).

Lemma 5.1.

$$k^2 M^2 + 2k M^2 i \omega_1 \mathrm{Cos}\,\alpha + \omega_1^2 (1 - M^2) + \omega_2^2 (1 - M^2 \mathrm{Sin}^2 \alpha)$$

is never equal to a negative number for any complex number k which is not pure imaginary.

Proof. If $M = 0$ or $k = 0$

$$k^2 M^2 + 2k M^2 i\omega_1 \text{Cos}\,\alpha + \omega_1^2(1 - M^2) + \omega_2^2(1 - M^2\text{Sin}^2\alpha)$$

$$= \omega_1^2(1 - M^2) + \omega_2^2(1 - M^2\text{Sin}^2\alpha) \geq 0.$$

Next let us consider $M \neq 0$. Let $k = \sigma + i\gamma$. We can calculate that the Im part $= 2M^2\sigma(\gamma + \omega_1\text{Cos}\,\alpha)$ and the real part $= (\sigma^2 - \gamma^2)M^2 - 2M^2\gamma\omega_1\text{Cos}\,\alpha + \omega_1^2(1 - M^2) + \omega_2^2(1 - M^2\text{Sin}^2\alpha)$.
We need to consider the case

$$\sigma(\gamma + \omega_1\text{Cos}\,\alpha) = 0,$$

and σ is not zero. But if $(\gamma + \omega_1\text{Cos}\,\alpha) = 0$, then the real part is

$$\sigma^2 M^2 + M^2\omega_1^2\text{Cos}^2\alpha + \omega_1^2(1 - M^2) + \omega_2^2(1 - M^2\text{Sin}^2\alpha) \geq 0. \qquad \square$$

Hence the usual principal value of the square root will be taken throughout.

As noted in Chap. 4, we can again show that the only solution that vanishes as $|z|$ goes to zero is given by:

$$\tilde{\hat{\phi}}(\lambda, i\omega_1, i\omega_2, z) = A_+ e^{r_1}, \quad z > 0,$$

$$= \frac{\hat{v}}{r_1} e^{r_1 z}, \quad z > 0,$$

$$= A_- e^{r_2 z}, \quad z < 0,$$

$$= \frac{\hat{v}}{r_2} e^{r_2 z}, \quad z < 0,$$

where we define

$$\hat{v} = \frac{\partial \tilde{\hat{\phi}}}{\partial z}\Big|z = 0,$$

which as we have noted is continuous at $z = 0$.

Now by (5.8)

$$\tilde{\hat{A}}(\lambda, i\omega_1, i\omega_2) = (\lambda + U_\infty i\omega_1\text{Cos}\,\alpha)\,\hat{v}\,(\lambda, i\omega_1, i\omega_2)\left(\frac{1}{r_1} - \frac{1}{r_2}\right). \qquad (5.16)$$

Hence

$$\hat{v}(\lambda, i\omega_1, i\omega_2) = \tilde{\hat{A}}(\lambda, i\omega_1, i\omega_2)\left(\frac{r_1 r_2}{r_2 - r_1}\right)\frac{1}{(\lambda + U_\infty i\omega_1\text{Cos}\,\alpha)}, \qquad (5.17)$$

where

$$\left(\frac{r_1 r_2}{r_2 - r_1}\right) = \frac{1}{2} \frac{1}{(\lambda + U_\infty i \omega_1 \text{Cos}\, \alpha)} \left(k^2 M^2 + 2k M^2 i \omega_1 \text{Cos}\, \alpha \right.$$

$$\left. + \omega_1^2(1 - M^2 \text{Cos}^2\alpha) + \omega_2^2\right) \Big/ \left(\sqrt{} \left(k^2 M^2 + 2k M^2 i \omega_1 \text{Cos}\, \alpha \right.\right.$$

$$\left.\left. + \omega_1^2(1 - M^2) + (1 - M^2 \text{Sin}^2\alpha)\omega_2^2\right)\right). \tag{5.18}$$

Let $\hat{P}(\lambda, x, y)$ denote the inverse Fourier Transform of

$$= \frac{1}{2} \frac{1}{(\lambda + U_\infty i \omega_1 \text{Cos}\, \alpha)} \left(k^2 M^2 + 2k M^2 i \omega_1 \text{Cos}\, \alpha \right.$$

$$\left. + \omega_1^2(1 - M^2 \text{Cos}^2\alpha) + \omega_2^2\right) \Big/ \left(\sqrt{} \left(k^2 M^2 + 2k M^2 i \omega_1 \text{Cos}\, \alpha \right.\right.$$

$$\left.\left. + \omega_1^2(1 - M^2) + (1 - M^2 \text{Sin}^2\alpha)\omega_2^2\right)\right).$$

Then taking inverse Fourier Transforms in (5.18) we obtain

$$\hat{v}(\lambda, x, y) = \int_{-b}^{b} \int_{0}^{\ell} \hat{P}(\lambda, x - \xi, y - \eta) \hat{A}(\lambda, \xi, \eta) d\xi d\eta, \quad -\infty < x, y < \infty$$

with the Kutta condition

$$\hat{A}(\lambda, \xi, \eta) \to 0 \text{ as } \xi \to b-,$$

where the kernel is specified in terms of its $L_p - L_q$ spatial double Fourier Transform

$$\int_{-\infty}^{\infty} \int_{-\infty}^{\infty} \hat{P}(\lambda, x, y) e^{-ix\omega_1 - iy\omega_2} dx dy = \frac{1}{2} \frac{1}{(\lambda + U_\infty i \omega_1 \text{Cos}\, \alpha)}$$

$$\times \left(k^2 M^2 + 2k M^2 i \omega_1 \text{Cos}\, \alpha + \omega_1^2 \left(1 - M^2 \text{Cos}^2\alpha\right) + \omega_2^2\right) \Big/ \left(\sqrt{} \left(k^2 M^2 \right.\right.$$

$$\left.\left. + 2k M^2 i \omega_1 \text{Cos}\, \alpha + \omega_1^2(1 - M^2) + (1 - M^2 \text{Sin}^2\alpha)\omega_2^2\right)\right), \tag{5.19}$$

which we denote by $\tilde{P}(\lambda, i\omega_1, i\omega_2)$.

In particular, specializing to Γ, and this is Possio's simple yet profound idea, we get the integral equation named after him:

The Linear Possio Integral Equation

$$\hat{v}(\lambda, x, y) = \int_{-b}^{b} \int_{0}^{\ell} \hat{P}(\lambda, x - \xi, y - \eta) \hat{A}(\lambda, \xi, \eta) d\xi \, d\eta, \ x, y \in \Gamma, \qquad (5.20)$$

where

$$\hat{A}(\lambda, \xi, \eta) \to 0 \text{ as } \xi \to b-,$$

and where

$$\int_{-\infty}^{\infty} \int_{-\infty}^{\infty} \hat{P}(\lambda, x, y) e^{-ix\omega_1 - iy\omega_2} dx \, dy = \frac{1}{2} \left(\frac{1}{\lambda + U_\infty i \omega_1 \text{Cos ff}} \right)$$

$$\left(k^2 M^2 + 2k M^2 i \omega_1 \text{Cos} \, \alpha + \omega_1^2 (1 - M^2 \text{Cos}^2 \alpha) + \omega_2^2 \right) \Big/ \left(\sqrt{\left(k^2 M^2 \right.} \right.$$

$$\left. \left. + 2k M^2 i \omega_1 \text{Cos} \, \alpha + \omega_1^2 (1 - M^2) + (1 - M^2 \text{Sin}^2 \alpha) \omega_2^2 \right) \right).$$

Or, in other words this is enough to determine $\hat{A}(\lambda, ., .)$.

In particular we note that if the Linear Aeroelastic problem has a solution, then the Laplace Transform of the Kussner Doublet Function defined in (5.8) will satisfy the Possio Equation (5.20).

And conversely, if the Possio equation has a solution, then the Laplace/ Fourier Transform of the solution of the potential equation (5.13) is given by:

$$\tilde{\phi}(\lambda, i\omega_1, i\omega_2, z) = \tilde{P}(\lambda, i\omega_1, i\omega_2) \tilde{\hat{A}}(\lambda, i\omega_1, i\omega_2) \frac{1}{r_1} e^{r_1 z} \quad z > 0, \qquad (5.21)$$

$$\tilde{P}(\lambda, i\omega_1, i\omega_2) \tilde{\hat{A}}(\lambda, i\omega_1, i\omega_2) \frac{1}{r_2} e^{r_2 z} \quad z < 0. \qquad (5.22)$$

Of course we need next to determine the solution as a function of the time/space coordinates that we started with. The Laplace/Fourier Transformation has to be inverted. But the Transforms by themselves—the Laplace Transforms in particular— are of independent interest, specifically in studying system stability assuming the Transform is invertible!—that it is the Transform of a time domain function which satisfies (5.3).

Existence and Uniqueness of Solution

The existence and uniqueness of the Possio equation (5.20) is of primary concern to us and we shall need to spread this over many stages.

First let us note that if the LDP–Laplace Domain Possio equation (5.20)–has two solutions and thus also two time-domain solutions of (5.3), then the difference will satisfy the equation with

$$w_a(t, x, y) = 0.$$

But then the difference will also yield a nonzero solution of the potential flow equation with zero pressure jump across the wing. Hence existence and uniqueness of solution of the Linear Euler equation with the Kutta–Joukowsky boundary condition is equivalent to the existence and uniqueness of solution to the LDP Equation with the extra condition that the solution is a Laplace Transform:

$$\hat{A}(\lambda, x, y) = \int_0^\infty e^{-\lambda t} A(t, x, y) dt \quad \text{Re.} \lambda > \sigma_a.$$

The Typical Section Case

Let us specialize now to the "Typical Section" or "Airfoil" case, where we set the y-coordinate to zero simplifying to 2D aerodynamics—which is equivalent to setting ω_2 to zero in the Fourier Transforms. Thus the 1D Possio Equation becomes:

$$\hat{w}_a(\lambda, x) = \int_{-b}^{b} \hat{P}(\lambda, x - \xi) \hat{A}(\lambda, \xi) d\xi, \quad |x| < b, \tag{5.23}$$

with the Kutta condition:

$$\hat{A}(\lambda, b-) = 0, \tag{5.24}$$

which may be weakened to $\hat{A}(\lambda, x)$ goes to a finite limit

$$\text{as } x \text{ goes to } b-, \tag{5.25}$$

where the $L_p - L_q$ Fourier Transform of the kernel is given by:

$$\tilde{P}(\lambda, i\omega) = \frac{1}{2} \frac{1}{(k + i\omega \text{Cos}\,\alpha)} \frac{\left(M^2 k^2 + 2k M^2 i\omega \text{Cos}\,\alpha + \omega^2 (1 - M^2 \text{Cos}\,\alpha) \right)}{\left(\sqrt{(M^2 k^2 + 2k M^2 i\omega \text{Cos}\,\alpha + \omega^2 (1 - M^2))} \right)},$$

$$-\infty < \omega < \infty. \tag{5.26}$$

And an immediate observation from (5.23) is that if $\hat{A}(\lambda, .)$ is the solution corresponding to $\hat{w}(\lambda, .)$, then $f(\lambda)\hat{A}(\lambda, .)$ is the solution corresponding to $f(\lambda)\hat{w}(\lambda, .)$.

In particular this implies that the solution has as many time derivatives as $w_a(t, .)$ has.

Typical Section/Zero Angle of Attack

If we set in addition

$$\alpha = 0$$

the kernel P becomes

$$\tilde{P}(\lambda, i\omega) = \frac{1}{2}\frac{1}{k+i\omega}\sqrt{\left(M^2k^2 + 2kM^2i\omega + \omega^2(1-M^2)\right)}. \qquad (5.27)$$

The next result is crucial and is the point of departure from the extant aeroelasticity literature.

Balakrishnan Formula

Theorem 5.2. *(Balakrishnan 2003)*
Let $\alpha = 0$.
Then in (5.27) $\tilde{P}(\lambda, \omega)$ can be expressed as

$$\tilde{P}(\lambda, \omega) = \frac{1}{2}\frac{|\omega|}{i\omega}\sqrt{1-M^2}\left(1 + \hat{\bar{B}}(k, i\omega)\right), \qquad (5.28)$$

where $\hat{\bar{B}}(k, i\omega)$ is the Fourier Transform

$$\hat{\bar{B}}(k, i\omega) = \int_{-\infty}^{\infty} \hat{B}(k, x)e^{-i\omega x}dx, \quad -\infty < \omega < \infty,$$

where

$$\hat{B}(k, x) = -k\frac{e^{-kx}}{\sqrt{1-M^2}} - k\int_0^{\alpha_1} e^{-kxs}a(M, s)ds \quad x > 0,$$

$$k\int_0^{\alpha_2} e^{kxs}a(M, -s)ds, \quad x < 0,$$

where

$$\alpha_1 = \frac{M}{1+M}; \; \alpha_2 = \frac{M}{1-M};$$

$$a(M, s) = \frac{1}{\pi}\left(\sqrt{((\alpha_1 - s)(\alpha_2 + s))}\right)/(1-s), \quad -\alpha_2 < s < \alpha_1,$$

and $\hat{B}(k, .)$ is in $L_p(R^1)$ for $p > 1$.

Proof. First let us prove that $\hat{B}(k, .)$ is in $L_p(R^1)$ for $p > 1$.

We note that $|a(M, s)|$ is bounded in $-\alpha_2 \leq s \leq \alpha_1$ for each M. Let $c(M)$ denote this bound. Since we have:

$$\hat{B}(k, x) = k\,\hat{B}(1, kx) \tag{5.29}$$

it is enough to consider $k = 1$. Then

$$\left| \int_0^{\alpha_1} e^{-xs} a(M, s) ds \right| \leq c(M) \frac{1 - e^{-x\alpha_1}}{x}, \quad x > 0$$

$$\left| \int_0^{\alpha_2} e^{xs} a(M, -s) ds \right| \leq c(M) \left(\left(1 - e^{-|x|\alpha_2} \right) \big/ (|x|) \right) \quad x < 0$$

and

$$\int_0^\infty \left(\frac{1 - e^{-x\alpha_1}}{x} \right)^P dx \;<\; \infty \text{ for } p > 1.$$

The representation (5.28) was discovered by the author [56] by contour integration (see [71]) but proved by direct integration, which we follow here.

First let us note that by (5.33) we may take

$$k = 1.$$

Let

$$-\alpha_1 < \text{Re. } z < \alpha_2 \quad z \neq 0.$$

Then

$$\int_{-\infty}^\infty e^{-zx} \hat{B}(1, x) dx = \frac{-1}{\sqrt{1 - M^2}} \frac{1}{1 + z} - \frac{1}{\pi} \int_0^{\alpha_1} \frac{1}{s + z} \left(\sqrt{((\alpha_1 - s)} \right.$$

$$\left. \times (\alpha_2 + s)) \right) \big/ (1 - s) ds + \frac{1}{\pi} \int_0^{\alpha_2} \frac{1}{s - z} \left(\sqrt{((\alpha_2 - s)} \right.$$

$$\left. \times (\alpha_1 + s)) \right) \big/ (1 + s) ds,$$

which by changing variables in the integrals (see [4] for details) can be expressed as:

$$= \frac{z}{z + 1} \sqrt{\left(\frac{1}{z^2}(z + \alpha_1)(z - \alpha_2) \right)} - 1$$

and for

$$z = i\omega$$

this is

$$= \frac{|\omega|}{i\omega} \frac{1}{1 + i\omega} \sqrt{((i\omega + \alpha_1)(\alpha_2 - i\omega))} - 1.$$

Hence

$$\tilde{B}(k, i\omega) = \frac{|\omega|}{i\omega} \frac{1}{k + i\omega} \sqrt{((ki\omega + \alpha_1)(\alpha_2 - ki\omega))} - 1$$

$$= \frac{|\omega|}{i\omega} \frac{1}{k + i\omega} \frac{1}{\sqrt{1 - M^2}} \sqrt{(M^2k^2 + 2kM^2i\omega + \omega^2(1 - M^2))} - 1$$

showing that(5.28) =(5.29) as required.
Finally let us note that for Typical Section and zero angle of attack, (5.22) and (5.23) specialize to

$$\tilde{\tilde{\phi}}(\lambda, i\omega, z) = -\frac{1}{k + i\omega} \tilde{\tilde{A}}(\lambda, i\omega) e^{rz} \quad z > 0$$

$$= +\frac{1}{k + i\omega} \tilde{\tilde{A}}(\lambda, i\omega) e^{-rz} \quad z < 0,$$

where

$$r = -\sqrt{(k^2 M^2 + 2kM^2i\omega + \omega^2(1 - M^2))}$$

so that, in particular

$$\phi(t, x, -z) = -\phi(t, x, z),$$

$$(\partial\phi(t, x, -z))/\partial z = (\partial\phi(t, x, z))/\partial z,$$

$$(\partial\phi(t, x, -z))/\partial x = -(\partial\phi(t, x, z))/\partial z.$$

Next we specialize to the most studied case $M = 0$ See [6].

The case $M = 0$ corresponds to "Incompressible Flow" for reasons explained below. The importance for us is that we can obtain an explicit solution for the Possio equation and a systematic use of which yields an alternate theory as compared with the classical theory in [6]. $M = 0$ is an "outlier" in that it is atypical in comparison to the solution for nonzero M.   □

Incompressible Flow

First we have:

Theorem 5.3. *For each α and $M, 0 \leq M \leq 1$, (5.27) defines an $L_p - L_q$ Mikhlin multiplier on $L_p(R^1)$ and hence a linear bounded operator on $L_p(R^1)$ into itself. Let $\Phi(\alpha, M, k)$ denote the operator. It is convenient to regard $L_p(-b, b)$ as a subspace of $L_p(-b, b)$ Let $\mathcal{P}$ denote the projection operator on $L_p(R^1)$ into $L_p(-b, b)$. Then the Possio Equation can be expressed in operator form as an equation in $L_p(-b, b)$, $1 \leq p < 2$:*

$$\hat{w}_a(\lambda, .) = \mathcal{P}\Phi(\alpha, M, k)\hat{A}(\lambda, .) \tag{5.30}$$

with the Kutta condition

$$\hat{A}(\lambda, b-) = 0.$$

Proof. By definition (see [28, 41, 56]), a scalar valued function $\mu(\cdot)$ defined on R^1 is called a "Mikhlin multiplier" if $\tilde{g}(\omega) = \mu(\omega)\tilde{f}(\omega)$, $-\infty < \omega < \infty$, excepting $\omega = 0$ is the Fourier Transform of a function in $L_p(R^1)$ whenever $\tilde{f}(.)$ is. A sufficient condition for this due to Mikhlin [1] is that $\mu(.)$ be continuously differentiable in $(-\infty, \infty)$ and

$$|\mu(\omega)| + |\omega\mu'(\omega)| \quad \leq c < \infty, \text{ omitting } \omega = 0.$$

This condition is readily verified for (5.27). The statement of the theorem follows from Mikhlin's theory [28] who shows that if T denotes the corresponding operator, then T is linear bounded with norm:

$$||T|| \leq cM_p \text{ where } M_p \text{ depends only on } p.$$

In our structure dynamics, the downwash is actually C_1 in $[-b, b]$.

That the space has to be L_p, $1 < p < 2$ can be seen by considering the special case where $k = 0$, which corresponds to the steady state case treated in Chap. 4. Let us look at some special cases where an explicit solution can be constructed.   □

Zero Angle of Attack

Note that for zero angle of attack, the kernel $\overset{\approx}{P}$ in (5.27) becomes:

$$\frac{1}{2}\sqrt{(M^2k^2 + 2kM^2i\omega + \omega^2(1 - M^2))}/(k + i\omega). \tag{5.31}$$

Recall that $\mathrm{Re}\, k > 0$ in what follows.

Theorem 5.4. *Let $\alpha = 0$ and $M = 0$ (Incompressible Case). Then the Possio Equation in this case:*

$$\hat{w}_a(\lambda, .) = \mathcal{P}\Phi(0, 0, k)\hat{A}(\lambda, .) \tag{5.32}$$

has a unique solution in L_p which we can express explicitly. Here we follow [4].

Proof. Let $\Phi(0, 0, k)$ be simplified to $\Phi(k)$. For $\alpha = 0 = M$ the multiplier is

$$\frac{1}{2}\frac{|\omega|}{k + i\omega},$$

which can be expressed as:

$$\frac{1}{2}\left(1 - \frac{k}{k + i\omega}\right)\frac{|\omega|}{i\omega}.$$

Let $\mathcal{R}(k)$ denote the operator corresponding to the multiplier

$$\frac{1}{k + i\omega}.$$

Then $\mathcal{R}(k)$ is given by:

$$\mathcal{R}(k)f = g; g(x) = \int_{-\infty}^{x} e^{-k(x-s)} f(s)ds, \quad -\infty < x < \infty. \tag{5.33}$$

Next the multiplier $\frac{|\omega|}{i\omega}$ is recognized as representing the Hilbert Transform, denoted $\mathcal{H}$, given by

$$\mathcal{H}f = g; \quad g(x) = \frac{1}{\pi}\int_{-\infty}^{\infty}\frac{f(\xi)}{x - \xi}d\xi, \quad -\infty < x < \infty. \tag{5.34}$$

And we note that:

$$\mathcal{H}\mathcal{R}(k) = \mathcal{R}(k)\mathcal{H}.$$

Then $\mathcal{P}\Phi(k)$ can be expressed:

$$\mathcal{P}\Phi(k)\mathcal{P}f = \frac{1}{2}\mathcal{P}(I - k\mathcal{R}(k))\mathcal{H}\mathcal{P}f \text{ for } f \text{ in } L_p(R^1).$$

Hence (5.33) becomes:

$$2\hat{w}_a(\lambda, .) = \mathcal{P}\mathcal{H}\mathcal{P}\hat{A}(\lambda, .) - k\mathcal{P}\mathcal{R}(k)\mathcal{H}\mathcal{P}\hat{A}(\lambda, .)$$

$$= \mathcal{P}\mathcal{H}\mathcal{P}\hat{A}(\lambda, .) - k\mathcal{P}\mathcal{H}\mathcal{R}(k)\mathcal{P}\hat{A}(\lambda, .)$$

$$= \mathcal{P}\mathcal{H}\mathcal{P}\hat{A}(\lambda, .) - k\mathcal{P}\mathcal{H}(\mathcal{P} + I - \mathcal{P})\mathcal{R}(k)\mathcal{P}\hat{A}(\lambda, .)$$

$$= \mathcal{P}\mathcal{H}\mathcal{P}(I - k\mathcal{R}(k)\mathcal{P})\hat{A}(\lambda, .) - k\mathcal{P}\mathcal{H}(I - \mathcal{P})\mathcal{R}(k)\mathcal{P}\hat{A}(\lambda, .). \tag{5.35}$$

Let us evaluate the function

$$\mathcal{P}\mathcal{H}(I - \mathcal{P})\mathcal{R}(k)\mathcal{P}\hat{A}(\lambda, .).$$

Denoting it by $q(.)$, we have:

$$q(x) = \frac{1}{\pi} \int_b^\infty \frac{1}{x-\xi} \int_{-b}^b e^{-k(\xi-s)} \hat{A}(\lambda,.) ds d\xi, \quad |x| < b$$

$$= \frac{1}{\pi} \int_0^\infty \frac{1}{x-b-\sigma} e^{-k\sigma} d\sigma \; L\left(k, \hat{A}(\lambda,.)\right),$$

where

$$L(k, f) = \int_{-b}^b e^{-k(b-s)} f(s) ds.$$

Hence we have

$$2\hat{w}(\lambda,.) = \mathcal{PHP}(I - k\mathcal{R}(k)\mathcal{P})\hat{A}(\lambda,.) - kq.$$

Hence

$$2\mathcal{T}\hat{w}(\lambda,.) = (I - k\mathcal{L}(k))\hat{A}(\lambda,.) - k\mathcal{T}q, \qquad (5.36)$$

where $\mathcal{L}(k)$ for each k is a Volterra operator on $L_p(-b, b)$ into itself, defined by:

$$\mathcal{L}(k) f = g; \quad g(x) = \int_{-b}^x e^{-k(x-s)} f(s) ds, \quad |x| < b$$

and $k\mathcal{T}q$ is the function

$$kL(k, \hat{A}(\lambda,.))h,$$

where

$$h(x) = \frac{1}{\pi} \sqrt{\frac{b-x}{b+x}} \left[\frac{1}{\pi} \int_{-b}^b \sqrt{\frac{b+s}{b-s}} \frac{1}{s-b-\sigma} \frac{1}{s-x} ds \int_0^\infty e^{-k\sigma} d\sigma \right],$$

where we can calculate that the integral

$$\frac{1}{\pi} \int_{-b}^b \sqrt{\frac{b+s}{b-s}} 1/((s-b-\sigma)(s-x)) ds = \sqrt{\left(\frac{2b+\sigma}{\sigma}\right)} \frac{1}{x-b-\sigma}. \qquad (5.37)$$

Here $\mathcal{T}$ is recognized as the "Tricomi Operator" which we treat more fully in the sequel.

Hence

$$h(x) = \frac{1}{\pi} \sqrt{\frac{b-x}{b+x}} \int_0^\infty e^{-k\sigma} \sqrt{\left(\frac{2b+\sigma}{\sigma}\right)} \frac{1}{x-b-\sigma} d\sigma \quad |x| < b. \qquad (5.38)$$

(Note that $(.)$ is NOT defined for k negative. An important point for us in the sequel, in Sect. 5.5.) Hence (5.37) becomes:

$$2\mathcal{T}\hat{w}(\lambda,.) = (I - k\mathcal{L}(k))\hat{A}(\lambda,.) - kL(\kappa, \hat{A}(\lambda,.))h. \tag{5.39}$$

$$\square$$

Lemma 5.5.
$$(I - k\mathcal{L}(k))^{-1} = (I + k\mathcal{L}(0)). \tag{5.40}$$

Proof. Either by the Volterra Expansion

$$(I - k\mathcal{L}(k))^{-1} = \sum_{n=0}^{\infty} (k\mathcal{L}(k))^n.$$

Or, "solving"

$$(I - k\mathcal{L}(k))^{-1} f = g,$$

$$f(x) = g(x) - k \int_{-b}^{x} e^{-k(x-s)} g(s) ds.$$

Hence

$$f'(x) = g'(x) - kg(x) + k(g(x) - f(x)),$$
$$g'(x) = f'(x) + kf(x).$$

Hence

$$g(x) = f(x) + k \int_{-b}^{x} f(s) ds, \text{ since } f(-b) = g(-b).$$

Hence

$$(I - k\mathcal{L}(k))^{-1} = (I + k\mathcal{L}(0)). \qquad \square$$

Hence

$$2(I + k\mathcal{L}(0))\mathcal{T}\hat{w}_a(\lambda,.) = \hat{A}(\lambda,.) - kL(k, \hat{A}(\lambda,.))(I + k\mathcal{L}(0))h \tag{5.41}$$

from which we see that all we need to evaluate is the functional:

$$L(k, \hat{A}(\lambda,.)).$$

For which we take the functional $L(k, .)$ on both sides of (5.41). $L(k, lhs) = L(k, rhs)$ which yields

$$L(k, \hat{A}(\lambda,.)) = (L(k, (I + k\mathcal{L}(0))2\mathcal{T}\hat{w}(\lambda,.)))/(1 - kL(k, (I + k\mathcal{L}(0))h)). \tag{5.42}$$

However we still need to verify that the denominator in (5.42) is not zero. Here we have a remarkable result due to Sears [110].

Sears Formula

Lemma 5.6. *(Sears)*
  For Re. $k > 0$:

$$(1 - kL(k, (I + k\mathcal{L}(0))h)) = k \int_0^\infty e^{-k\sigma} \sqrt{\left(\frac{2b + \sigma}{\sigma}\right)} d\sigma.$$

Proof. First we calculate
$$kL(k, (I + k\mathcal{L}(0))h).$$

First, integrating by parts in

$$L(k, k\mathcal{L}(0)h) = \int_{-b}^b e^{-k(b-x)} k \int_{-b}^x h(s) ds dx$$

yields

$$= \int_{-b}^b h(s) ds - L(k, h).$$

Hence

$$kL(k, (I + k\mathcal{L}(0))h)) = k \int_0^\infty e^{-k\sigma} \sqrt{\left(\frac{2b + \sigma}{\sigma}\right)} \left(\frac{1}{\pi} \int_{-b}^b \sqrt{\frac{b - x}{b + x}} \frac{1}{x - b - \sigma} dx\right) d\sigma,$$

where the integral in parenthesis

$$= \sqrt{\left(\frac{\sigma}{2b + \sigma}\right)} - 1.$$

Hence we obtain that

$$1 - kL(k, (I + k\mathcal{L}(0))h) = k \int_0^\infty e^{-k\sigma} \sqrt{\left(\frac{2b + \sigma}{\sigma}\right)} d\sigma. \qquad (5.43)$$

That it is analytic in k in the entire plane except for $k \leq 0$ where it has a logarithmic singularity—typifying the analyticity properties functions of k we shall use. This can be seen from rewriting (5.43) in the form—by a change of variable:

$$\int_0^\infty e^{-t} \sqrt{\left(\frac{2bk + t}{t}\right)} dt. \qquad (5.44)$$

Or expressing it in terms of Bessel K functions:

$$k \int_0^\infty e^{-k\sigma} \sqrt{\left(\frac{2b + \sigma}{\sigma}\right)} d\sigma = bk[K_0(bk) + K_1(bk)]e^{bk}. \qquad (5.45)$$

But now the interesting question is: what is

$$1 \bigg/ \left(k \int_0^\infty e^{-kt} \sqrt{\left(\frac{2b+t}{t}\right)} \, dt \right). \tag{5.46}$$

Is this a Laplace transform? Is (5.46) expressible as

$$\int_0^\infty c_1(t) e^{-kt} dt, \qquad c_1(.) \geq 0.$$

This is answered in the affirmative by Sears [110]—as we have seen—who actually gives an approximation to the function $c_1(.)$, which is referred to as the Sears function. This is necessary to obtain the time domain structure response. But it is easier of course to show just that (5.44) is bounded away from zero.

In the form (5.44) we see that as k goes to zero in the right-half-plane, the function has the limit equal to 1, and goes to infinity as $\text{Re}\, k$ goes to infinity. In fact we have

$$\int_0^\infty e^{-t} \sqrt{\left(\frac{2bk+t}{t}\right)} \, dt - 1 = \int_0^\infty e^{-t} \left(\sqrt{\left(\frac{2bk}{t}+1\right)} - 1 \right) dt \geq 0 \quad \text{for } k \geq 0$$

and $\text{Re}\, \sqrt{1+z} - 1 \geq 0 \quad \text{for } \text{Re}\, z \geq 0.$ □

Solution of Possio Equation in Laplace Domain: $M = 0$

Hence finally we have the solution:

$$\hat{A}(\lambda,.) = (I + k\mathcal{L}(0)) \left[2T\hat{w}_a(\lambda,.) + h \frac{(L(k,(I+k\mathcal{L}(0))2T\hat{w}_a(\lambda,.)))}{\int_0^\infty e^{-t} \sqrt{\frac{2bk+t}{t}} dt} \right].$$

Or

$$\hat{A}(\lambda,.) = (I + k\mathcal{L}(0)) \left[2T\hat{w}_a(\lambda,.) + h \frac{(L(k,(I+k\mathcal{L}(0))2T\hat{w}_a(\lambda,.)))}{bk(K_0(bk) + K_1(bk))e^{bk}} \right]. \tag{5.47}$$

To show that

$$\hat{A}(\lambda, b-) = 0,$$

we make a change of variable

$$\sigma = t^2(b-x)$$

in

$$h(x) = \frac{1}{\pi}\sqrt{\frac{b-x}{b+x}}\int_0^\infty e^{-k\sigma}\sqrt{\frac{2b+\sigma}{\sigma}}\frac{1}{x-b-\sigma}d\sigma$$

yielding

$$= \frac{1}{\pi}(-1)\sqrt{\frac{1}{b+x}}\int_0^\infty e^{-kt^2(b-x)}\sqrt{(2b+t^2(b-x))}\frac{2dt}{1+t^2},$$

and in this form it follows readily that $h(x) \to -1$ as $x \to b-$. Now from (5.34), evaluating the limit as $x \to b-$, we see that the left side goes to zero, while the right side

$$= (I - k\mathcal{L}(k))\hat{A}(\lambda,.) - kL(\kappa,\hat{A}(\lambda,.))h$$

$$0 = \hat{A}(\lambda,b-) - k\int_{-b}^b e^{-k(b-s)}\hat{A}(\lambda,s)ds + k\int_{-b}^b e^{-k(b-s)}\hat{A}(\lambda,s)ds.$$

Or $\hat{A}(\lambda,b-) = 0$.   □

Remarks: It is of interest in the sequel to determine the range of values of k beyond $\operatorname{Re} k > 0$ for which the solution continues to be valid.

Here we use the solution in the form (5.47) involving the Bessel K functions. They are defined and analytic and nonzero except for $k < 0$; and we have already covered $k = 0$. Hence the solution is valid for all k omitting the negative real axis where we have a line of singularity. What about the kernel in the equation? Here the problem is that $1/(k + i\omega)$ is no longer the Laplace transform of a function in the positive time domain for $\operatorname{Re} k$ negative. This ends the incompressible case where we have presented a closed-form solution to the Possio equation in the Laplace domain. We consider the time–domain solution later. Next we go on to consider $M > 0$ where we no longer have the luxury of a closed-form solution of the Possio equation (not as yet, of course).

5.4   Linear Possio Equation: Compressible Flow: $M > 0$

Now we go on to consider nonzero M. Unfortunately there is no longer a closed-form solution available (except for $M = 1$). Nor does the technique of solution for $M = 0$ carry over.

In fact we show that $M = 0$ is typical and may be labeled "singular." We call the flow "compressible" if the far field Mach number is bigger than 0. There is a further qualification: transonic is somewhat vague, for $M >\sim\sim 0.8$ (this depends on the Mach number of the flow in the interior; see [49]). This further distinction plays no role in our theory and we do not invoke it. Our first objective is to develop an abstract version of the equation.

Abstract Version of Possio Equation $M > 0$

We use the language of semigroup theory of operators [16, 28]. We assume throughout that

$$w_a(t, .) \text{ is in } C_1[-b, b] \text{ for every } t$$

and

$$\int_0^\infty e^{-\sigma t} \|w_a(t, .)\| dt < \infty \, \sigma > \sigma_a > 0.$$

For most if not all of our purposes we may take the abscissa σ_a to be zero. The C_1 condition is indeed true for the downwash.

Let $\mu(\omega)$ denote the $L_p - L_q$ Mikhlin multiplier

$$\mu(\omega) = \frac{1}{2} \frac{|\omega|}{i\omega} \sqrt{1 - M^2} \left(1 + \bar{\bar{B}}(k, i\omega)\right),$$

where

$$\bar{\bar{B}}(k, i\omega) = k \left(\frac{-1}{\sqrt{1 - M^2}} \frac{1}{k + i\omega} - \int_{-\alpha_2}^{\alpha_1} \frac{1}{ks + i\omega} a(M, s) ds\right).$$

Let us convert this multiplier to the corresponding operator on $L_p(-b, b)$,

$$\frac{1}{2} \sqrt{1 - M^2} \, \mathbb{H}(I + \mathbb{B}(k))$$

so that the Possio equation (5.23) in abstract form is:

$$\hat{w}_a(\lambda, .) = \frac{1}{2} \sqrt{1 - M^2} \, \mathbb{P}\mathbb{H}(I + \mathbb{B}(k))\mathbb{P}\hat{A}(\lambda, .), \tag{5.48}$$

where $\mathbb{H}$ is the Hilbert transform

$$\mathbb{H}f = g; \quad g(x) = \int_{-\infty}^{\infty} \frac{1}{\pi(x - s)} f(s) ds, \quad -\infty < x < \infty$$

and $\mathbb{B}(k)$ can be expressed:

$$\mathbb{B}(k)A = \frac{-k\mathbb{R}(k)A}{\sqrt{1 - M^2}} - \int_{-\alpha_2}^{\alpha_1} k\mathbb{R}(ks)Aa(M, s) ds, \quad A \in L_p(-b, b),$$

where $\mathbb{R}(k)$ is the bounded linear operator on $L_p(R^1)$ into itself corresponding to the multiplier

$$\frac{1}{k + i\omega},$$

and is the resolvent of the generator $\mathbb{D}$, defined for all k off the imaginary axis.

$$\mathbb{D}f = -f'$$

generating the negative shift group $\mathbb{S}(.)$ on $L_p(R^1)$:

$$\mathbb{S}(t)f = f(.-t) \qquad t \geq 0$$

$R(k)$ is short for $R(k, \mathbb{D})$, the resolvent of $\mathbb{D}$ defined for λ off the imaginary axis:

$$R(\lambda, \mathbb{D})f = \int_0^\infty e^{-\lambda t} \mathbb{S}(t)f\,dt, \qquad \operatorname{Re}\lambda > 0,$$

$$R(-\lambda, \mathbb{D})f = -\int_0^\infty e^{-\lambda t} \mathbb{S}(-t)f\,dt, \qquad \operatorname{Re}\lambda > 0.$$

Again $\mathbb{P}$ denotes the projection operator on $L_p(R^1)$ into $L_p[-b, b]$ $1 < p < 2$. The Hilbert transform can be expressed [10] as

$$\mathbb{H}f = \frac{1}{\pi} \int_0^\infty \frac{(\mathbb{S}(t)f - \mathbb{S}(-t)f)}{t}\,dt,$$

where

$$\mathbb{H}R(\lambda, D) = R(\lambda, \mathbb{D})\mathbb{H}$$

and we note that

$$\mathbb{P}R(\lambda, D)\mathbb{P}f = \int_0^{2b} e^{-\lambda t} \mathbb{P}\mathbb{S}(t)\mathbb{P}f\,dt$$

is an entire function of λ. In particular:

Lemma 5.7. *As* $\operatorname{Re}k$ *goes to infinity,* $\mathbb{B}(k)$ *converges strongly over* $L_p(R^1)$ *to*

$$-I - \frac{M}{\sqrt{1 - M^2}}\mathbb{H},$$

which we denote as $\mathbb{B}(\infty)$.

Proof. $\mathbb{B}(\infty)$ is the operator corresponding to the multiplier:

$$-1 - \frac{M}{\sqrt{1 - M^2}}\frac{|\omega|}{i\omega}.$$

The multiplier corresponding to $\mathbb{B}(k)$ is

$$\tilde{B}(k, i\omega) = \frac{i\omega}{k + i\omega}\sqrt{1 + \frac{kM^2(k + 2i\omega)}{\omega^2(1 - M^2)}} - 1,$$

which, as $\operatorname{Re} k \to \infty$, converges pointwise to the multiplier

$$-1 - \frac{M}{\sqrt{1 - M^2}} \frac{|\omega|}{i\omega}.$$

This is enough but we establish this directly using elementary semigroup theory:

$$\mathbb{B}(k)f = \frac{-kR(k, \mathbb{D})f}{\sqrt{1 - M^2}} - \int_{-\alpha_2}^{\alpha_1} kR(ks, \mathbb{D})fa(M, s)ds.$$

We note that

$$kR(k, \mathbb{D})f \to 0 \quad \text{as } \operatorname{Re} k \to \infty.$$

Hence the first term goes to zero. But $\|kR(k, \mathbb{D})\|$ does not. $kR(k, \mathbb{D})$ converges only strongly. However, we do have that

$$\|ksR(ks, \mathbb{D})\| \leq \frac{|k|}{\operatorname{Re} k}, \quad \operatorname{Re} k > 0, \ -\alpha_2 < s < \alpha_1.$$

And the integral can be expressed:

$$\int_{-\alpha_2}^{\alpha_1} kR(ks, \mathbb{D})fa(M, s)ds \, a(M, 0) \int_{-\alpha_2}^{\alpha_1} kR(ks, \mathbb{D})f \, ds$$

$$+ \int_{-\alpha_2}^{\alpha_1} kR(ks, \mathbb{D})f(a\,(M, s) - a(M, 0))ds$$

where

$$a(M, 0) = \frac{M}{\sqrt{1 - M^2}}$$

and

$$\int_{-\alpha_2}^{\alpha_1} kR(ks, \mathbb{D})f \, ds = \int_0^\infty \int_0^{\alpha_1} ke^{-kst} ds \, \mathbb{S}(t)f \, dt$$

$$- \int_0^\infty \int_0^{\alpha_2} ke^{-kst} ds \, \mathbb{S}(-t)f \, dt$$

$$= \int_0^\infty \int_0^{\alpha_1} e^{-kst} ds \, k(\mathbb{S}(t)f - \mathbb{S}(-t)f)dt$$

$$- \int_0^\infty \int_{\alpha_1}^{\alpha_2} ke^{-kst} ds \, \mathbb{S}(-t)f \, dt$$

$$= \int_0^\infty (1 - e^{-k\alpha_1 t}) \frac{(\mathbb{S}(t)f - \mathbb{S}(-t)f)}{t} dt$$

$$- \int_0^\infty \frac{(e^{-k\alpha_1 t} - e^{-k\alpha_2 t})}{t} \mathbb{S}(-t)f \, dt,$$

which converges strongly to

$$\int_0^\infty \frac{(\mathbb{S}(t)f - \mathbb{S}(-t)f)}{t} dt = \mathbb{H} f.$$

Now

$$-\int_{-\alpha_2}^{\alpha_1} ks \mathbb{R}(ks, \mathbb{D}) f \frac{a(M,s) - a(M,0)}{s} ds - \frac{-k\mathbb{R}(k,\mathbb{D})f}{\sqrt{1-M^2}}$$

converges boundedly in any sector of opening less than $\pi/2$ to

$$-f \int_{-\alpha_2}^{\alpha_1} \left(\frac{a(M,s) - a(M,0)}{s} \right) ds - \frac{f}{\sqrt{1-M^2}}.$$

To evaluate this, let us note that

$$\int_{-\alpha_2}^{\alpha_1} \frac{k}{ks+i\omega} a(M,s) ds = \int_{-\alpha_2}^{\alpha_1} \frac{k}{ks+i\omega} (a(M,s) - a(M,0)) ds$$
$$+ a(M,0) \int_{-\alpha_2}^{\alpha_1} \frac{k}{ks+i\omega} ds,$$

where the second term goes to the multiplier corresponding to the Hilbert transform:

$$\frac{M}{\sqrt{1-M^2}} \frac{|\omega|}{i\omega}.$$

But

$$\bar{\bar{B}}(k,i\omega) = \frac{-k}{\sqrt{1-M^2}} \frac{1}{k+i\omega} - \int_{-\alpha_2}^{\alpha_1} \frac{k}{ks+i\omega} (a(M,s) - a(M,0)) ds$$
$$+ \frac{M}{\sqrt{1-M^2}} \int_{-\alpha_2}^{\alpha_1} \frac{k}{ks+i\omega} ds - \int_{-\alpha_2}^{\alpha_1} \frac{k}{ks+i\omega} (a(M,s) - a(M,0)) ds$$
$$= \bar{\bar{B}}(k,i\omega) + \frac{k}{\sqrt{1-M^2}} \frac{1}{k+i\omega} - \frac{M}{\sqrt{1-M^2}} \left[\int_{-\alpha_2}^{\alpha_1} \frac{k}{ks+iw} ds \right].$$

Hence as $\operatorname{Re} k$ goes to infinity in the sector, we have

$$-\int_{-\alpha_2}^{\alpha_1} (a(M,s) - a(M,0)) \frac{1}{s} ds - \frac{1}{\sqrt{1-M^2}} = -1,$$

which establishes the result. But we emphasize that the convergence is only strong. □

Using this result we can express the Possio equation in abstract form as an equation in $L_p[-b, b]$:

$$2\hat{w}_a(\lambda, .) = \sqrt{1 - M^2}\, \mathbb{P}\mathbb{H}(I + \mathbb{B}(k))\hat{A}(\lambda, .)$$

$$= \sqrt{1 - M^2}\, \mathbb{P}\mathbb{H}(I + \mathbb{B}(k) - \mathbb{B}(\infty))\hat{A}(\lambda, .)$$

$$+ \sqrt{1 - M^2}\, \mathbb{P}\mathbb{H}\left[-I - \frac{M}{\sqrt{1 - M^2}}\mathbb{H} \right]\hat{A}(\lambda, .).$$

Or

$$\frac{2}{M}\hat{w}_a(\lambda, .) = \hat{A}(\lambda, .) + \frac{\sqrt{1 - M^2}}{M}\mathbb{P}\mathbb{H}(\mathbb{B}(k) - \mathbb{B}(\infty))\hat{A}(\lambda, .), \quad 0 < M < 1,$$

$$(5.49)$$

which is defined for all λ off the imaginary axis where $R(k, \mathbb{D})$ is defined. It is important to note that in (5.49), we are restricted to $0 < M < 1$. The cases $M = 0$ and $M = 1$ are treated separately below. Note that (5.49) does not impose the Kutta condition.

The Kutta Condition

It is not evident from (5.49) that the solution will satisfy the Kutta condition: For this we go to an equivalent but different version of the LDP (Laplace domain Possio) by invoking the Tricomi operator.

Theorem 5.8. *The Possio equation can be expressed in the form:*

$$\frac{2\mathcal{T}\hat{w}_a(\lambda, .)}{\sqrt{1 - M^2}} = ((I + \mathbb{P}\mathbb{B}(k)\mathbb{P}) + \mathcal{T}\mathbb{P}\mathbb{H}(I - \mathbb{P})\mathbb{B}(k)\mathbb{P})\hat{A}(\lambda, .), \qquad (5.50)$$

where $\mathcal{T}$ is the Tricomi operator.

Proof. For nonzero M we multiply both sides of (5.48) by M to obtain:

$$2\hat{w}_a(\lambda, .) = M\hat{A}(\lambda, .) + \left(\sqrt{1 - M^2} \right)\mathbb{P}\mathbb{H}(\mathbb{B}(k) - \mathbb{B}(\infty))\hat{A}(\lambda, .),$$

where

$$\sqrt{1 - M^2}\mathbb{P}\mathbb{H}(\mathbb{B}(k) - \mathbb{B}(\infty)) = \sqrt{1 - M^2}\mathbb{P}\mathbb{H}(\mathbb{B}(k) + I) - MI.$$

Hence

$$\frac{2\hat{w}_a(\lambda, .)}{\sqrt{1 - M^2}} = \mathbb{P}\mathbb{H}\hat{A}(\lambda, .) + \mathbb{P}\mathbb{H}\mathbb{B}(k)\hat{A}(\lambda, .),$$

where the right side can be expressed:

$$= \mathbb{PHP}\hat{A}(\lambda, .) + \mathbb{PHPB}(k)\mathbb{P}\hat{A}(\lambda, .) + \mathbb{PH}(I - \mathbb{P})\mathbb{B}(k)\hat{A}(\lambda, .). \quad (5.51)$$

□

Lemma 5.9. *Suppose f is in $L_p[-b, b], 1 < p < 2$ and*

$$\mathcal{T}f = 0.$$

Then

$$f = 0.$$

Proof. Let

$$\mathcal{T}f = g,$$

$$g(x) = \frac{1}{\pi} \sqrt{\frac{b-x}{b+x}} \int_{-b}^{b} \sqrt{\frac{b+s}{b-s}} \frac{f(s)}{s-x} ds,$$

which is given to be

$$= 0, \quad |x| < b.$$

Hence

$$\int_{-b}^{b} \sqrt{\frac{b+s}{b-s}} \frac{f(s)}{s-x} ds = 0, \quad |x| < b.$$

Hence (see [11]):

$$\sqrt{\frac{b+s}{b-s}} f(s) = \frac{\text{const}}{\sqrt{b^2 - s^2}},$$

$$f(s) = \frac{\text{const}}{b+s}, \quad |s| < b,$$

which leads to a contradiction, because this function is not in $L_p(-b, b)$, $1 < p < 2$. □

Hence

$$\mathcal{T}\left[\frac{2\hat{w}_a(\lambda, .)}{\sqrt{1 - M^2}} - \left(\mathbb{PH}\hat{A}(\lambda, .) + \mathbb{PHB}(k)\hat{A}(\lambda, .)\right)\right] = 0$$

is equivalent to

$$\frac{2\hat{w}_a(\lambda, .)}{\sqrt{1 - M^2}} - \left(\mathbb{PH}\hat{A}(\lambda, .) + \mathbb{PHB}(k)\hat{A}(\lambda, .)\right) = 0,$$

proving the theorem. □

Lemma 5.10. $\mathbb{P}\mathbb{B}(k)\mathbb{P}$ is a Volterra Operator on $L_p(-b, b)$ into itself, and its range is contained in $C_1[-b, b]$. It is defined for all k and is an entire function.

Proof.

$$\mathbb{P}\mathbb{B}(k)\mathbb{P}A = \frac{-k\mathbb{P}\mathbb{R}(k)\mathbb{P}A}{\sqrt{1 - M^2}} - \int_{-\alpha_1}^{\alpha_1} k\mathbb{P}\mathbb{R}(ks)\mathbb{P}A\, a(M, s)ds$$
$$- \int_{-\alpha_2}^{-\alpha_1} k\mathbb{P}\mathbb{R}(ks)\mathbb{P}A\, a(M, s)ds,$$

where

$$\mathbb{P}\mathbb{R}(k)\mathbb{P}f = g; \qquad g = \int_0^\infty e^{-kt}\mathbb{P}\mathbb{S}(t)\, P\, f\, dt.$$

And because

$$\mathbb{P}\mathbb{S}(t)P = 0 \quad \text{for } t > b,$$

we have:

$$g(x) = \int_{-b}^x e^{-k(x-s)} f(s)\, ds, \quad |x| < b.$$

Let us use the notation:

$$\mathcal{L}(k) = \mathbb{P}\mathbb{R}(k)\mathbb{P}$$

$\mathcal{L}(k)$ is then an entire function of k. And

$$\mathbb{P}\mathbb{B}(k)\mathbb{P} = \frac{-k\mathcal{L}(k)}{\sqrt{1 - M^2}} - \int_{-\alpha_2}^{\alpha_1} k\mathcal{L}(ks)\, a\, (M, s)ds$$

is a Volterra operator on $L_p(-b, b)$ into itself, and $\mathcal{L}(k)$ is defined for every k in the plane. In fact we have for

$$A = \mathbb{P}\mathbb{B}(k)\mathbb{P}f$$

$$A(x) = \frac{-k\int_{-b}^x e^{-k(x-\xi)} f(\xi)d\xi}{\sqrt{1 - M^2}}$$
$$- k\int_{-b}^x f(\xi) \int_{-\alpha_2}^{\alpha_1} e^{-ks(x-\xi)} a(M, s)\, ds\, d\xi, \quad |x| < b.$$

Furthermore, we see that A $(.)$ is in $C_1[-b, b]$, so that $\mathbb{P}\mathbb{B}(k)\mathbb{P}$ maps $L_p[-b, b]$ into $C_1[-b, b]$. □

Getting back to (5.51), we see that by the Tricomi theorem

$$\mathcal{T}\left[\mathbb{P}\mathbb{H}\mathbb{P}\hat{A}(\lambda, .) + \mathbb{P}\mathbb{H}\mathbb{P}\mathbb{B}(k)\mathbb{P}\hat{A}(\lambda, .)\right] = (I + \mathbb{P}\mathbb{B}(k)\mathbb{P})\hat{A}(\lambda, .)$$

and is in $L_p[-b, b]$. Next note that the third term on the right in (5.51),

$$\mathbb{P}\mathbb{H}(I - \mathbb{P})\mathbb{B}(k)\hat{A}(\lambda, .),$$

is in $L_p[-b,b]$, $1 < p < 2$. Hence by the Tricomi theorem [11],

$$\mathcal{T}(\mathbb{P}\mathbb{H}(I - \mathbb{P})\mathbb{B}(k)\hat{A}(\lambda, .))$$

is in $L_p[-b,b]$ for $1 < p < 4/3$. We can prove more by actual calculation.

Lemma 5.11. *Let* $W(k) = \mathcal{T}\mathbb{P}\mathbb{H}(I - \mathbb{P})\mathbb{B}(k)\mathbb{P}$ *which by (5.50) is defined for every* k *off the imaginary axis. However for* $\operatorname{Re} k > 0$ *by [5, 8] for A in* $L_p[-b,b]$, $1 < p < 2$,

$$W(k)A = \frac{k}{\sqrt{1 - M^2}} h_-(k)L_-(k, A) + k \int_0^{\alpha_1} h_-(ks)L_-(ks, A)a(M, s)\mathrm{d}s$$

$$+ k \int_0^{\alpha_2} (h_+(ks) + j(ks))L_+(ks, A)a\,(M, -s)\mathrm{d}s, \qquad (5.52)$$

where the functionals are

$$L_+\,(k, A) = \int_{-b}^{b} e^{-k(b+\xi)} A(\xi)\mathrm{d}\xi,$$

$$L_-(k, A) = \int_{-b}^{b} e^{-k(b-\xi)} A(\xi)\mathrm{d}\xi$$

and the functions are

$$h_+(k, x) = \frac{1}{\pi} \sqrt{\frac{b - x}{b + x}} \int_0^{\infty} \sqrt{\frac{\sigma}{2b + \sigma}} \frac{1}{b + \sigma + x} e^{-k\sigma}\mathrm{d}\sigma,$$

$$h_-(k, x) = \frac{1}{\pi} \sqrt{\frac{b - x}{b + x}} \int_0^{\infty} \sqrt{\frac{2b + \sigma}{\sigma}} \frac{1}{b + \sigma - x} e^{-k\sigma}\mathrm{d}\sigma,$$

$$j(k, x) = e^{k(b+x)}, \qquad |x| < b.$$

Each function on the right in (5.52) is in $L_p[-b,b]$, $1 < p < 2$. Furthermore,

$$h_+(k, b-) = 0,$$

$$h_-(k, b-) = 1.$$

Proof.

$$W(k)A = T\mathbb{P}\mathbb{H}(I - \mathbb{P})\mathbb{B}(k)\mathbb{P}A(.)$$

$$= T\mathbb{P}\mathbb{H}(I - \mathbb{P}) \left[\frac{-kR(k)A(.)}{\sqrt{1 - M^2}} - \int_{-\alpha_2}^{\alpha_1} kR(ks)A(.)a(M, s)ds \right].$$

Computing the first term we have

$$- T\mathbb{P}\mathbb{H}(I - \mathbb{P})kR(k)A(.) = \frac{-k}{\pi} \sqrt{\frac{b - x}{b + x}} \int_{-b}^{b} \sqrt{\frac{b + \xi}{b - \xi}} \frac{1}{\xi - x}$$

$$\times \frac{1}{\pi} \left[\int_{-\infty}^{\infty} \int_{-b}^{b} \right] \frac{1}{\xi - s} \int_{-\infty}^{s} e^{-k(s-\eta)}(PA)(\eta)d\eta \, ds \, d\xi.$$

We therefore have

$$-kT\mathbb{P}\mathbb{H}(I - \mathbb{P})R(k)A(.) = \frac{-k}{\pi} \sqrt{\frac{b - x}{b + x}} \int_{-b}^{b} \sqrt{\frac{b + \xi}{b - \xi}} \frac{1}{\xi - x}$$

$$\times \frac{1}{\pi} \left[\int_{-\infty}^{-b} + \int_{b}^{\infty} \right] \frac{1}{\xi - s} \int_{-b}^{s} e^{-k(s-\eta)}(PA)(\eta) \, d\eta \, ds \, d\xi$$

$$= \frac{-k}{\pi} \sqrt{\frac{b - x}{b + x}} \int_{-b}^{b} \sqrt{\frac{b + \xi}{b - \xi}} \frac{1}{\xi - x}$$

$$\times \frac{1}{\pi} \int_{b}^{\infty} \frac{1}{\xi - s} \int_{-b}^{b} e^{-k(s-\eta)}(A)(\eta)d\eta \, ds \, d\xi.$$

Changing variables using $s = \sigma + b$, and changing the order of integration we obtain

$$T\mathbb{P}\mathbb{H}(I - \mathbb{P})kR(k)A(.) = \frac{-k}{\pi} \sqrt{\frac{b - x}{b + x}} \int_{-b}^{b} \sqrt{\frac{b + \xi}{b - \xi}} \frac{1}{\xi - x}$$

$$\times \frac{1}{\pi} \int_{0}^{\infty} \frac{e^{-k\sigma}}{\xi - (\sigma + b)} \int_{-b}^{b} e^{-k(b-\eta)} A(\eta)d\eta \, d\sigma \, d\xi$$

$$= \frac{-k}{\pi} \int_{0}^{\infty} \int_{-b}^{b} e^{-k(b-s)} A(s)ds \, e^{-k\sigma}$$

$$\times \frac{1}{\pi} \int_{-b}^{b} \sqrt{\frac{b + \xi}{b - \xi}} \frac{1}{\xi - x} \frac{1}{\xi - (\sigma + b)} d\xi \, d\sigma.$$

Evaluating the innermost integral

$$I = \frac{1}{\pi} \int_{-b}^{b} \sqrt{\frac{b+\xi}{b-\xi}} \frac{1}{\xi - x} \frac{1}{\xi - (\sigma + b)} d\xi,$$

we get that

$$I = \frac{1}{\pi} \int_{-b}^{b} \sqrt{\frac{b+\xi}{b-\xi}} \frac{1}{x - (\sigma + b)} \left(\frac{1}{\xi - x} - \frac{1}{\xi - (\sigma + b)} \right) d\xi.$$

We next use the fact that

$$\frac{1}{\pi} \int_{-b}^{b} \sqrt{\frac{b+\xi}{b-\xi}} \frac{1}{\xi - x} = 1 \quad \text{for } |x| < b \text{ and the integration identity,}$$

$$\frac{1}{\pi} \int_{-b}^{b} \sqrt{\frac{b+\xi}{b-\xi}} \frac{1}{\xi - z} d\xi = \frac{b+z}{\sqrt{z^2 - b^2}} + 1 \quad \text{for } z > b \text{ and therefore,}$$

$$I = \frac{1}{x - (\sigma + b)} \left[1 - \frac{2b + \sigma}{\sqrt{\sigma^2 + 2b\sigma}} - 1 \right]$$

$$= \frac{1}{(\sigma + b) - x} \frac{\sqrt{2b + \sigma}}{\sqrt{\sigma}}.$$

Therefore, returning to the expression for $-\mathcal{T}\mathbb{P}\mathbb{H}(I - \mathbb{P})kR(k)A(\,.\,)$ we have

$$-\mathcal{T}\mathbb{P}\mathbb{H}(I - \mathbb{P})kR(k)A(\cdot) = \frac{-k}{\pi} \sqrt{\frac{b-x}{b+x}} \int_{0}^{\infty} \frac{e^{-k\sigma}}{(\sigma + b) - x} \frac{\sqrt{2b + \sigma}}{\sqrt{\sigma}}$$

$$\times \int_{-b}^{b} e^{-k(b-\eta)} A(\eta) d\eta \, d\sigma$$

$$= -kh_{-}(k, x)L_{-}(k, A).$$

We now utilize this expression to compute

$$- \mathcal{T}\mathbb{P}\mathbb{H}k \left[\int_{0}^{\alpha_1} (I - \mathbb{P})R(ks)A(\,.\,)a(M, s)ds \right]$$

$$= \int_{0}^{\alpha_1} -kh_{-}(ks, x)L_{-}(ks, A)a(M, s)ds \,].$$

We finally compute the term

$$\mathcal{T}\mathbb{P}\mathbb{H}k \left[\int_{0}^{\alpha_2} (I - \mathbb{P})R(-ks)A(\,.\,)a(M, -s)ds \right]$$

starting with

$$T\mathbb{PH}k\left[\int_0^{\alpha_2} R(-ks)A(.)a(M,-s)\mathrm{d}s\right]$$

$$= \frac{k}{\pi}\sqrt{\frac{b-x}{b+x}}\int_{-b}^b \sqrt{\frac{b+\xi}{b-\xi}}\frac{1}{\xi-x}\frac{1}{\pi}\left[\int_{-\infty}^{-b}+\int_{-b}^b+\int_b^\infty\right]\frac{1}{\xi-z}$$

$$\times\left[\int_0^{\alpha_2}\int_z^\infty e^{ks(z-\eta)}(PA)(\eta)\mathrm{d}\eta\,a(M,-s)\mathrm{d}s\right]\mathrm{d}z\,\mathrm{d}\xi,$$

where we used the fact that $R(k)$ is the operator corresponding to the multiplier $\frac{-1}{-k-i\omega}$ which when $\operatorname{Re}k < 0$ is

$$R(k)f = -\int_z^\infty e^{-k(z-\eta)}f(\eta)\mathrm{d}\eta$$

and hence we have

$$T\mathbb{PH}k\left[\int_0^{\alpha_2} R(-ks)A(.)a(M,-s)\mathrm{d}s\right]$$

$$= \frac{-k}{\pi}\sqrt{\frac{b-x}{b+x}}\int_{-b}^b \sqrt{\frac{b+\xi}{b-\xi}}\frac{1}{\xi-x}\frac{1}{\pi}\left[\int_{-\infty}^{-b}+\int_{-b}^b\right]\frac{1}{\xi-z}$$

$$\times\left[\int_0^{\alpha_2}\int_z^b e^{ks(z-\eta)}A(\eta)\mathrm{d}\eta\,a(M,-s)\mathrm{d}s\right]\mathrm{d}z\,\mathrm{d}\xi. \qquad (5.53)$$

Now considering the first integral $I1$, we have

$$I1 = \frac{-k}{\pi}\sqrt{\frac{b-x}{b+x}}\int_{-b}^b \sqrt{\frac{b+\xi}{b-\xi}}\frac{1}{\xi-x}\frac{1}{\pi}\int_{-\infty}^{-b}\frac{1}{\xi-z}$$

$$\times\left[\int_0^{\alpha_2}\int_{-b}^b e^{ks(z-\eta)}A(\eta)\mathrm{d}\eta\,a(M,-s)\mathrm{d}s\right]\mathrm{d}z\,\mathrm{d}\xi$$

$$= \frac{-k}{\pi}\int_0^{\alpha_2}\sqrt{\frac{b-x}{b+x}}\int_0^\infty\left[\int_{-b}^b \sqrt{\frac{b+\xi}{b-\xi}}\frac{1}{\xi-x}\frac{1}{\pi}\frac{1}{\xi+\sigma+b}\mathrm{d}\xi\right]$$

$$\times\left[\int_{-b}^b e^{-ks(\sigma+b+\eta)}A(\eta)\mathrm{d}\eta\,a(M,-s)\right]\mathrm{d}\sigma\,\mathrm{d}s.$$

Evaluating

$$\frac{1}{\pi} \int_{-b}^{b} \sqrt{\frac{b+\xi}{b-\xi}} \frac{1}{\xi - x} \frac{1}{\xi + (\sigma + b)} d\xi$$

$$= \frac{1}{\pi} \int_{-b}^{b} \sqrt{\frac{b+\xi}{b-\xi}} \frac{1}{(\sigma + b) + x} \left(\frac{1}{\xi - x} - \frac{1}{\xi + (\sigma + b)} \right) d\xi$$

$$= \frac{1}{(\sigma + b) + x} \left[1 - \frac{-\sigma}{\sqrt{\sigma^2 + 2b\sigma}} - 1 \right]$$

$$= \frac{1}{x + (\sigma + b)} \frac{\sqrt{\sigma}}{\sqrt{\sigma + 2b}},$$

where we have used the fact that

$$\frac{1}{\pi} \int_{-b}^{b} \sqrt{\frac{b+\xi}{b-\xi}} \frac{1}{\xi - x} = 1, \quad \text{for } |x| < b \text{ and the integration identity,}$$

$$\frac{1}{\pi} \int_{-b}^{b} \sqrt{\frac{b+\xi}{b-\xi}} \frac{1}{\xi - z} d\xi = \frac{b - z}{\sqrt{z^2 - b^2}} + 1 \quad \text{for } |z| > 1.$$

Thus, $I1$ becomes

$$I1 = \frac{-k}{\pi} \int_{0}^{\alpha_2} \sqrt{\frac{b-x}{b+x}} \int_{0}^{\infty} \int_{0}^{\infty} \frac{1}{x + (\sigma + b)} \frac{\sqrt{\sigma}}{\sqrt{\sigma + 2b}} d\xi \right]$$

$$\times \left[\int_{-b}^{b} e^{-ks(\sigma + b + \eta)} A(\eta) d\eta \, a(M, -s) \right] d\sigma \, ds$$

$$= -k \int_{0}^{\alpha_2} h_+(ks, x) L_+(ks, A) a(M, -s) ds.$$

We now compute the second integral $I2$ in (5.53) to get

$$I2 = \frac{-k}{\pi} \sqrt{\frac{b-x}{b+x}} \int_{-b}^{b} \sqrt{\frac{b+\xi}{b-\xi}} \frac{1}{\xi - x} \frac{1}{\pi} \int_{-b}^{b} \frac{1}{\xi - z}$$

$$\times \left[\int_{0}^{\alpha_2} \int_{z}^{b} e^{ks(z - \eta)} A(\eta) d\eta \, a(M, -s) ds \right] dz \, d\xi$$

$$= -k \int_{0}^{\alpha_2} T\mathbb{PHP} \left[\int_{.}^{b} e^{ks(. - \eta)} A(\eta) d\eta \right] a(M, -s) ds$$

$$= -k \int_0^{\alpha_2} \int_x^b e^{ks(x-\eta)} A(\eta) d\eta \, a(M, -s) ds.$$

It now remains to subtract the term

$$I3 = k \int_0^2 T\mathbb{P}\mathbb{H}\mathbb{P}R(-ks)A(.)a(M, -s)ds$$

$$= k \int_0^{\alpha_2} \mathcal{L}(ks)\mathbb{P}A(.)a(M, -s)ds$$

$$= k \int_0^{\alpha_2} \int_{-b}^x e^{ks(x-\eta)} A(\eta) d\eta \, a(M, -s)ds.$$

Therefore, we get

$$I1 + I2 - I3 = -k \int_0^{\alpha_2} h_+(ks, x)L_+(ks, A)a(M, -s)ds$$

$$- k \int_0^{\alpha_2} \int_{-b}^b e^{ks(x-\eta)} A(\eta) d\eta \, a(M, -s)ds$$

$$= -k \int_0^{\alpha_2} h_+(ks, x)L_+(ks, A)a(M, -s)ds$$

$$- k \int_0^{\alpha_2} e^{ks(x+b)} \int_{-b}^b e^{-ks(b+\eta)} A(\eta) d\eta a(M, -s)ds$$

$$= -k \int_0^{\alpha_2} (h_+(ks, x) + e^{ks(x+b)})L_+(ks, A)a(M, -s)ds$$

as desired. Next we show that $h_-(ks, .)$ is in $L_p[-b, b]$. We have

$$h_-(ks, x) = \frac{1}{\pi} \sqrt{\frac{b-x}{b+x}} \int_0^\infty \sqrt{\frac{2b+\sigma}{\sigma}} \frac{1}{b+\sigma-x} e^{-ks\sigma} d\sigma, \qquad s > 0, \quad (5.54)$$

where the integral:

$$\left| \int_0^\infty \sqrt{\frac{2b+\sigma}{\sigma}} \frac{1}{b+\sigma-x} e^{-ks\sigma} d\sigma \right|, \qquad s > 0$$

is

$$\leq \frac{1}{b-x} \int_0^\infty \sqrt{\frac{2b+\sigma}{\sigma}} e^{-\mathrm{Re}\, ks\sigma} d\sigma$$

and

$$\sqrt{\frac{b-x}{b+x}} \frac{1}{b-x} = \frac{1}{\sqrt{b^2-x^2}}, \qquad |x| < b$$

and is in $L_p[-b,b]$, $1 < p < 2$. Next

$$h_+(ks,x) = \frac{1}{\pi} \sqrt{\frac{b-x}{b+x}} \int_0^\infty \sqrt{\frac{\sigma}{2b+\sigma}} \frac{1}{b+\sigma+x} e^{-ks\sigma} d\sigma, \qquad s > 0,$$

$$|h_+(ks,x)| \le \frac{1}{\pi} \sqrt{\frac{b-x}{b+x}} \int_0^\infty \sqrt{\frac{\sigma}{2b+\sigma}} \frac{1}{\sigma} e^{-\operatorname{Re} ks\sigma} d\sigma, \qquad s > 0, \qquad (5.55)$$

and hence $h_+(ks,.)$ is in $L_p[-b,b]$, $1 < p < 2$ for $s > 0$, $\operatorname{Re} k > 0$, and

$$h_+(ks,b-) = 0. \qquad (5.56)$$

Hence

$$\left| \int_0^{\alpha_1} h_-(ks,x) L_-(ks,A) a(M,s) ds \right| \le \frac{\int_{-b}^b |A(\xi)| d\xi}{\pi} \sqrt{\frac{1}{b^2-x^2}}$$

$$\times \int_0^\infty \sqrt{\frac{\sigma}{2b+\sigma}} \frac{1}{\sigma} \left(\int_0^{\alpha_1} a(M,s) e^{-\operatorname{Re} ks\sigma} ds \right) d\sigma, \qquad (5.57)$$

verifying that the second term in (5.52) is in $L_p[-b,b]$, $1 < p < 2$. Similarly

$$\left| \int_0^{\alpha_2} (h_+(ks,x) + j(ks,x)) L_+(ks,A) a(M,-s) ds \right|$$

$$\le \frac{\int_{-b}^b |A(\xi)| d\xi}{\pi} \sqrt{\frac{b-x}{b+x}} \int_0^\infty \sqrt{\frac{\sigma}{2b+\sigma}} \frac{1}{\sigma} \left(\int_0^{\alpha_2} a(M,-s) e^{-\operatorname{Re} ks\sigma} ds \right) d\sigma$$

$$+ \int_{-b}^b |A(\xi)| d\xi \int_0^{\alpha_2} e^{\operatorname{Re} ksx} a(M,-s) ds,$$

and is thus in $L_p[-b,b]$, $1 < p < 2$. Finally

$$h_+(k,b-) = 0 \text{ follows from (5.56)}$$

and

$$h_-(k,b-) = 0 \text{ follows from (5.57).} \qquad \square$$

Theorem 5.12. *Suppose the (Possio) equation given by (5.50) for any λ has a unique solution. Then it is given by*

$$\hat{A}(\lambda,.) = ((I + \mathbb{P}B(k)\mathbb{P}) + W(k))^{-1}\frac{2T\hat{w}_a(\lambda,.)}{\sqrt{1-M^2}}$$

and $\hat{A}(\lambda,.)$ satisfies the Kutta condition:

$$\hat{A}(\lambda,x) \to 0 \quad \text{as } x \to b.$$

Lemma 5.13. *The operator $W(k)$ is compact on $L_p[-b,b]$ into itself $1 < p < 2$.*

Proof. Follows immediately from (5.51) because it is defined in terms of the continuous linear functionals $L_+(k,A)$ and $L_-(k,A)'$.   □

It follows from the Lemma 5.13 that

$$I + \mathbb{P}B(k)\mathbb{P} + T\mathbb{P}\mathbb{H}(I - \mathbb{P})B(k)\mathbb{P} = I + C(k),$$

where $C(k)$ is compact on $L_p[-b,b]$ into itself, $1 < p < 2$. Hence either there is a nonzero A such that

$$(I + C(k))A = 0$$

or

$$(I + C(k))$$

has a bounded inverse. Hence if the solution is unique, there is no nonzero A such that

$$(I + C(k))A = 0$$

and the theorem follows.

Theorem 5.14. *Suppose (5.50) has a unique solution. Then the solution will satisfy the Kutta condition.*

Proof. Let

$$g = ((I + \mathbb{P}B(k)\mathbb{P}) + T\mathbb{P}\mathbb{H}(I - P)B(k)P)\hat{A}(\lambda,.).$$

Then

$$g(x) = \hat{A}(\lambda,x) + \frac{-k\int_{-b}^{x}e^{-k(x-\xi)}\hat{A}(\lambda,\xi)d\xi}{\sqrt{1-M^2}}$$

$$- k\int_{-b}^{x}\hat{A}(\lambda,\xi)\int_{-\alpha_2}^{\alpha_1}e^{-ks(x-\xi)}a(M,s)d\xi ds, \quad |x| < b$$

$$+ \frac{k}{\sqrt{1-M^2}}h_-(k,x)L_-(k,A) + k\int_{0}^{\alpha_1}h_-(ks,x)L_-(ks,A)a(M,s)ds$$

$$+ k\int_{0}^{\alpha_2}(h_+(ks,x) + j(ks,x))L_+(ks,A)a(M,-s)ds.$$

Hence, by Lemma 5.6

$$g(b-) = \hat{A}(\lambda, b-) + \frac{-k \int_{-b}^{b} e^{-k}(b - \xi) \hat{A}(\lambda, \xi) d\xi}{\sqrt{1 - M^2}}$$

$$- k \int_{-b}^{b} \hat{A}(\lambda, \xi) \int_{-\alpha_2}^{\alpha_1} e^{-ks(b-\xi)} a(M, s) d\xi ds, \qquad |x| < b$$

$$+ \frac{k}{\sqrt{1 - M^2}} + L_-(k, A) + k \int_{0}^{\alpha_1} L_-(ks, A) a(M, s) ds$$

$$+ k \int_{0}^{\alpha_2} j(ks, b) L_+(k, A) a(M, -s) ds$$

$$= \hat{A}(\lambda, b-).$$

But the right side by Theorem 5.16 $= 0$.   □

Remarks: What is new is the introduction of the operator $W(k)$ which is defined for all k, except for the negative real axis where it has a logarithmic singularity. See below on the latter.

Existence Theorem

In lieu of a blanket existence theorem for every M and every k, we offer two results, each with some limitation.
Existence Theorem for k in Bounded Vertical Strip

Theorem 5.15. *For values of k in a bounded vertical strip, that is,*

$$|\text{Re. } k| < k_b < \infty,$$

we can find M_0 such that the LDP has a unique solution for all $M < M_0$, $M_0 > 0$.

Proof. We begin with the equation in the form:

$$\frac{2T \hat{w}_a(\lambda, .)}{\sqrt{1 - M^2}} = (I + \mathbb{P}B(k)\mathbb{P} + W(k)) \hat{A}(\lambda, .).$$

To make explicit the dependence on M, we decompose

$$\mathbb{P}B(k)\mathbb{P} \text{ as } = \mathbb{P}\mathbb{B}_0(k)\mathcal{P} + \mathcal{P}\mathcal{B}_M(k)\mathcal{P},$$

where

$$\mathbb{P}\mathbb{B}P_0(k)\mathcal{P} = -k\mathcal{L}(k),$$

$$\mathcal{P}\mathcal{B}_M(k)\mathcal{P} = -k\,\mathcal{L}(k)\left(\frac{1}{\sqrt{1-M^2}} - 1\right) - \int_{-\alpha_2}^{\alpha_1} k\,\mathcal{L}(ks)a(M,s)ds,$$

$$W(k) = W_0(k) + W_M(k),$$

where

$$W_0(k)A = kh_-(k)L_-(k,A),$$

$$W_M(k)A = (kh_-(k)L_-(k,A))\left(\frac{1}{\sqrt{1-M^2}} - 1\right)$$

$$+ k\int_0^{\alpha_1} h_-(ks)L_-(ks,A)a(M,s))ds$$

$$+ k\int_0^{\alpha_2} (h_+(ks) + j(ks))L_+(ks,A)a(M,-s)ds,$$

so that

(Note $\mathcal{P} \to \mathbb{P}$ $\mathcal{B} \to \mathbb{B}$ here and below.)

$$\frac{2\mathcal{T}\hat{w}_a(\lambda,.)}{\sqrt{1-M^2}} = (I + \mathcal{P}\mathcal{B}_0(k)\mathcal{P} + W_0(k) + \mathcal{P}\mathcal{B}_M(k)\mathcal{P} + W_M(k))\hat{A}(\lambda,.).$$

For $M = 0$, we have already proved that

$$(I + \mathcal{P}\mathcal{B}_0(k)\mathcal{P} + w_0(k))$$

has a bounded inverse. Hence

$$(I + (I + \mathcal{P}\mathcal{B}_0(k)\mathcal{P} + W_0(k)))^{-1}(\mathcal{P}\mathcal{B}_M(k)\mathcal{P} + W_M(k))\hat{A}(\lambda,.)$$

$$= (I + \mathcal{P}\mathcal{B}_0(k)\mathcal{P} + W_0(k))^{-1}\frac{2\mathcal{T}\hat{w}_a(\lambda,.)}{\sqrt{1-M^2}},$$

where the right side is the solution for $M = 0$.
  Next we show that

$$(I + \mathcal{P}\mathcal{B}_0(k)\mathcal{P} + W_0(k))^{-1}(\mathcal{P}\mathcal{B}_M(k)\mathcal{P} + W_M(k))$$

can be made less than 1 in operator norm for all k, Re $k > \sigma_a$ for $0 < M < M_0$ for some M_0.

We calculate

$$(\mathcal{P}\mathcal{B}_M(k)\mathcal{P} + \mathcal{W}_M(k))A$$

$$= -k\mathcal{L}(k)A\left(\frac{1}{\sqrt{1-M^2}} - 1\right) - \int_{-\alpha_2}^{\alpha_1} k\mathcal{L}(ks)Aa(M,s)\mathrm{d}s$$

$$+(kh_-(k)L_-(k,A))\left(\frac{1}{\sqrt{1-M^2}} - 1\right)$$

$$+k\int_0^{\alpha_1} h_-(ks)L_-(ks,A)\, a(M,s)\mathrm{d}s$$

$$+k\int_0^{\alpha_2} (h_+(ks) + j(ks))L_+(ks,A)a(M,-s)\mathrm{d}s$$

$$= \left(\frac{1}{\sqrt{1-M^2}} - 1\right)(-k\mathcal{L}(k)A + kh_-(k)L_-(k,A))$$

$$-\int_{-\alpha_2}^{\alpha_1} k\mathcal{L}(ks)Aa(M,s)\mathrm{d}s + k\int_0^{\alpha_1} h_-(ks)L_-(ks,A)a(M,s)\mathrm{d}s$$

$$+k\int_0^{\alpha_2} (h_+(ks) + j(ks))L_+(ks,A)\, a(M,-s)\mathrm{d}s,$$

where

$$-\int_{-\alpha_2}^{\alpha_1} k\mathcal{L}(ks)Aa(M,s)\mathrm{d}s$$

can be expressed

$$= -\int_{-\alpha_2}^{\alpha_1} sk\mathcal{L}(ks)A\left(\frac{(a(M,s) - a(M,0))}{s}\right)\mathrm{d}s - \int_{-\alpha_2}^{\alpha_1} k\mathcal{L}(ks)Aa(M,0)\mathrm{d}s.$$

On a bounded strip continuous functions are bounded.

$$\left(\frac{1}{\sqrt{1-M^2}} - 1\right) \to 0 \quad \text{as} \quad M \to 0$$

and

$$a(M,s) = \frac{1}{\pi}\frac{\sqrt{(\alpha_1 - s)(\alpha_2 + s)}}{1-s}, \quad -\alpha_2 < s < \alpha_1$$

$$= \frac{1}{\pi}\frac{\sqrt{-s^2 + \frac{2M^2}{1-M^2}s + \frac{M^2}{1-M^2}}}{1-s},$$

$$a(M,0) = \frac{1}{\pi}\sqrt{\frac{M^2}{1-M^2}}.$$

Hence

$$\frac{a(M,s) - a(M,0)}{s} = \frac{1}{\pi s}\left[\frac{\sqrt{-s^2 + -\frac{2M^2}{1-M^2}s + \frac{M^2}{1-M^2}}}{1 - s} - \frac{M}{\sqrt{1 - M^2}}\right]$$

$$= \frac{1}{\pi\sqrt{1 - M^2}}\left[\frac{\sqrt{-s^2(1 - M^2) - 2M^2s + M^2} - M(1 - s)}{(1 - s)s}\right]$$

$$= \frac{1}{\pi\sqrt{1 - M^2}}\left[\frac{\sqrt{M^2(1 - s^2) - s^2} - M(1 - s)}{(1 - s)s}\right]$$

$$= \frac{-M}{\pi\sqrt{1 - M^2}}\frac{\left(1 - \sqrt{1 - \frac{s^2}{M(1-s)}}\right)}{s}.$$

Hence for $|s| < \sqrt{M^2 + 4M} - M$, this is in absolute magnitude

$$< \pi\sqrt{1 - M^2}\frac{1}{2}\left(\frac{s}{(1 - s)}\right).$$

And hence it follows that

$$\int_{-\frac{M}{1-M}}^{\frac{M}{1+M}} |a(M,s)|ds \to 0 \quad \text{with } M,$$

$$\int_{-\frac{M}{1-M}}^{\frac{M}{1+M}} |a(M,0)|ds \to 0 \quad \text{with } M,$$

$$\int_{-\frac{M}{1-M}}^{\frac{M}{1+M}} \left|\frac{a(M,s) - a(M,0)}{s}\right|ds \to 0 \quad \text{with } M.$$

Hence we can find M_0 such that

$$\|(I + \mathcal{P}B_0(k)\mathcal{P} + W_0(k))^{-1}(\mathcal{P}B_M(k)\mathcal{P} + W_M(k))\| < 1$$

for k in the strip.
Hence

$$(I + (I + \mathcal{P}B_0(k)\mathcal{P} + W_0(k))^{-1}(\mathcal{P}B_M(k)\mathcal{P} + W_M(k)))^{-1}A$$

$$= \sum_{n=0}^{\infty}(-1)^{n-1}((I + \mathcal{P}B_0(k)\mathcal{P} + W_0(k))^{-1}(\mathcal{P}B_M(k)\mathcal{P} + W_M(k)))^n A,$$

for any element A in $L_p[-b, b]$, and finally

$$\hat{A}(\lambda, .) = (I + (I + PB_0(k)P + w_0(k))^{-1}(PB_M(k)P + W_M(k)))^{-1}$$

$$\times (I + PB_0(k)P + W_0(k))^{-1}\frac{2T\hat{w}_a(\lambda, .)}{\sqrt{1 - M^2}}$$

$$= (I + PB_0(k)P + W_0(k))^{-1}(I - PB_M(k)P + W_M(k))\frac{2T\hat{w}_a(\lambda, .)}{\sqrt{1 - M^2}}$$

up to the first two terms.   □

Existence Theorem for Small k

Theorem 5.16. *The Possio equation has a unique solution for small enough k, given by the Neumann expansion:*

$$\hat{A}(\lambda, .) = \sum_{n=0}^{\infty}(-1)^n(PB(k)P + W(k))^n\frac{2T\hat{w}_a(\lambda, .)}{\sqrt{1 - M^2}}.$$

In particular for $k = 0$, we have:

$$\hat{A}(0, .) = \frac{2T\hat{w}_a(0, .)}{\sqrt{1 - M^2}}$$

and satisfies the Kutta condition if

$$\hat{w}_a(0, .) \text{ is in } C_1[-b, b].$$

Proof. We don't need to invoke the fact that $W(k)$ is compact. We prove that $\mathbb{PB}(k)\mathbb{P} + W(k)$ goes to zero in operator norm with k.

Now $||(I + \mathbb{PB}(k)\mathbb{P})^{-1}$ -I $||$ goes to zero as k goes to zero, because $||\mathbb{PB}(k)\mathbb{P}||$ goes to zero as k goes to zero. Hence it is enough to show that $||W(k)||$ is small for small k.

Now

$$h_+(k, x) = \frac{1}{\pi}\sqrt{\frac{b - x}{b + x}}\int_0^\infty \sqrt{\frac{\sigma}{2b + \sigma}}\frac{1}{b + \sigma + x}e^{-k\sigma}d\sigma$$

$$= \frac{1}{\pi}\sqrt{\frac{b - x}{b + x}}\int_0^\infty \sqrt{\frac{\sigma}{2bk + \sigma}}\frac{1}{\sigma + k(b + x)}e^{\sigma}d\sigma,$$

which is then defined for every k except $k \leq 0$.

For $k = i\omega$, this yields

$$h_+(i\omega, x) = \frac{1}{\pi}\sqrt{\frac{b-x}{b+x}}\int_0^\infty \sqrt{\frac{\sigma}{2bi\omega + \sigma}}\frac{1}{\sigma + i\omega(b+x)}e^{-\sigma}\,d\sigma,$$

$$|h_+(i\omega, x)| \leq \frac{1}{\pi}\frac{1}{\sqrt{2b|\omega|}}\sqrt{\frac{b-x}{b+x}}\int_0^\infty \sqrt{\frac{1}{\sigma}}e^{-\sigma}\,d\sigma.$$

And for $k > 0$,

$$\leq \frac{1}{\pi}\sqrt{\frac{b-x}{b+x}}\int_0^\infty \sqrt{\frac{\sigma}{2bk + \sigma}}\frac{1}{\sigma}e^{-\sigma}\,d\sigma$$

$$\leq \frac{1}{\pi}\sqrt{\frac{b-x}{b+x}}\int_0^\infty \sqrt{\frac{\sigma}{2bk}}\frac{1}{\sigma}e^{-\sigma}\,d\sigma$$

$$\leq \sqrt{\frac{1}{2bk}}\frac{1}{\pi}\sqrt{\frac{b-x}{b+x}}\int_0^\infty \frac{1}{\sqrt{\sigma}}e^{-\sigma}\,d\sigma.$$

Hence for $k = |k|e^{i\theta}$, we have

$$\|kh_+(k)\| = 0\left(\frac{\sqrt{|k|}}{\sqrt{|\cos\theta|}}\right)$$

$$= 0\left(\sqrt{|k|}\right) \quad \text{if } \theta = \frac{\pi}{2}.$$

We have a similar estimate for $\|kh_-(k)\|$.

Hence the theorem holds as k goes to zero in such a way that the real part is bounded away from zero, or equal to zero. We note that the larger the M, the smaller the k we will need. The Neumann expansion follows as soon as k is small enough so that the operator norm

$$\|(I + \mathbb{P}\mathbb{B}(k)\mathbb{P})^{-1}W(k)\| < 1. \qquad \square$$

Existence Theorem for Large k

The Neumann expansion does not converge for large Re k.

Here we go back to the Possio equation in the form (5.50):

$$\frac{2}{M}\hat{w}_a(\lambda,.) = \hat{A}(\lambda,.) + \frac{\sqrt{1-M^2}}{M}\mathbb{PH}(\mathbb{B}(k) - \mathbb{B}(\infty))\hat{A}(\lambda,.)$$

whereas Re k goes to infinity. Here $\mathbb{PH}(\mathbb{B}(k)-\mathbb{B}(\infty))$ goes to zero strongly only, not in operator norm! Hence, for example, if $\hat{w}_a(\lambda,.) = f(.)$ and if $\hat{A}(\lambda,.)$ converges then the limit will be $(2/M)f(.)$.

Note this cannot hold for $M = 0$. The limit blows up as we show in that case. We show that it does hold for $M = 1$. Note also that in (5.50) we cannot assert that

$$\mathbb{PH}(\mathbb{B}(k) - \mathbb{B}(\infty))$$

is compact. Indeed we cannot assert that

$$I + \frac{\sqrt{1-M^2}}{M}\mathbb{PH}(\mathbb{B}(k) - \mathbb{B}(\infty))$$

has a bounded inverse even for large Re k.

Theorem 5.17. *Suppose the Possio equation has a unique solution. Then*

$$I + \frac{\sqrt{1-M^2}}{M}\mathbb{PH}(\mathbb{B}(k) - \mathbb{B}(\infty))\mathbb{P}$$

has a bounded inverse on $L_p(-b,b)$.

Proof.

$$\hat{A}(\lambda,.) = ((I + \mathbb{PB}(k)\mathbb{P}) + \mathcal{T}\mathbb{PH}(I - \mathbb{P})\mathbb{B}(k)\mathbb{P})^{-1}\frac{2\mathcal{T}\hat{w}_a(\lambda,.)}{\sqrt{1-M^2}}.$$

Hence by (5.50)

$$\left(I + \frac{\sqrt{1-M^2}}{M}\mathbb{PH}(\mathbb{B}(k) - \mathbb{B}(\infty))\right) \cdot (I + \mathbb{PB}(k)\mathbb{P}$$

$$+ \mathcal{T}\mathbb{PH}(I - \mathbb{P})B(k)\mathbb{P})^{-1}\frac{2\mathcal{T}\hat{w}_a(\lambda,.)}{\sqrt{1-M^2}} = \frac{2}{M}\hat{w}_a(\lambda,.).$$

Here $\hat{w}_a(\lambda,.)$ is any element of $C_1[-b,b]$, therefore we have (for nonzero M):

$$\left(I - \frac{\sqrt{1-M^2}}{M}\mathbb{PH}(\mathbb{B}(k) - \mathbb{B}(\infty))\right) \cdot (I + \mathbb{PB}(k)\mathbb{P}$$

$$+ \mathcal{T}\mathbb{PH}(I - \mathbb{P})\mathbb{B}(k)\mathbb{P})^{-1}\frac{M}{\sqrt{1-M^2}}\mathcal{T} = I,$$

where I is the identity operator on $C_1[-b, b]$. Hence

$$\left[I + \frac{\sqrt{1 - M^2}}{M} \mathbb{P}\mathbb{H}(\mathbb{B}(k) - \mathbb{B}(\infty)) \right]^{-1} f$$

$$= (I + \mathbb{P}\mathbb{B}(k)\mathbb{P} + \mathcal{T}\mathbb{P}\mathbb{H}(I - \mathbb{P})\mathbb{B}(k)\mathbb{P})^{-1} \frac{M}{\sqrt{1 - M^2}} \mathcal{T} f$$

for f in $C_1[-b, b]$.   □

The convergence for large Re k being only in the strong sense, we cannot assert that either side converges. However, assuming both sides converge strongly, we can check whether:

$$f = (I + \mathbb{P}\mathbb{B}(k)\mathbb{P} + \mathcal{T}\mathbb{P}\mathbb{H}(I - \mathbb{P})\mathbb{B}(k)\mathbb{P})^{-1} \frac{M}{\sqrt{1 - M^2}} \mathcal{T} f.$$

Now for f in $C_1[-b, b]$

$$(I + \mathbb{P}\mathbb{B}(\infty)\mathbb{P}) f = I - \mathbb{P}\left(I + \frac{M\mathbb{H}}{\sqrt{1 - M^2}} \right) \mathbb{P} f$$

$$= -\frac{M}{\sqrt{1 - M^2}} \mathbb{P}\mathbb{H}\mathbb{P} f,$$

$$\mathcal{T}\mathbb{P}\mathbb{H}(I - \mathbb{P})\mathbb{B}(\infty)\mathbb{P} f = \mathcal{T}\mathbb{P}\mathbb{H}(I - \mathbb{P})\left(-\mathbb{P} - \frac{M}{\sqrt{1 - M^2}} \mathbb{H}\mathbb{P} \right) f$$

$$= \frac{-M}{\sqrt{1 - M^2}} \mathcal{T}\mathbb{P}\mathbb{H}(I - \mathbb{P})\mathbb{H}\mathbb{P} f$$

$$= \frac{M}{\sqrt{1 - M^2}} (\mathcal{T} + \mathcal{T}\mathbb{P}\mathbb{H}\mathbb{P}\mathbb{H}) f.$$

Because

$$\mathcal{T}\mathbb{P}\mathbb{H}\mathbb{P}\mathbb{H} f = \mathbb{P}\mathbb{H} f, \quad \text{for } f \text{ in } C_1[-b, b],$$

we have

$$(I + \mathbb{P}\mathbb{B}(k)\mathbb{P} + \mathcal{T}\mathbb{P}\mathbb{H}(I - \mathbb{P})\mathbb{B}(k)\mathbb{P})^{-1} \frac{M}{\sqrt{1 - M^2}} \mathcal{T} f$$

$$\mathcal{T}^{-1}\mathcal{T} f = f.$$

Hence at least under the assumption that the Possio equation has a solution we do have that

$$\left(I + \frac{\sqrt{1 - M^2}}{M} \mathbb{P}\mathbb{H}(\mathbb{B}(k) - \mathbb{B}(\infty)) \right)^{-1} \qquad f = f$$

for f in $C_1[-b, b]$ as Re k goes to infinity.

This is as far as we go in terms of the question of existence of solution of the LDP.

We turn next to the time domain solutions.

Time Domain Solution: Incompressible Flow

So far we have followed the traditional (and Possio's own original formulation in the frequency domain where the main concern is to calculate the aeroelastic modes. However, as we delve more into the problem of stability—our main concern—we need the time-domain solution.

For incompressible flow we simply set $M = 0$ in the Possio equation. Let us see what the corresponding linear potential equation is.

Going back to (5.3) we set $M = 0$ therein. This yields

$$\frac{\partial^2 \phi}{\partial t^2} + 2U_\infty \left(\cos \alpha \frac{\partial^2 \phi}{\partial t \, \partial x} + \sin \alpha \frac{\partial^2 \phi}{\partial t \, \partial z} \right) - a_\infty^2 \left[\frac{\partial^2 \phi}{\partial x^2} + \frac{\partial^2 \phi}{\partial z^2} + \frac{\partial^2 \phi}{\partial y^2} \right] = 0.$$

Next we note that $M = 0$ for any U is equivalent to $a_\infty \to \infty$.

Hence dividing both sides by a_∞^2 and taking the limit, we obtain

$$\frac{\partial^2 \phi}{\partial x^2} + \frac{\partial^2 \phi}{\partial z^2} + \frac{\partial^2 \phi}{\partial y^2} = 0,$$

which is then our field equation for $M = 0$. This is of course the celebrated Laplace equation and time does not appear in it but does go into the boundary conditions:

$$\text{Far Field Velocity} = i U_\infty.$$

Flow Tangency

$$\frac{\partial \phi_1(t, x, y, 0\pm)}{\partial z} = w_a(t, x)$$

$$= (-1)[\dot{h}(t, y) + (x - a)\dot{\theta}(t, y) + U_\infty \cos \alpha \, \theta(t, y)],$$

$$x, y \in \Gamma. \tag{5.58}$$

Kutta–Joukowsky

With the Kussner–Doublet function

$$A_1(t, x, y) = -\frac{\delta \psi_1(t, x, y)}{U_\infty} \tag{5.59}$$

$$= 0 \text{ off the wing} \to 0 \text{ as } x \to b - \text{ as defined before.}$$

These are then the boundary dynamics.

We note that we can go over to the finite plane model, with the nonzero angle of attack making no difference except for replacing U_∞ by $U_\infty \cos \alpha$, and small α at that.

The Mikhlin multiplier (5.14) becomes

$$\frac{1}{2} \frac{1}{\lambda + i \omega_1 U_\infty \cos \alpha} \sqrt{\omega_1^2 + \omega_2^2}. \tag{5.60}$$

At the present time we have no solution technique for the corresponding Possio equation.

Note that we get the beam or high aspect ratio solution if we set $\omega_2 = 0$ which we may consider as a first approximation and try a power series expansion

$$\frac{1}{2} \frac{1}{\lambda + i \omega_1 U_\infty \cos \alpha} \sqrt{\omega_1^2 + \omega_2^2} = \frac{1}{2} \frac{|\omega_1|}{\lambda + i \omega_1 U_\infty \cos \alpha} \left(1 + \frac{1}{2} \frac{\omega_2^2}{\omega_1^2} \cdots \right).$$

See [55] for more on this.

Hence we specialize to the following:

Time Domain Solution Incompressible Flow: Typical Section Theory/Zero Angle of Attack

We can rewrite the potential field equation (5.43) now as simply that the divergence of the flow is zero:

$$\nabla . q = 0,$$

where q is the velocity vector and this is the defining property of incompressible flow.

The potential flow is now 2D:

$$\Delta \phi(t, x, z) = 0, \qquad x, z \in R^2, \qquad \text{except } z = 0; \qquad |x| < b,$$

which is then the boundary; we call attention to the fact that it is a singular boundary which makes all the difference and characterizes aeroelasticity.

We start with the Laplace Fourier transforms we have calculated. Thus

$$\hat{\phi}_1(\lambda, i\omega, z) = -\frac{1}{2}\frac{1}{k + i\omega}\hat{A}(\lambda, i\omega)e^{-|z|\omega},$$

$$\hat{A}(\lambda, i\omega) = (I + k\mathcal{L}(0))(2T\hat{w}_a(\lambda, .))$$

$$+\frac{L(k, (I + k\mathcal{L}(0))2T\hat{w}_a(\lambda, .))}{\int_0^\infty e^{-kt}\sqrt{\frac{2b+t}{t}}dt}\hat{h}\Bigg],$$

$$\hat{h}(x) = \frac{1}{\pi}\sqrt{\frac{b - x}{b + x}}\int_0^\infty e^{-k\sigma}\sqrt{\frac{2b + \sigma}{\sigma}}\frac{1}{x - b - \sigma}d\sigma,$$

$$|x| < bk = \lambda/U_\infty. \tag{5.61}$$

To get the time domain response we need to find the inverse Laplace/Fourier transforms.

Possio Equation Time Domain Solution: Incompressible Flow

Theorem 5.18. Let $u(t, .) = 2Tw_a(t, .)$, where $w_a(t, .)$ is given by (5.54) and is in $C_1(-b, b)$, as before.

We assume that the initial structure state is zero so that

$$w_a(0, .) = 0.$$

Assume the Sears inversion formula:

$$\frac{1}{k\int_0^\infty e^{-k\sigma}\sqrt{\frac{2b+\sigma}{\sigma}}d\sigma} = \int_0^\infty e^{-\lambda(t/U_\infty)}c_1(t)dt$$

$$= \int_0^\infty e^{-\lambda t}U_\infty c_1(tU_\infty)dt.$$

Then for $|x| < b$

$$A(t, x) = \gamma(t, x) + \frac{1}{U_\infty}\int_{-b}^x \dot{\gamma}(t, s)ds,$$

where

$$\gamma(t, x) = u(t, x) + \int_0^t h_c(t - \sigma, x)\left(\int_{-b}^b (\dot{u}(\sigma, s)ds\right)d\sigma,$$

where

$$\hat{h}_c = \frac{\hat{h}}{k \int_0^\infty e^{-k\sigma} \sqrt{\frac{2b+\sigma}{\sigma}} \, d\sigma} = \frac{\int_0^\infty e^{-k\sigma} \frac{1}{\pi} \sqrt{\frac{b-x}{b+x}} \frac{\sqrt{2b+\sigma}}{\sigma} \frac{1}{x-b-\sigma} \, d\sigma}{k \int_0^\infty e^{-k\sigma} \sqrt{\frac{2b+\sigma}{\sigma}} \, d\sigma} \tag{5.62}$$

and the inverse transform:

$$h_c(t,x) = \int_0^t U_\infty h(U_\infty \sigma, x) U_\infty c_1(U_\infty(t-\sigma)) d\sigma,$$

where

$$h(\sigma, x) = \frac{1}{\pi} \sqrt{\frac{b-x}{b+x}} \sqrt{\frac{2b+\sigma}{\sigma}} \frac{1}{x-b-\sigma}, \qquad |x| < b.$$

Proof.

$$\hat{A}(\lambda, i\omega) = (I + k\mathcal{L}(0)) \left[\hat{u}(\lambda, .) + \int_{-b}^b \hat{u}(\lambda, x) dx \frac{\hat{h}}{\int_0^\infty e^{-kt} \sqrt{\frac{2b+t}{t}} dt} \right],$$

which follows from (5.61) because we have:

$$L(k, (I + k\mathcal{L}(0))\hat{u}(\lambda, .)) = \int_{-b}^b e^{-k(b-x)} \left(\hat{u}(\lambda, x) + k \int_{-b}^x \hat{u}(\lambda, \xi) d\xi \right) dx,$$

where

$$\int_{-b}^b e^{-k(b-x)} k \int_{-b}^x \hat{u}(\lambda, \xi) d\xi dx = \int_{-b}^b \hat{u}(\lambda, \xi) d\xi - \int_{-b}^b e^{-k(b-x)} \hat{u}(\lambda, x) dx.$$

Hence

$$L(k, (I + k\mathcal{L}(0))\hat{u}(\lambda, .)) = \int_{-b}^b \hat{u}(\lambda, \xi) d\xi.$$

Hence

$$\hat{A}(\lambda, i\omega) = (I + k\mathcal{L}(0)) \left(\hat{u}(\lambda, .) + k \int_{-b}^b \hat{u}(\lambda, x) dx \hat{h}_c \right).$$

Let

$$\hat{\gamma}(\lambda, .) = \hat{u}(\lambda, .) + k \int_{-b}^b \hat{u}(\lambda, x) dx \frac{\hat{h}_c}{k}.$$

Then

$$kL(0)\hat{\gamma}(\lambda,.) = \frac{1}{U_\infty} \int_{-b}^{x} \lambda \hat{\gamma}(\lambda,\xi)d\xi.$$

Hence taking inverse transforms we have:

$$\gamma(t,.) = u(t,.) + \int_{-b}^{b} \int_{0}^{t} u(t-\sigma,x)dx h_c(\sigma,.)d\sigma,$$

which is (5.62). and $kL(0)\hat{\gamma}(\lambda,.)$ is the transform of $(1/U_\infty) \int_{-b}^{x} \dot{\gamma}(t,x)dx$, which proves the theorem.   □

Next using this result, we can derive the solution to the potential equation obtained a different way in [6].

Time Domain Potential

Theorem 5.19. *The time domain potential:*

$$\phi_1(t,x,z)\frac{z}{\pi} \int_{-b}^{b} \int_{0}^{t} A(t-\sigma,y)\frac{1}{((x-y-U_\infty\sigma)^2 + z^2)}d\sigma dy. \qquad (5.63)$$

Proof. Use

$$(k+i\omega)^{-1} = U_\infty \int_{0}^{\infty} e^{-\lambda t - i\omega t U_\infty}dt,$$

$$e^{-|\omega|z} = \frac{1}{2\pi} \int_{-\infty}^{\infty} e^{-i\omega x}\frac{2z}{x^2 + z^2}dx, \quad z > 0,$$

$$\frac{e^{-|\omega|z}}{(k+i\omega)} = \int_{0}^{\infty} e^{-\lambda t} \int_{-\infty}^{\infty} e^{-iwx}\left(\frac{1}{\pi}\frac{z}{(x-U_\infty t)^2 + z^2}\right)dxdt$$

yielding

$$\hat{\phi}_1(\lambda,i\omega,z) = \int_{-\infty}^{\infty} e^{-ix\omega}$$

$$\int_{0}^{\infty} e^{-\lambda t}dt \int_{0}^{t} \frac{z}{\pi} \int_{-b}^{b} \frac{1}{((x-y-U_\infty\sigma)^2 + z^2)}A(t-\sigma,x-y)d\sigma dydx.$$

Hence:

$$\phi_1(t, x, z) = \frac{z}{\pi} \int_{-b}^{b} \int_{0}^{t} A(t - \sigma, y) \frac{1}{((x - y - U_\infty \sigma)^2 + z^2)} d\sigma dy. \qquad \square$$

Remarks: The formula (5.63) actually holds for every M. So do:

$$\phi_1(t, x, -z) = -\phi_1(t, x, z),$$

$$\frac{\partial \phi_1(t, x, -z)}{\partial x} = -\frac{\partial \phi_1(t, x, z)}{\partial x},$$

$$\frac{\partial \phi_1(t, x, -z)}{\partial z} = \frac{\partial \phi_1(t, x, z)}{\partial z}.$$

Also the Fourier transform of

$$\frac{\partial \phi_1(t, x, z)}{\partial x} = i\omega.$$

Our next case where we are able to obtain a closed-form solution to the Possio equation is for $M = 1$, which is referred to as the sonic case.

The Sonic Case: Laplace Domain Typical Section; Zero Angle of Attack

For $M = 1$, the Possio multiplier equation becomes:

$$\hat{v}(\lambda, i\omega) = \frac{1}{2} \sqrt{\frac{k^2 + 2ki\omega}{k + i\omega}} \hat{A}(\lambda, i\omega), \qquad (5.64)$$

where the multiplier can be expressed in the form:

$$\Phi(1, k) = \frac{1}{2} \sqrt{\frac{k^2 + 2ki\omega}{k + i\omega}} = \frac{1}{2}(\sqrt{k}) \left(2 - \frac{k}{k + i\omega}\right) \left(\frac{1}{\sqrt{k + 2i\omega}}\right). \qquad (5.65)$$

Then let as before $\mathcal{L}(\lambda)$ denote the operator corresponding to the multiplier $1/\lambda + i\omega$ and $\mathcal{L}_2(k)$ correspond to the multiplier $\left(1/(\sqrt{k + 2i\omega})\right)$

$$\mathcal{L}_2(k) f = g; \qquad g(x) = \int_{-b}^{x} e^{-(k/2)(x-s)} \frac{f(s)}{\sqrt{2\pi(x - s)}} ds, \qquad |x| < b. \quad (5.66)$$

This is also a Volterra operator on $L_p(-b,b)$ into itself.

Thus we have

$$P\Phi(k,1)PA(\hat{\lambda},.) = \sqrt{k}(2I - k(k))\mathcal{L}_2(k)\hat{A}(\lambda,.)$$
$$= \hat{w}(\lambda,.). \qquad (5.67)$$

The Tricomi operator plays no role at all! $(2\,I - k\mathcal{L}(k))$ has a bounded inverse but not of course $\mathcal{L}_2(k)$. Thus we have

$$\mathcal{L}_2(k)\hat{A}(\lambda,.) = \frac{1}{\sqrt{k}}(2I - k\mathcal{L}(k))^{-1}\hat{w}(\lambda,.), \qquad (5.68)$$

where

$$(2I - k\mathcal{L}(k))^{-1} = \frac{1}{2}\left(I + \frac{k}{2}\mathcal{L}\left(\frac{k}{2}\right)\right).$$

But

$$\mathcal{L}_2(k)f = g$$

is essentially the Abel integral equation to find f given g. But we have to restrict g. Thus following Tricomi [11] we require that g have an L_p $(-b,b)$ derivative also. This defines a Sobolev space; we denote it $\mathcal{W}_p^1$. We can verify that the right side of (5.68) is indeed contained in this space in as much as

$$\hat{w}(\lambda,.) \qquad \text{is in } C_1(-b,b).$$

Hence we have the following.

Theorem 5.20. *The Possio equation (5.67) has the unique solution*

$$\hat{A}(\lambda,.) = \frac{2}{\sqrt{k}}\hat{w}_a(\lambda,-b) + \frac{2}{\sqrt{k}}\mathcal{L}_2(k)(\hat{w}_{a'}(\lambda,.) + k\hat{w}_a(\lambda,.)). \qquad (5.69)$$

Lemma 5.21. *For g in $\mathcal{W}_p^1$ the integral equation $\mathcal{L}_2(k)f = g$ has a unique solution in $L_p(-b,b)$ given explicitly by:*

$$f(x) = f_1(x) + f_2(x) + f_3(x), \qquad |x| < b, \qquad (5.70)$$

where

$$f_1(x) = \frac{k}{2\sqrt{\pi}}\int_{-b}^{x} e^{-((k(x-s))/2)}\frac{g(s)}{\sqrt{x-s}}ds,$$

$$f_2(x) = \frac{1}{\sqrt{\pi}}\int_{-b}^{x} e^{-((k(x-s))/2)}\frac{g'(s)}{\sqrt{x-s}}ds,$$

$$f_3(x) = \frac{1}{\sqrt{\pi}}g(-b)\frac{e^{-k((x+b)/2)}}{\sqrt{x+b}}.$$

We note that

$$f_1 + f_2 = \sqrt{2}\mathcal{L}_2(k)\left(g' + \frac{k}{2}g\right).$$

Proof. First we show that zero is not in the point spectrum of $\mathcal{L}_2(k)$. For if $\mathcal{L}_2(k)f = 0$ for f in $L_p(-b, b)$ by (5.61) we must have

$$\int_{-b}^x \frac{h(s)}{\sqrt{x-s}}ds = 0, \qquad |x| < b, \quad \text{where } h(x) = e^{k(x/2)}f(x).$$

Next we follow the ingenious argument by Tricomi [11] to show that $h(.)$ must be zero.

$$\int_{-b}^x \frac{1}{\sqrt{x-y}}\int_{-b}^y \frac{h(s)}{\sqrt{y-s}}dsdy = \int_{-b}^x h(\sigma)\int_\sigma^x \frac{1}{\sqrt{(x-y)(y-\sigma)}}dyd\sigma,$$

where the inner integral is $= \pi$ from which the result follows.

Next (5.61) is also proved by Tricomi [11] and we can use the notation:

$$f = \mathcal{L}_2(k)^{-1}g$$

proving the lemma.

Next we go back to (5.68):

$$\hat{A}(\lambda, .) = \mathcal{L}_2(k)^{-1}\frac{1}{2\sqrt{k}}\left(I + \frac{k\mathcal{L}(\frac{k}{2})}{2}\right)\hat{w}_a(\lambda, .).$$

To simplify this further, we note that if

$$g = \left(I + \frac{k\mathcal{L}(\frac{k}{2})}{2}\right)f, \text{ then}$$

$$f(-b) = g(-b)$$

and

$$g' + \frac{k}{2}g = f' + kf$$

from which (5.69) follows. $\square$

It is interesting to look at some special cases first.

Special Cases

1. Case 1: Solution for

$$\hat{w}_a(\lambda, x) = f_1(x) = 1 \quad |x| < b$$

is

$$\hat{A}_1(\lambda, .) = \frac{2}{\sqrt{k}} \frac{e^{-k((b+x)/2)}}{\sqrt{2\pi(b+x)}} + 2\sqrt{k} \int_{-b}^{x} e^{-(k/2)(x-s)} \frac{1}{\sqrt{2\pi(x-s)}} ds, \quad |x| < b.$$

Note that the solution does not converge as $k \to 0$, and goes to the value 2, as $\mathrm{Re}\, k \to \infty (2/M$ more generally).

This is consistent with our general theory for nonzero M.

2. Case 2:

$$\hat{w}_a(\lambda, x) = f_2(x) = x, \quad |x| < b,$$

left as an exercise.

Time Domain Solution $M = 1$

Next let us look at the time domain solution.

To get the field equation we have only to make $M = 1$ in the general form (5.3):

$$\frac{\partial^2 \phi}{\partial t^2} + 2a_\infty \frac{\partial^2 \phi}{\partial t \partial x} - a_\infty^2 \frac{\partial^2 \phi}{\partial z^2} = 0. \tag{5.71}$$

The far field is now

$$\phi_\infty = x a_\infty \quad \text{because } U_\infty = a_\infty.$$

The boundary dynamics are the same as for every M, except that now $U_\infty = a_\infty$.

Hence to get the time domain response we have only to take inverse Laplace transforms in (5.70) which here unlike the case $M = 0$, where we needed the Sears theorem, is more straightforward.

Here we merely state it, referring to [19] for more details as necessary.

$$A(t, .) = g_1(t, .) + g_2(t, .) + g_3(t, .),$$

$$g_1(t, x) = 2 \int_0^t w_a(t - \sigma, -b) G(\sigma, x) d\sigma,$$

$$G(t,x) = \frac{1}{\sqrt{2(b+x)}} \left(t - \frac{(b+x)}{2} \right)^{-(1/2)}, \qquad t > \frac{(b+x)}{2},$$

$$g_2(x) = 2 \int_0^t \int_{-b}^x \frac{1}{\sqrt{2(x-\xi)}} \left(\sigma - \frac{(x-\xi)}{2} \right)^{-(1/2)} w_a'(t-\sigma,\xi) d\sigma d\xi,$$

$$g_3(t,x) = 2 \frac{d}{dt} \int_0^t \int_{-b}^x \frac{1}{\sqrt{2(x-\xi)}} \left(\sigma - \frac{(x-\xi)}{2} \right)^{-(1/2)} w_a(t-\sigma,\xi) d\sigma d\xi$$

$$+ 2w_a(t,x).$$

5.5   Linear Time Domain Airfoil Dynamics

The Convolution/Evolution Equation

Having calculated explicitly the pressure jump across the wing, for $M = 0$ and $M = 1$ we now go on to consider the time domain structure response to the aerodynamic lift and moment.

We begin with incompressible flow: $M = 0$.

It turns out that we can save a lot of tedious analysis by first taking the Laplace transforms of the structure dynamics equations and waiting until the end before inverting to the time domain. Here we follow [21] and consider Re $\lambda > 0$, unless otherwise specified.

First note that now for us

$$\hat{w}_a(\lambda,x) = -[\lambda \hat{h}(\lambda,s) + (x-ab)\lambda\hat{\theta}(\lambda,s) + U_\infty\hat{\theta}(\lambda,s)] \qquad |x| < b. \quad (5.72)$$

To find the corresponding lift and moment it is convenient (and customary) to normalize the half-chord "b" to be equal to 1. Thus we redefine k as

$$k = \frac{b\lambda}{U_\infty}$$

and define

$$\hat{w}(\lambda,x) = \hat{w}_a(\lambda,bx), \qquad |x| < 1,$$

which can be expressed as

$$= -U_\infty[k\hat{h}(\lambda,s) + (k(x-a)+1)\hat{\theta}(\lambda,s)], \qquad |x| < 1. \quad (5.73)$$

And the normalized Possio equation, cf. (5.23), becomes

$$\hat{w}(\lambda, x) = \int_{-1}^{1} \hat{P}(\lambda, x - \xi)\hat{A}(\lambda, \xi)d\xi, \qquad |x| < 1,$$

where continuing the normalization,

$$\hat{P}(\lambda, x) = bP(\lambda, bx), \quad |x| < 1,$$

$$\hat{A}(\lambda, x) = A(\lambda, bx), \quad |x| < 1.$$

So we may proceed with using our formulae for the solution just taking $b = 1$, with k now being taken as $b(\lambda/U_\infty)$. In particular let $\hat{A}_i(\lambda, x)$ be the solution corresponding to $\hat{w}(\lambda, x) = f_i(x), i = 1, 2$ where

$$f_1(x) = 1; \qquad f_2(x) = x, \qquad |x| < 1$$

and

$$w_{ij} = \int_{-1}^{1} f_i(x)\hat{A}_j(\lambda, x)dx, \qquad i = 1, 2; \qquad j = 1, 2. \tag{5.74}$$

To calculate the lift and moment we take:

$$w_a(\lambda, x) = -[\lambda\hat{h}(\lambda, s) + (x - ab)\lambda\hat{\theta}(\lambda, s) + U_\infty\hat{\theta}(\lambda, s)] \qquad |x| < b \tag{5.75}$$

$$= -U_\infty[k\hat{h}(\lambda, s) + (k(x - a) + 1)\hat{\theta}(\lambda, s)], \qquad |x| < 1, \tag{5.76}$$

so that the transform of the lift

$$\hat{L}(\lambda, s) = \rho bU_\infty \int_{-1}^{1} \hat{A}(\lambda, x)dx$$

$$= -\rho bU_\infty^2 \left[\frac{k}{b}w_{11}\hat{h}(\lambda, s) + (kw_{12} + (1 - ka)w_{11})\hat{\theta}(\lambda, s)\right] \tag{5.77}$$

and the transform of the moment

$$\hat{M}(\lambda, s) = \rho b^2 U_\infty \int_{-1}^{1} (x - a)\hat{A}(\lambda, x)dx$$

$$= -\rho b^2 U_\infty^2 \left[\frac{k}{b}(w_{21} - aw_{11})\hat{h}(\lambda, s)\right.$$

$$\left. + (kw_{22} + (1 - ak)w_{21} - akw_{12} - a(1 - ak)w_{11})\hat{\theta}(\lambda, s)\right]. \tag{5.78}$$

Next let us calculate the w_{ij} for $M = 0; \alpha = 0$.

Thus for $\hat{A}_1(\lambda, .)$, we go back to (5.61) and note that where now

$$u(\lambda, .) = 2T\hat{w}_a(\lambda, .)$$

$$u(\lambda, x) = \sqrt{\frac{1-x}{1+x}} \frac{2}{\pi} \int_{-1}^{1} \sqrt{\frac{1+\xi}{1-\xi}} \frac{1}{\xi - x} d\xi$$

$$= 2\sqrt{\frac{1-x}{1+x}}, \qquad |x| < 1$$

and hence

$$\int_{-1}^{1} u(\lambda, x) dx = 2\pi.$$

Hence

$$\hat{A}_1(\lambda, .) = (I + k\mathcal{L}(0))(u(\lambda, .) + 2\pi \hat{h}_c(\lambda, .))$$

and in turn

$$w_{11} = 2\pi + 2\pi \int_{-1}^{1} \hat{h}_c(\lambda, x) dx + k \int_{-1}^{1} (1-x)(u(\lambda, x) + 2\pi \hat{h}_c(\lambda, x)) dx,$$

where

$$\int_{-1}^{1} \hat{h}_c(\lambda, x) dx = \frac{1}{k \left(\int_0^\infty e^{-k\sigma} \sqrt{\frac{2+\sigma}{\sigma}} d\sigma \right)} - 1$$

$$= \frac{1}{k} \frac{1}{(K_0(k) + K_1(k))e^k} - 1,$$

where $K_0(.)$. $K_1(.)$ are Bessel K functions of order zero and one.

$$k \int_{-1}^{1} (1-x) u(\lambda, x) dx = 2k \int_{-1}^{1} (1-x) \sqrt{\frac{1-x}{1+x}} dx$$

$$= 3k\pi.$$

And

$$k \int_{-1}^{1} (1-x) \hat{h}_c(\lambda, x) dx = \left(-\frac{1}{k} - e^k K_1(k) \right) \Big/ ((K_0(k) + K_1(k))e^k) - 1.$$

Using

$$\int_{-1}^{1} \frac{1}{\pi}(1-x)^{3/2}\frac{1}{\sqrt{1+x}(x-1-\sigma)}dx$$

$$= \left(-1-\sigma\left(\sqrt{\frac{\sigma}{2+\sigma}}-1\right)\right)\left(\sqrt{\frac{2+\sigma}{\sigma}}\right)$$

and

$$\int_{0}^{\infty} e^{-k\sigma}\left(\sqrt{\sigma(2+\sigma)}-\sigma-\sqrt{\frac{2+\sigma}{\sigma}}\right)d\sigma \Bigg/ \left(\int_{0}^{\infty} e^{-kt}\sqrt{\frac{2+t}{t}}dt\right)$$

$$= \frac{\int_{0}^{\infty} e^{-kt}(\sqrt{\sigma(2+\sigma)}-\sigma)d\sigma}{\int_{0}^{\infty} e^{-kt}\sqrt{\frac{2+t}{t}}dt} - \frac{1}{k},$$

where

$$\int_{0}^{\infty} e^{-k\sigma}(\sqrt{\sigma(2b+\sigma)}-\sigma)d\sigma$$

$$= \int_{0}^{\infty} e^{-k\sigma}\sigma\left(\sqrt{\frac{1}{\sigma}(2b+\sigma)}-1\right)d\sigma$$

$$= -\frac{d}{dk}\int_{0}^{\infty} e^{-k\sigma}\left(\sqrt{\frac{1}{\sigma}(2b+\sigma)}-1\right)d\sigma, \quad b=1$$

$$= -\frac{d}{dk}\left\{(e^{k}(K_0(k)+K_1(k))-\frac{1}{k}\right\}$$

$$= -\left[e^{k}(K_0(k)+K_1(k))+(e^{k})\right.$$

$$\left.\left(-K_1(k)-K_0(k)-\frac{K_1(k)}{k}\right)+\frac{1}{k^2}\right]$$

$$= e^{k}\frac{K_1(k)}{k}-\frac{1}{k^2}.$$

Hence

$$w_{11} = 2\pi + 2\pi \left(\frac{1}{k} \frac{1}{(K_0(k) + K_1(k))e^k} - 1 \right) + 3k\pi$$

$$+ 2\pi \left(e^k K_1 - \frac{1}{k} \right) ((K_0(k) + K_1(k))e^k)^{-1} - 2\pi k$$

$$= 2\pi T(k) + \pi k,$$

where

$$T(k) = \frac{K_1(k)}{K_0(k) + K_1(k)}$$

is called the Theodorsen function [6].

We may note also that

$$\hat{A}_1(\lambda, x) = 2\sqrt{\frac{1-x}{1+x}}$$

$$+ \left(2\sqrt{\frac{1-x}{1+x}} \int_0^\infty \sqrt{\frac{2+\sigma}{\sigma}} e^{-k\sigma} \frac{d\sigma}{x-1-\sigma} \right) \Big/ e^k (K_0(k) + K_1(k))$$

$$+ 2k \int_{-1}^x \sqrt{\frac{1-s}{1+s}} ds + 2k \int_{-1}^x \sqrt{\frac{1-s}{1+s}} ds$$

$$\times \int_0^\infty \sqrt{\frac{2+\sigma}{\sigma}} e^{-k\sigma} \frac{d\sigma}{s-1-\sigma} \Big/ (e^k (K_0(k) + K_1(k))).$$

This is another example of an explicit solution to the Possio equation.

Note that in this form we can show

$$Ltx \to 1 \quad \hat{A}(\lambda, x) = 0.$$

Next

$$w_{21} = \int_{-1}^1 x \hat{A}_1(\lambda, x) dx = -\pi T(k)$$

omitting the details of the calculation, which are similar to the previous case.

We need next

$$\hat{A}_2(\lambda, x).$$

Here

$$u(\lambda, x) = 2\sqrt{1-x^2},$$

$$\int_{-1}^{1} u(\lambda, x)dx = \pi.$$

Hence

$$\hat{A}_2(\lambda, x) = 2\sqrt{1 - x^2}$$

$$+ \left(\pi \sqrt{\frac{1-x}{1+x}} \int_0^\infty \sqrt{\frac{2+\sigma}{\sigma}} e^{-k\sigma} \frac{d\sigma}{x - 1 - \sigma} \right) \bigg/ e^k (K_0(k) + K_1(k))$$

$$+ 2k \int_{-1}^{x} \sqrt{1 - s^2} ds$$

$$+ k\pi \left[\int_{-1}^{x} \sqrt{\frac{1-s}{1+s}} ds \int_0^\infty \sqrt{\frac{2+\sigma}{\sigma}} e^{-k\sigma} \frac{d\sigma}{s - 1 - \sigma} \bigg/ e^k (K_0(k) + K_1(k)) \right].$$

Hence exploiting the previous calculations of integrals we obtain:

$$w_{12} = \pi T(k),$$

$$w_{22} = \frac{\pi}{2} \left(1 + \frac{k}{4} - T(k) \right),$$

$$w_{11} = k\pi + 2\pi T(k),$$

$$w_{21} = -\pi T(k). \tag{5.79}$$

These hold for any b, with

$$k = \frac{\lambda b}{U}.$$

Next let us calculate the lift:

$$\hat{L}(\lambda, s) = - \rho b U_\infty^2 \left[\frac{k}{b}(k\pi + 2\pi T(k))\hat{h}(\lambda, s) \right.$$

$$\left. + (k\pi T(k) + (1 - ak)(k\pi + 2\pi T(k)))\hat{\theta}(\lambda, s) \right]$$

$$= - \rho b U_\infty^2 \left(k^2 \frac{\pi}{b} \hat{h}(\lambda, s) + (1 - ak)k\pi\hat{\theta}(\lambda, s) \right)$$

$$+ T(k)(-2\pi\rho b U_\infty) \left(\lambda\hat{h}(\lambda, s) + b \left(\frac{1}{2} - a \right) \lambda\hat{\theta}(\lambda, s)U_\infty\hat{\theta}(\lambda, s) \right)$$

$$= -\lambda^2 \rho \pi b^2 \hat{h}(\lambda, s) + \lambda^2 \pi \rho b^3 a \hat{\theta}(\lambda, s) - \lambda \pi \rho b^2 U_\infty \hat{\theta}(\lambda, s)$$

$$- T(k) 2\pi \rho b U_\infty \left(\lambda \hat{h}(\lambda, s) + b \left(\frac{1}{2} - a \right) \lambda \hat{\theta}(\lambda, s) + U_\infty \hat{\theta}(\lambda, s) \right).$$

$$(5.80)$$

Note the appearance of terms containing λ^2, and the term containing the Theodorsen function. We note first that

$$K_0(k) = \int_1^\infty e^{-kt} \frac{dt}{\sqrt{t^2 - 1}},$$

$$K_1(k) = -K_0'(k)$$

$$= \int_1^\infty e^{-kt} \frac{t\,dt}{\sqrt{t^2 - 1}}.$$

Hence

$$K_0(k) + K_1(k) = \int_1^\infty e^{-kt} \sqrt{\frac{t+1}{t-1}}\,dt,$$

$$e^k(K_0(k) + K_1(k)) = \int_0^\infty e^{-kt} \sqrt{\frac{t+2}{t}}\,dt,$$

$$e^k K_0(k) = \int_0^\infty e^{-kt} \sqrt{(1/(t(t+2)))}\,dt,$$

$$e^k K_1(k) = \int_0^\infty e^{-kt} ((t+1)/(\sqrt{(t(t+2))}))\,dt$$

and

$$\frac{T(k)}{k} = e^k K_1(k)/k e^k(K_0(k) + K_1(k))$$

$$= \int_0^\infty e^{-kt} L(t)\,dt,$$

where

$$L(t) = \int_0^t c_1(t - \sigma) \frac{\sigma + 1}{\sqrt{\sigma(\sigma + 2)}}\,d\sigma \qquad t > 0,$$

known as the Wagner function [6], where

$$c_1(t) \text{ is nonnegative and hence so is } L(.)$$

and

$$\int_0^\infty c_1(t)dt = 1$$

because

$$1/(k(K_0(k) + K_1(k))e^k) \to 1 \quad \text{as} \quad k \to 0.$$

Here it is known also (Kussner–Sears [6]) that

$$c_1(t) = O(t^{-(1/2)}) \quad \text{for small } t$$

and we see so is

$$\frac{t+1}{\sqrt{t(t+2)}}.$$

Hence $L(t)$ does not go to zero as t goes to zero!
To find the limit let us note that $T(k)$ has the following expansion.
   At zero

$$\begin{aligned}
T(k) =& 1 + (\text{EulerGamma} - \log[2] + \log[k])k \\
&+ (\text{EulerGamma}^2 - 2\text{EulerGamma} \log[2] \\
&+ \log[2]^2 + 2\text{EulerGamma} \log[k] - 2\log[2]\log[k] \\
&+ \log[k]^2)k^2 + O[k]^3
\end{aligned}$$

showing the logarithmic essential singularity along the negative real line.
   Also

$$kT'(k) \to 0 \quad \text{as } k \to 0.$$

At infinity:

$$T(k) = \frac{1}{2} + \frac{1}{8k} - \frac{1}{16k^2} + O\left[\frac{1}{k}\right]^3.$$

Note that

$$kT'(k) \to 0 \quad \text{at infinity.}$$

The inverse Laplace transform of $T(k)$ contains a delta function at the origin and hence

$$k\left(\frac{T(k)}{k}\right) \quad \text{goes to} \quad \frac{1}{2} = L(0+) \text{ as } \quad \text{Re } k \to \infty.$$

Now

$$T(k) - \frac{1}{2} = \frac{1}{2}(K_1(k) - K_0(k))/(K_1(k) + K_0(k))$$

$$= \frac{1}{2}k(e^k K_1(k) - e^k K_0(k))/(ke^k(K_1(k) + K_0(k)))$$

and

$$k(\mathrm{e}^k K_1(k) - \mathrm{e}^k K_0(k)) = \int_0^\infty k\mathrm{e}^{-kt}\sqrt{\frac{t}{t+2}}\mathrm{d}t$$

$$= \int_0^\infty \mathrm{e}^{-kt}\frac{\mathrm{d}}{\mathrm{d}t}\sqrt{\frac{t}{t+2}}\mathrm{d}t$$

$$= \int_0^\infty \mathrm{e}^{-kt}1/(\sqrt{(t(t+2)^3)})\,\mathrm{d}t.$$

Hence

$$T(k) - \frac{1}{2} = \int_0^\infty \mathrm{e}^{-kt}\ell(t)\mathrm{d}t,$$

where

$$\ell(t) = \int_0^t c_1(t-\sigma)1/(\sqrt{(\sigma(\sigma+2)^3)})\mathrm{d}\sigma,$$

where

$$\ell(t) \geq 0; \quad \int_0^\infty \ell(t)\mathrm{d}t = \frac{1}{2}.$$

Hence we have that

$$L(t) = \frac{1}{2} + \int_0^t \ell(t)\mathrm{d}t,$$

$$L(\infty) = 1 = T(0+),$$

$$\int_0^\infty \mathrm{e}^{-kt}\ell(t)\mathrm{d}t = k\int_0^\infty \mathrm{e}^{-kt}L(t)\mathrm{d}t - \frac{1}{2}. \tag{5.81}$$

Hence

$$\text{as} \quad |k| \to 0, \qquad \operatorname{Re} k > 0,$$

$$k\int_0^\infty \mathrm{e}^{-kt}L(t)\mathrm{d}t \to 0. \tag{5.82}$$

Next the moment:

$$\hat{M}(\lambda, s) = -\rho U_\infty^2 b^2 \left\{ \left[\frac{k}{b}(-\pi T(k) - a(k\pi + 2\pi T(k))\right]\hat{h}(\lambda, s) \right.$$

$$+ \left[k\frac{\pi}{2}\left(1 + \frac{k}{4} - T(k)\right) - (1-ak)\pi T(k) - ak\pi T(k)\right.$$

$$\left. -a(1-ak)(k\pi + 2\pi T(k))\right]\hat{\theta}(\lambda, s)\right\}$$

$$= \rho b^3 \lambda^2 a \pi \hat{h}(\lambda, s) - \pi \rho b^4 \left(a^2 + \frac{1}{8} \right) \lambda^2 \hat{\theta}(\lambda, s)$$

$$+ (-\rho U_\infty b^3) \left(\frac{\pi}{2} - a\pi \right) \lambda \hat{\theta}(\lambda, s) + T(k)[-\rho U_\infty^2 b^2]$$

$$\times \left\{ \left(-\pi \frac{k}{b} - 2\pi a \frac{k}{b} \right) \hat{h}(\lambda, s) + \left(-k \frac{\pi}{2} - \pi - a(1 - ak)2\pi \right) \hat{\theta}(\lambda, s) \right\},$$

where the last term containing the Theodorsen function can be expressed as

$$T(k)[-\rho U_\infty^2 b^2] \left(\left(-\pi \frac{k}{b} \hat{h}(\lambda, s) + \left(-k \frac{\pi}{2} - \pi \right) \right) \hat{\theta}(\lambda, s) \right.$$

$$- 2\pi a \, \hat{\theta}(\lambda, s) + a^2 k 2\pi \hat{\theta}(\lambda, s) - 2\pi a \frac{k}{b} \hat{h}(\lambda, s))$$

$$= T(k)[\rho U_\infty b^2] \left(\frac{1}{2} + a \right) \left(2\pi \lambda \hat{h}(\lambda, s) + b\pi(1 - 2a)\lambda \hat{\theta}(\lambda, s) + 2\pi U_\infty \hat{\theta}(\lambda, s) \right).$$

It is convenient now to define a new function

$$\hat{w}(\lambda, s) = 2\pi \lambda \hat{h}(\lambda, s) + b\pi(1 - 2a)\lambda \hat{\theta}(\lambda, s) + 2\pi U_\infty \hat{\theta}(\lambda, s)$$

with the inverse Laplace transform

$$w(t, s) = 2\pi \dot{h}(t, s) + b\pi(1 - 2a)\dot{\theta}(t, s) + 2\pi U_\infty \theta(t, s), \qquad t > 0.$$

Hence using this function:

$$\hat{L}(\lambda, s) = -\lambda^2 \rho \pi b^2 \hat{h}(\lambda, s) + \lambda^2 \pi \rho b^3 a \hat{\theta}(\lambda, s) - \lambda \pi \rho b^2 U_\infty \hat{\theta}(\lambda, s)$$
$$- T(k)\rho b U_\infty \hat{w}(\lambda, s)$$

$$\hat{M}(\lambda, s) = \rho b^3 \lambda^2 a \pi \hat{h}(\lambda, s) - \pi \rho b^4 \left(a^2 + \frac{1}{8} \right) \lambda^2 \hat{\theta}(\lambda, s) + (-\rho U_\infty b^3)$$

$$\times \left(\frac{\pi}{2} - a\pi \right) \lambda \hat{\theta}(\lambda, s) + T(k)[\rho U_\infty b^2] \left(\frac{1}{2} + a \right) \hat{w}(\lambda, s)$$

$$= \rho b^3 \lambda^2 a \pi \hat{h}(\lambda, s) - \pi \rho b^4 \left(a^2 + \frac{1}{8} \right) \lambda^2 \hat{\theta}(\lambda, s)$$

$$- \frac{\rho b^2 U_\infty}{2} \hat{w}(\lambda, s) + \rho b^2 U_\infty \pi \lambda \hat{h}(\lambda, s) + \pi \rho b^2 U_\infty^2 \hat{\theta}(\lambda, s)$$

$$+ T(k)[\rho U_\infty b^2] \left(\frac{1}{2} + a \right) \hat{w}(\lambda, s).$$

Using

$$(-\rho U_\infty b^3 \pi) \left(\frac{1}{2} - a \right) \lambda \hat{\theta}(\lambda, s)$$

$$= -\frac{\rho b^2 U_\infty}{2} (2\pi \lambda \hat{h}(\lambda, s) + \pi b(1 - 2a)\lambda \hat{\theta}(\lambda, s)$$

$$+ 2\pi U_\infty \hat{\theta}(\lambda, s)) + \pi \rho b^2 U_\infty^2 \hat{\theta}(\lambda, s)$$

$$= -\frac{\rho b^2 U_\infty}{2} \hat{w}(\lambda, s) + \frac{\rho b^2 U_\infty}{2} 2\pi \lambda \hat{h}(\lambda, s) + \pi \rho b^2 U_\infty^2 \hat{\theta}(\lambda, s).$$

We recall now our Hilbert space formulation of the structure dynamics in Chap. 2 with the Hilbert space H as therein and M and A defined as therein along with the notation:

$$x(t) = \begin{pmatrix} h(t) \\ \theta(t) \end{pmatrix},$$

$$M\ddot{x}(t) + Ax(t) = \begin{pmatrix} L(t) \\ M(t) \end{pmatrix}.$$

We take Laplace transforms on both sides, with $x(0) = 0$ and using the notation $\hat{x}(\lambda)$ for the Laplace transform we have

$$\lambda^2 M\hat{x}(\lambda) + A\hat{x}(\lambda) = \begin{pmatrix} \hat{L}(\lambda, .) \\ \hat{M}(\lambda, .) \end{pmatrix}.$$

Substituting for

$$\begin{pmatrix} \hat{L}(\lambda, .) \\ \hat{M}(\lambda, .) \end{pmatrix}$$

and collecting terms containing λ^2 and λ we have

$$\lambda^2 \mathcal{M}\hat{x}(\lambda) + \lambda \mathcal{D}\hat{x}(\lambda) + \mathcal{K}\hat{x}(\lambda) + A\hat{x}(\lambda) = \hat{\mathcal{F}}(\lambda)\hat{x}(\lambda, .), \qquad (5.83)$$

where

$$\mathcal{M} = \begin{pmatrix} m + \pi \rho b^2 & s - \pi \rho b^3 a \\ S - \rho b^3 a \pi & I_\theta + \pi \rho b^4 (a^2 + \frac{1}{8}) \end{pmatrix},$$

which is again positive definite because

$$\pi\rho b^2 \pi\rho b^4 \left(a^2 + \frac{1}{8} \right) - (\pi\rho b^3 a)^2 = \frac{1}{8} \left(\pi\rho b^3 \right)^2.$$

Next

$$\mathcal{D} = \begin{pmatrix} 0 & \pi\rho b^2 \ U_\infty \\ -\pi\rho b^2 \ U_\infty & 0 \end{pmatrix},$$

$$\mathcal{K} = \begin{pmatrix} 0 & 0 \\ 0 & -\pi\rho b^2 \ U_\infty^2 \end{pmatrix},$$

$$\hat{\mathcal{F}}(\lambda) = \begin{pmatrix} & \frac{T(k)}{k}\rho b U_\infty \\ \frac{T(k)}{k} \ \rho U_\infty \ b^2\left(\frac{1}{2} + a\right) - & \frac{\rho b^2 U_\infty}{2} \end{pmatrix}.$$

We have thus obtained the structure dynamics under aerodynamic loads in the Laplace transform domain.

The next step is to take the inverse Laplace transforms. Here the only slight complication is that the inverse transform of $T(\cdot)$ contains a delta function. Thus in taking the inverse transform of

$$T(k)\hat{w}_a(\lambda, s),$$

we note that we are already assuming that $w_a(t, .)$ is differentiable in t and hence we may write:

$$T(k)\hat{w}(\lambda, s) = \left(\frac{T(k)}{k} \right) k \hat{w}_a(\lambda, s)$$

$$= \frac{b}{U_\infty} \left(\frac{T(k)}{k} \right) (\lambda \hat{w}_a(\lambda, s))$$

and hence the inverse transform can be expressed

$$\frac{b}{U_\infty} \int_0^t L(t - \sigma)\dot{w}_a(\sigma, s)d\sigma,$$

where $L(\cdot)$ is given by:

$$L(t) = \int_0^t c_1(t - \sigma)(\sigma + 1)/(\sqrt{(\sigma(\sigma + 2))})d\sigma.$$

Or as

$$\frac{b}{U_\infty} \left(\frac{1}{2} w_a(t,s) + \int_0^t \ell(t-\sigma) w(\sigma,s) \mathrm{d}\sigma \right).$$

The first form is preferred traditionally [6] and so we use it for the time domain version.

Hence with

$$Y(t) = \begin{pmatrix} x(t) \\ \dot{x}(t) \end{pmatrix}$$

with range in $H \times H$ and defining

$$B = \begin{Bmatrix} 0 \\ 2\pi U_\infty \\ 2\pi \\ b\,(1-2a)\,\pi \end{Bmatrix}$$

$$= \begin{pmatrix} B_1 \\ B_2 \end{pmatrix},$$

we have

$$w_a(t,.) = B^* Y(t) = B_1^* x(t) + B_2^* \dot{x}(t).$$

We can now rewrite the structure equations in the time domain as

$$\mathcal{M}\ddot{x}(t) + \mathcal{D}\dot{x}(t) + Ax(t) + \mathcal{K}x(t) = \int_0^t \mathcal{F}(t-\sigma) B^* \dot{Y}(\sigma)\mathrm{d}\sigma,$$

where

$$\mathcal{F}(t) = \begin{pmatrix} -\rho b U_\infty L(U_\infty t) \\ \rho b^2 \left(a + \frac{1}{2}\right) U_\infty (L(U_\infty t) - \frac{1}{2}) \end{pmatrix}.$$

As usual we convert this to a first-order equation in time:

$$\dot{Y}(t) = A_S Y(t) + \mathcal{T}Y(t) + \int_0^t \mathcal{L}(t-\sigma)\dot{Y}(\sigma)\mathrm{d}\sigma$$

recognized as a convolution/evolution equation, where

$$A_s = \begin{pmatrix} 0 & I \\ -\mathcal{M}^{-1}A & 0 \end{pmatrix}$$

$$\mathcal{T} = \begin{Bmatrix} 0 & 0 \\ -\mathcal{M}^{-1}K & -\mathcal{M}^{-1}\mathcal{D} \end{Bmatrix}$$

(not to be confused with the Tricomi operator) the subscript S denoting structure. The operator $\mathcal{T}$ reflects the so-called non-circulatory terms in the literature; see [6].

$$\begin{pmatrix} \mathcal{L}(t)Y = & \begin{array}{c} 0 \\ 0 \\ \mathcal{M}^{-1}\mathcal{F}(t)B^*Y \end{array} \end{pmatrix}.$$

We now modify the energy space $\mathcal{H}$ using $\mathcal{M}$ in place of M. Thus modifying (5.3) we have:

The energy space $\mathcal{H}$ consists of elements of the form:

$$Y = \begin{pmatrix} x_1 \\ x_2 \end{pmatrix},$$

$$x_1 \in D(\sqrt{A}), \qquad x_2 \in H$$

endowed with the energy inner product:

$$[Y, Z]_E = [\sqrt{A}x_1, \sqrt{A}z_1] + [\mathcal{M}x_2, \quad z_2], \tag{5.84}$$

where

$$z = z_1$$

$$z_2,$$

$$z_1 \in D(\sqrt{A}), \qquad z_2 \in H.$$

Note that if we set U_∞ to be zero, we get back to the pure en vacuo structure equations of Chap. 2 without the aerodynamic forcing terms.

Hereinafter we simply continue to use $[,]$ and drop the subscript E, because the context should make clear which inner product is being used. Also we use:

$$Y = \begin{pmatrix} x_1 \\ x_2 \end{pmatrix}; \qquad x_1 = \begin{pmatrix} h_1 \\ \theta_1 \end{pmatrix}; \qquad x_2 = \begin{pmatrix} h_2 \\ \theta_2 \end{pmatrix}.$$

$\mathcal{A}_s$ is the generator of a C_0 semigroup, therefore so is

$$\mathcal{A} = \mathcal{A}_s + \mathcal{T}$$

as is well known [11, 16]. Let us denote the semigroup by $\mathcal{S}(t), t \geq 0$. Unlike the semigroup generated by $\mathcal{A}_s$, this semigroup is not necessarily a contraction. In fact we have:

$$\text{Re}[\mathcal{A}Y, Y] = \text{Re}[\mathcal{T}Y, Y],$$

because

$$\text{Re}[\mathcal{A}_s Y, Y] = 0; \qquad \text{Re}[\mathcal{D}x_2, x_2] = 0$$

and

$$\mathrm{Re}[\mathcal{T}Y, Y] = -\mathrm{Re}[\mathcal{K}x_1, x_2] = +\pi\rho U_\infty^2 b^2 \mathrm{Re}[\theta_1, \theta_2],$$

which shows that the semigroup $\mathcal{S}(\cdot)$ may not be a contraction.

But the growth bound is of course finite because the generator is a perturbation of $\mathcal{A}_s$ by a bounded operator; and the precise growth bound is not needed here anyway.

What is important, however, is that the resolvent of $\mathcal{A}$, denoted $\mathcal{R}(\lambda, \mathcal{A})$ is compact because the resolvent of $\mathcal{A}_s$ is compact.

What can we say about the convolution part?

The properties of $\mathcal{L}(\cdot)$ are readily obtained from those of the numerical function $L(\cdot)$. We have seen that

$$L(0) = \frac{1}{2}; \qquad L(\infty) = 1$$

and the derivative of $L(t)$ is given by $\ell(t)$ which is continuous and is integrable $[0, \infty]$. $L(\cdot)$ is nonnegative and

$$\int_0^\infty e^{-kt} L(t) dt < \infty \ \mathrm{Re}\, k > 0.$$

Hence we define

$$\hat{\mathcal{L}}(\lambda) = \int_0^\infty e^{-\lambda t} \mathcal{L}(t) dt \qquad \mathrm{Re}\, \lambda > 0.$$

Again thus defined, its properties depend upon $(T(\lambda))/\lambda$ which is not defined along the negative real axis, including zero, and can be extended analytically to the whole plane, except the negative real axis and $\lambda = 0$. On the other hand $\lambda\hat{\mathcal{L}}(\lambda)$ which involves only $T(\lambda)$ is now defined at $\lambda = 0$ by continuity. Hence $\lambda\hat{\mathcal{L}}(\lambda)$ is defined on the whole finite plane excepting $\lambda < 0$.

The calculation of the aeroelastic modes is the central part of Flutter analysis, which is really part of the spectral theory of the stability operator.

Spectral Theory: Aeroelastic Modes

$$M = 0; \quad \alpha = 0.$$

But before we prove existence and uniqueness of the solution of the convolution/evolution equation we need to first study spectral theory, which also plays a crucial role in stability theory.

Thus we go back to the Laplace transform theory.

We have: for every λ except the negative real axis:

$$\lambda^2 \mathcal{M}\hat{x}(\lambda) + \lambda\mathcal{D}\hat{x}(\lambda) + \mathcal{K}x(\lambda) + A\hat{x}(\lambda) = \hat{\mathcal{F}}(\lambda)\hat{x}(\lambda, .),$$

which by taking Laplace transforms of

$$\dot{Y}(t) = A_S Y(t) + TY(t) + \int_0^t \mathcal{L}(t - \sigma)\dot{Y}(\sigma)d\sigma$$

takes the form:

$$(\lambda I - A - \lambda\hat{\mathcal{L}}(\lambda))\hat{Y}(\lambda) = (I - \hat{\mathcal{L}}(\lambda))Y(0),$$

where $\hat{Y}(\lambda)$ is the Laplace transform:

$$\hat{Y}(\lambda) = \int_0^\infty e^{-\lambda t} Y(t)dt$$

and

$$\hat{\mathcal{L}}(\lambda) = \mathcal{M}^{-1}\hat{\mathcal{F}}(\lambda)B^*,$$

which is defined in the whole plane except for Re λ < 0. Hence, following semigroup theory techniques, we consider the generalized resolvent equation:

$$(\lambda I - A - \lambda\hat{\mathcal{L}}(\lambda))Y = Z$$

in $\mathcal{H}$.

Exploiting the compactness of $\mathcal{R}(\lambda, A_S)$ we have the next theorem.

Theorem 5.22. *Given any λ in the resolvent set of A_s (not an en vacuo structure mode in other words), and omitting the negative half-line, either*

$$(\lambda I - A - \lambda\hat{\mathcal{L}}(\lambda))Y = 0 \qquad Y \neq 0$$

for some nonzero Y or $(\lambda I - A - \lambda\hat{\mathcal{L}}(\lambda))$ has a bounded inverse, which we call the generalized resolvent.

Proof.
$$\lambda I - A - \lambda\hat{\mathcal{L}}(\lambda) = \lambda I - A_s - (T + \lambda\hat{\mathcal{L}}(Y)).$$

Hence

$$\mathcal{R}(\lambda, A_S)(\lambda I - A - \lambda\hat{\mathcal{L}}(\lambda))Y = (I - \mathcal{R}(\lambda, A_S)(\tau + \lambda\hat{\mathcal{L}}(\lambda)))Y,$$

where

$$\mathcal{R}(\lambda, A_s)(T + \lambda\hat{\mathcal{L}}(\lambda))$$

is compact because $\mathcal{R}(\lambda, A_s)$ is.

Hence either

$$(I - \mathcal{R}(\lambda, A_S)(T + \lambda\hat{L}(\lambda)))Y = 0$$

for some nonzero Y in which case

$$(\lambda I - A - \lambda \hat{\mathcal{L}}(\lambda))Y = 0$$

or

$$I - \mathcal{R}(\lambda, A_S)(\mathcal{T} + \lambda \hat{\mathcal{L}}(\lambda))$$

and equivalently

$$(\lambda I - A - \lambda \hat{\mathcal{L}}(\lambda))$$

has a bounded inverse. □

We call the bounded operator

$$(\lambda I - A - \lambda \hat{\mathcal{L}}(\lambda))^{-1}$$

the generalized resolvent, denoted $\mathcal{R}(\lambda)$.

Note, however, that $\mathcal{R}(\lambda)$ does not satisfy the resolvent equation [10, 41].

Aeroelastic Modes

The zeros of $\lambda I - A - \lambda \hat{\mathcal{L}}(\lambda)$ are called the "aeroelastic modes." They are not identified as eigenvalues at the expense of changing the space, as we show later.

The zeros of the adjoint:

$$(\lambda I - A - \lambda \hat{\mathcal{L}}(\lambda))^* Y = 0$$

are called the adjoint modes.

The main objective in flutter analysis is to calculate these modes and the corresponding "mode shapes" $Y(.)$.

Calculating Aeroelastic Modes: $M = 0$

We begin with $M = 0$. Here we recall the analysis in Chap. 2, including now the aerodynamic loading. Thus solving

$$(\lambda I - A - \lambda \hat{\mathcal{L}}(\lambda))Y = 0$$

leads us back to where we began:

$$\lambda^2 M \hat{x}(\lambda) + A \hat{x}(\lambda) = \begin{pmatrix} \hat{L}(\lambda, .) \\ \hat{M}(\lambda, .) \end{pmatrix},$$

which unravels to

$$\hat{h}''''(\lambda, s) = -\frac{1}{EI}\left[\lambda^2(m\hat{h}(\lambda, s) + S\hat{\theta}(\lambda, s)) - \hat{L}(\lambda, s)\right],$$

$$\hat{\theta}''(\lambda, s) = \frac{1}{GJ}\left[\lambda^2(I_\theta + S\hat{h}(\lambda, s)) - \hat{M}(\lambda, s)\right], \quad 0 < s < \ell$$

with the end conditions:

$$\hat{h}(\lambda, 0) = 0; \qquad \hat{h}'''(\lambda, \ell) = 0; \qquad \hat{h}''(\lambda, \ell) = 0,$$

$$\hat{\theta}(\lambda, 0) = 0; \qquad \hat{\theta}(\lambda, \ell) = 0.$$

Note immediately that if λ is a mode, so is the conjugate.

This is recognized as a two-point boundary value problem for ordinary differential equations for each λ.

Two-Point Boundary Value Problem

To solve this let us define the 6×1 vector

$$Y(\lambda, s) = \begin{pmatrix} \hat{h}(\lambda, s) \\ \hat{h}'(\lambda, s) \\ \hat{h}''(\lambda, s) \\ \hat{h}'''(\lambda, s) \\ \hat{\theta}(\lambda, s) \\ \hat{\theta}'(\lambda, s) \end{pmatrix}.$$

Then the equation becomes:

$$\frac{d}{ds}y(\lambda, s) = A(\lambda)Y(\lambda, s),$$

where now (cf. (2.13))

$$A(\lambda) = \begin{pmatrix} 0 & 1 & 0 & 0 & 0 & 0 \\ 0 & 0 & 1 & 0 & 0 & 0 \\ 0 & 0 & 0 & 1 & 0 & 0 \\ w_1 & 0 & 0 & 0 & w_2 & 0 \\ 0 & 0 & 0 & 0 & 0 & 1 \\ w_3 & 0 & 0 & 0 & w_4 & 0 \end{pmatrix},$$

where now:

$$w_1 = -\frac{1}{EI}(m\lambda^2 + \lambda\rho U_\infty b w_{11}),$$

$$w_2 = -\frac{1}{EI}\Big[\lambda^2 S + \rho b U_\infty^2((1-ak)w_{11} + kw_{12})\Big],$$

$$w_3 = \frac{1}{GJ}\Big[\lambda^2 S + \lambda\rho b U_\infty^2(w_{21} - aw_{11})\Big],$$

$$w_4 = \frac{1}{GJ}\Big[\lambda^2 I_\theta + \rho b^2 U_\infty^2(w_{21} + kw_{22} - a(1-ak)w_{11} - ak(w_{21} + w_{12}))\Big]$$

and is defined for every λ except the negative half-line. The w_{ij} are given in (5.79) for $M = 0$.

The solution is given by

$$y(\lambda, s) = e^{sA(\lambda)}y(\lambda, 0),$$

where to satisfy the end conditions (CF) we must have:

$$y(\lambda, 0) = Qz(\lambda),$$

and

$$Py(\lambda, \ell) = Pe^{\ell A(\lambda)}Qz(\lambda), \quad z(\lambda) \neq 0. \tag{5.85}$$

Hence we must have

$$d(0, \lambda, U_\infty) = \det\big[Pe^{\ell A(\lambda)}Q\big] = 0,$$

where as before in Chap. 2:

$$P = \begin{pmatrix} 0\,0\,1\,0\,0\,0 \\ 0\,0\,0\,1\,0\,0 \\ 0\,0\,0\,0\,0\,1 \end{pmatrix},$$

$$Q = \begin{pmatrix} 0\,0\,0 \\ 0\,0\,0 \\ 1\,0\,0 \\ 0\,1\,0 \\ 0\,0\,0 \\ 0\,0\,1 \end{pmatrix}$$

$$= P^*.$$

The main result here is the following

Theorem 5.23. *The function*

$$d(0, \lambda, U)$$

for each U positive is an analytic function of λ except for a branch cut along the negative real axis, logarithmic singularity along Re $\lambda \leq 0$, with at most a countable number of isolated zeros in a half-plane that do not have an accumulation in the finite part of the plane.

The aeroelastic modes are the roots of

$$d(0, \lambda, U_\infty) = 0.$$

Proof. The analyticity properties are inherited from those of the matrix function $A(\lambda)$ and in turn by those of $T(k)$, and ultimately the analyticity properties of the Bessel K functions. Hence the stated analyticity properties follow.   □

Let us explore the aeroelastic modes.
    Let

$$Y = \begin{pmatrix} x_1 \\ x_2 \end{pmatrix}, \qquad x_1 = \begin{pmatrix} h_1 \\ \theta_1 \end{pmatrix}, \qquad x_2 = \begin{pmatrix} h_2 \\ \theta_2 \end{pmatrix}; \qquad [Y.Y] = 1.$$

Then

$$(\lambda I - \mathcal{A} - \lambda \hat{\mathcal{L}}(\lambda))Y = 0$$

or, equivalently:

$$\lambda x_1 = x_2,$$

$$\lambda \mathcal{M} x_2 = -A x_1 - \mathcal{K} x_1 - \lambda \mathcal{D} x_1$$

$$+ \begin{pmatrix} T(k)\rho U_\infty^2 \\ T(k)\rho b(a+\frac{1}{2})U_\infty^2 - \dfrac{\lambda \rho b^2 U_\infty}{2} \end{pmatrix} (2\pi U_\infty \theta_1 + 2\pi h_2 + \pi b(1-2a)\theta_2),$$

where

$$h_2 = \lambda h_1; \qquad \theta_2 = \lambda \theta_1.$$

We begin with the special case $\lambda = 0$, of interest because it is an unstable mode even though it is not a structure mode.

Theorem 5.24. *Zero is an aeroelastic mode for a sequence U_n of values of U_∞ given by*

$$U_n = (2n + 1)(\sqrt{\pi GJ})/(2b\ell \sqrt{(\rho(1 + 2a))}).$$

Proof. We have:

$$x_2 = 0,$$

$$0 = \begin{pmatrix} 2\pi\rho U_\infty^3 \theta_1 \\ 2\pi\rho b \left(a + \tfrac{1}{2}\right) U_\infty 3\theta_1 \end{pmatrix} - Ax_1.$$

Or

$$\begin{pmatrix} EI\, h_1''''(s) \\ -GJ\, \theta_1''(s) \end{pmatrix} = \begin{pmatrix} 2\pi\rho U_\infty^2 \theta_1(s) \\ 2\pi\rho b(a + \tfrac{1}{2})\, U_\infty^2 \theta_1(s) \end{pmatrix},$$

where we see that

$$-GJ\,\theta_1''(s) = 2\pi\rho b \left(a + \frac{1}{2} \right) U_\infty^2 \theta_1(s) \qquad 0 < s < \ell$$

with the CF end conditions:

$$\theta_1(0) = 0; \qquad \theta_1'(\ell) = 0$$

is an eigenvalue problem and has solution only for a sequence of values of U_∞ given by

$$U_n = (2n + 1)\left(\sqrt{\pi GJ}\right) \Big/ (2b\ell\sqrt{(\rho(1 + 2a))}),$$

$$\theta_1(s) = \text{const} \times \sin\frac{\pi s}{2\ell} \qquad 0 < s < \ell \text{ for all } n. \qquad \square$$

The minimal value is called the divergence speed. See Sect. 5.5; of course the speeds are also determined from solving

$$d(0, 0, U) = 0,$$

where $d(0, 0, U)$ is an entire function of U of finite order and has a countable number of zeros.

Theorem 5.25. *The aeroelastic modes are confined to a finite vertical strip in the complex plane; the real part is bounded.*

Proof. Let λ be an aeroelastic mode and Y the mode shape, where we normalize so that

$$[Y, Y] = 1.$$

Hence

$$\lambda[Y, Y] = \lambda = [AY, Y] + \lambda[\hat{\mathcal{L}}(\lambda)Y, Y]$$
$$= -2\mathrm{Re}\lambda[x_1, Ax_1] - |\lambda|^2[\mathcal{D}x_1, x_1] - |\lambda|^2[\mathcal{K}x_1, x_1] + [\lambda\hat{\mathcal{L}}(\lambda)Y, Y],$$

where:

$$Y = \begin{pmatrix} x_1 \\ \lambda x_1 \end{pmatrix},$$

$$[Y, Y] = [x_1, Ax_1] + |\lambda|^2[\mathcal{M}x_1, x_1] = 1$$
$$[\mathcal{D}x_1, x_1] = 0; \quad [\mathcal{K}x_1, x_1] = \pi\rho b^2 U_\infty^2[\theta_1, \theta_1]$$
$$\mathrm{Re}\lambda(1 + 2[x_1, Ax_1] + \pi\rho b^2 U_\infty^2[\theta_1, \theta_1])$$
$$= \mathrm{Re}[\lambda\hat{\mathcal{L}}(\lambda)Y, Y],$$

where the right-hand side goes to zero as $\mathrm{Re}\lambda \to \infty$. Hence $\mathrm{Re}\lambda$ is bounded. The aeroelastic modes are confined to a vertical strip in the complex plane. Furthermore there can only be a finite number of zeros with positive real part [8]. Indeed if we have a sequence of the form: $\lambda_k = \sigma + i\omega_k$ and $\omega_k \to \infty$ we would contradict the asymptotic properties of the roots.   □

Mode Shapes

By the "mode shape" corresponding to an aeroelastic mode λ_k, we mean the component $x_{k,1}$ of $Y_k(.)$:

$$Y_k = \begin{pmatrix} x_{k,1} \\ x_{k,2} \end{pmatrix},$$

$$x_{k,1} = \begin{pmatrix} h_k(s) \\ \theta_k(s) \end{pmatrix}, \quad 0 < s < \ell.$$

Let us calculate Y_k.

$$\lambda_k Y_k - AY_k - \lambda_k \hat{\mathcal{L}}(\lambda_k) = 0$$

and hence we have:

$$x_{k,2} = \lambda_k x_{k,1},$$

$$\lambda_k^2 \mathcal{M}x_{k,1} = -Ax_{k,1} - \mathcal{K}x_{k,1} - \lambda_k \mathcal{D}x_{k,1}$$

$$+ \begin{pmatrix} T\left(\dfrac{\lambda_k b}{U_k}\right)\rho U_k^2 \\ T\left(\dfrac{\lambda_k b}{U_k}\right)\rho b \left(a+\dfrac{1}{2}\right) U_k^2 - \dfrac{\lambda_k \rho b^2 U_k}{2} \end{pmatrix}$$

$$\times (2\pi U_\infty \theta_k + 2\pi h_k + \pi b(1-2a)\lambda_k \theta_k).$$

Here we follow [56].

Recall that

$$\text{Det.} P e^{\ell A(\lambda_k)} Q = 0. \tag{5.86}$$

Hence there is a nonzero z_k such that

$$P e^{\ell A(\lambda_k)} Q z_k = Q^* e^{\ell A(\lambda_k)} Q z_k = 0. \tag{5.87}$$

Note that the dimension of the nullspace of the matrix is at most two. Define

$$z_k(s) = e^{sA(\lambda_k)} Q z_k \qquad 0 < s < \ell$$

$$= \text{col}[h_k(s), h_k'(s), h_k''(s), h_k'''(s), \theta_k(s), \theta_k'(s)],$$

$$z_k(0) = Q z_k.$$

Thus defined $z_k(s)$ is an analytic function of s.

Next define

$$Y_k(s) = \text{col}[h_k(s), \theta_k(s), \lambda_k h_k(s), \lambda_k \theta_k(s)], \qquad 0 < s < \ell.$$

Note that we can express $Y_k(.)$ as

$$Y_k(s) = T_{46}(\lambda_k) e^{sA(\lambda_k)} Q z_k \qquad 0 < s < \ell,$$

$$P e^{\ell A(\lambda_k)} Q z_k = 0,$$

where $T_{46}(\lambda_k)$ is the 4×6 matrix given by:

$$T_{46}(\lambda) = \begin{pmatrix} 1 & 0 & 0 & 0 & 0 & 0 \\ 0 & 0 & 0 & 0 & 1 & 0 \\ \lambda & 0 & 0 & 0 & 0 & 0 \\ 0 & 0 & 0 & 0 & \lambda & 0 \end{pmatrix}.$$

And we have thus determined the mode shapes

$$Y_k = \begin{pmatrix} x_{k,1} \\ x_{k,2} \end{pmatrix}.$$

Generalized Resolvent

We now derive a closed-form expression for the generalized resolvent. The generalized resolvent equation

$$\left(\lambda I - \mathcal{A} - \lambda \hat{\mathcal{L}}(\lambda)\right) Y = Y_g,$$

where

$$Y_g = \begin{pmatrix} x_{1,g} \\ x_{2,g} \end{pmatrix}; \qquad x_{2,g} = \begin{pmatrix} h_g(.) \\ \theta_g(.) \end{pmatrix},$$

$$Y = \begin{pmatrix} x_1 \\ x_2 \end{pmatrix} \tag{5.88}$$

reduces to the nonhomogeneous equations

$$\lambda x_1 - x_2 = x_{1,g},$$

$$\hat{h}''''(\lambda, s) = -\frac{1}{EI}\left[\lambda^2\left(m\hat{h}(\lambda, s) + S\hat{\theta}(\lambda, s)\right) - \hat{L}(\lambda, s)\right] + \frac{h_g(s)}{EI}, \tag{5.89}$$

$$\hat{\theta}''(\lambda, s) = \frac{1}{GJ}\left[\lambda^2\left(I_\theta + S\hat{h}(\lambda, s)\right) - \hat{M}(\lambda, s)\right] + \frac{\theta_g(s)}{GJ},$$

$$0 < s < \ell, \tag{5.90}$$

$$\frac{d}{ds}y(\lambda, s) = A(\lambda)y(\lambda, s) + y_g(s), \tag{5.91}$$

$$y_g(s) = \mathrm{col}\left[0, 0, 0, \frac{h_g(s)}{EI}, 0, \frac{\theta_g(s)}{GJ}\right],$$

$$y(\lambda, 0) = Qz(\lambda),$$

$$y(\lambda, s) = e^{sA(\lambda)}Qz(\lambda) + \int_0^s e^{(s-\sigma)A(\lambda)}y_g(\sigma)d\sigma, \tag{5.92}$$

$$P\left(e^{\ell A(\lambda)}Qz(\lambda) + v(\lambda)\right) = 0; \qquad v(\lambda) = \int_0^\ell e^{(\ell-\sigma)A(\lambda)}y_g(\sigma)d\sigma,$$

or,

$$D(\lambda)z(\lambda) = -Pv(\lambda); \quad D(\lambda) = Pe^{\ell A(\lambda)}Q.$$

So, omitting the aeroelastic modes this has the solution:

$$z(\lambda) = -D(\lambda)^{-1}Pv(\lambda)$$

and in turn:

$$y(\lambda, s) = -e^{sA(\lambda)} Q D(\lambda)^{-1} P v(\lambda) + \int_0^s e^{(s-\sigma)A(\lambda)} y_g(\sigma) d\sigma,$$

$$0 < s < \ell \tag{5.93}$$

and finally the generalized resolvent

$$\mathcal{R}(\lambda) Y_g = \begin{pmatrix} x_1(\lambda) \\ x_2(\lambda) \end{pmatrix},$$

$$x_1(\lambda) = T_{26} y(\lambda, .); \qquad x_2(\lambda) = \lambda x_1(\lambda) - x_{1,g}, \tag{5.94}$$

where

$$T_{26} y(\lambda, s) = \begin{pmatrix} h(\lambda, s) \\ \theta(\lambda, s) \end{pmatrix} \qquad 0 < s < \ell.$$

Modal Expansion of Solution

Let us now get back to the solution to the initial value problem:

$$\left(\lambda I - (A) - \lambda \hat{\mathcal{L}}(\lambda) \right) \hat{Y}(\lambda) = \left(I - \hat{\mathcal{L}}(\lambda) \right) Y(0),$$

which has the solution

$$\hat{Y}(\lambda) = \mathcal{R}(\lambda) \left(I - \hat{\mathcal{L}}(\lambda) \right) Y(0), \quad \lambda \neq \lambda_k.$$

Because $D(\lambda)$ has only simple zeros at λ_k, it follows that $\mathcal{R}(\lambda)$ has only isolated simple poles at λ_k, confined to a finite strip.

Hence the inverse Laplace transform

$$W(t) = \lim_{L \to \infty} \frac{1}{2\pi} \int_{-L}^{L} e^{t(\sigma + i\omega)} \mathcal{R}(\sigma + i\omega) d\omega, \quad t > 0.$$

Following [33, 44] we deform the contour of the inverse Laplace transform to obtain [8]

$$W(t)Y = \left(\sum_{k=1}^{\infty} P_k Y e^{\lambda_k t} + \int_0^{\infty} e^{-rt} \mathcal{J}(r) Y dr \right)$$

(where we define the integral term presently),

$$P_k Y = \frac{1}{2\pi i} \oint \mathcal{R}(\lambda) Y \, d\lambda, \qquad \lambda = \lambda_k + re^{i\theta}, \quad r \text{ small enough}$$

$$\left(\lambda_k I - \mathcal{A} - \lambda_k \hat{\mathcal{L}}(\lambda_k) \right) P_k Y = 0.$$

Or P_k is an idempotent mapping into the null space of

$$\left(\lambda_k I - \mathcal{A} - \lambda_k \hat{\mathcal{L}}(\lambda_k) \right).$$

The null space is of dimension at most 2, but numerical calculations indicate it is only 1 and we may assume this for simplicity.

Next note that the generalized resolvent is defined for every λ except $\{\lambda_k\}$ and the negative half-line. We also exclude $\lambda_k = 0$.

Then with $r > 0$ we have to consider the integral along the bounding lines:

$$\lambda = -r + i\omega \quad \text{and} \quad \lambda = -r - i\omega,$$

where we note the jump:

$$\mathcal{R}(-r, -i\omega) - \mathcal{R}(-r, +i\omega) = \mathcal{J}(r, \omega)$$

can be calculated in many ways, our main interest being in the limit as r goes to zero.

We now define

$$\mathcal{J}(r)Y = \frac{1}{2\pi i} \lim_{\omega \to 0} [\mathcal{R}(-r - i\omega) - \mathcal{R}(-r + i\omega)]Y, \qquad r > 0,$$

where we need to first show that the limits exist. Now

$$R(-r, -i\omega) = ((-r + i\omega)I - \mathcal{A} - (-r + i\omega)\mathcal{L}(-r + i\omega))^{-1} Y,$$

where

$$\mathcal{L}(-r + i\omega)Y = \begin{pmatrix} 0 \\ 0 \\ \mathcal{M}^{-l} \hat{\mathcal{F}}(-r + i\omega) B^* Y \end{pmatrix},$$

where

$$\hat{\mathcal{F}}(-r + i\omega) = \begin{pmatrix} \dfrac{T(k)}{k} bU\rho \\ \dfrac{T(k)}{k} b^2 U\rho \left(\dfrac{1}{2} + a \right) - \dfrac{\rho b^2 U}{2} \end{pmatrix},$$

$$k = (-r + i\omega).$$

Note the appearance of the "scale" factor b/U for the first time. We need next to calculate the jump. Now

$$K_0(z) = -I_0(z)\log[z] + I_0(z)\log 2 + \sum_{m=0}^{\infty} \frac{\left(\frac{z}{2}\right)^{2m}}{(m!)^2}.$$

So in the definition we need to go to the Riemann sheet where the value of the logarithm jumps by $2\pi i$ every time we cross the negative half-line. Using for $r > 0, \omega > 0$,

$$\log[-r + i\omega] = \log\sqrt{r^2 + \omega^2} + i(\pi - \theta), \qquad \text{Tan}\theta = \frac{\omega}{r},$$

$$\log[-r - i\omega] = \log\sqrt{r^2 + \omega^2} - i(\pi - \theta),$$

$$\log[-r + i\omega] - \log[-r + i\omega] = 2\pi i - 2i\theta,$$

which leads to the jump in the Bessel K functions:

$$K_0(-r + i\omega) = -I_0(-r + i\omega)\log[-r + i\omega] + I_0(-r + i\omega)\log 2$$

$$+ \sum_{k=0}^{\infty} (-r + i\omega)^{2k} \bigg/ \left(2^{2k}(k!)^2\right) \Psi(k + 1),$$

$$\lim \omega \to 0 \ (K_0(-r + i\omega) - K_0(-r - i\omega)) = -I_0(r)2\pi i.$$

We can define:

$$K_{0+}(-r) = \text{limit } \omega \to 0 \ K_0(-r + i\omega) = K_0(r) - i\pi I_0(r),$$

$$K_{0-}(-r) = \text{limit } \omega \to 0 \ K_0(-r - i\omega) = K_0(r) + i\pi I_0(r).$$

Similarly:

$$K_{1+}(-r) = \text{limit } \omega \to 0 \ K_1(-r + i\omega) = K_1(r) + i\pi I_1(r),$$

$$K_{1-}(-r) = \text{limit } \omega \to 0 \ K_1(-r - i\omega) = K_1(r) - i\pi I_1(r).$$

Let

$$J_T(r) = \lim \omega \to 0(T(-r - i\omega) - T(-r + i\omega))$$

$$= \frac{K_1(r) + i\pi I_1(r)}{K_1(r) + i\pi I_1(r) + K_0(r) + i\pi I_0(r)}$$

$$- \frac{K_1(r) - i\pi I_1(r)}{K_1(r) - i\pi I_1(r) + K_0(r) - i\pi I_0(r)}.$$

And

$$\lim \omega \to 0 \; T(-r + i\omega) = \frac{K_1(r) - i\pi I_1(r)}{K_1(r) - i\pi I_1(r) + K_0(r) - i\pi I_0(r)} = T_+(r),$$

$$\lim \omega \to 0 \; T(-r - i\omega) = \frac{K_1(r) - i\pi I_1(r)}{K_1(r) - i\pi I_1(r) + K_0(r) - i\pi I_0(r)} + J_T(r).$$

Then

$$\mathcal{J}(r)Y = \frac{1}{2\pi i} \lim_{\omega \to 0}[\mathcal{R}(-r - i\omega) - \mathcal{R}(-r + i\omega)]Y, \qquad r > 0$$

$$= \frac{1}{2\pi i} \lim_{\omega \to 0}[((-r - i\omega)I - \mathcal{A} - (-r - i\omega)\mathcal{L}(-r - i\omega))^{-1}Y$$

$$-((-r + i\omega)I - \mathcal{A} - (-r + i\omega)\mathcal{L}(-r + i\omega))^{-1}Y]$$

$$= \frac{1}{2\pi i}\left[(-rI - \mathcal{A} + r\mathcal{L}_-(r))^{-1}(r(\mathcal{L}_-(r)\right.$$

$$\left. -\mathcal{L}_+(r)))(-rI - \mathcal{A} + r\mathcal{L}_+(r))^{-1}\right]Y,$$

where

$$\mathcal{L}_-(r) = \lim \omega \to 0 \qquad \mathcal{L}(-r - i\omega),$$

$$\mathcal{L}_+(r) = \lim \omega \to 0 \qquad \mathcal{L}(-r + i\omega).$$

We are most interested in the limit as $r \to 0$, because of the familiar relation for Laplace transforms: following (5.20)

$$\lim t \to \infty \; t \int_0^\infty e^{-rt} \mathcal{J}(r)Y \, dr = \int_0^\infty e^{-R} \mathcal{J}(R/t)Y \, dR$$

$$\to \mathcal{J}(0+)Y \quad \text{as } t \to \infty$$

and

$$\mathcal{J}(0+)Y = 0.$$

Also we have:

$$\mathcal{R}(\lambda)Y = \sum_{k=1}^{\infty} \frac{P_k Y}{\lambda - \lambda_k} + \int_0^\infty \frac{\mathcal{J}(r)}{\lambda + r} Y \, dt.$$

The idempotents P_k are not orthogonal but $P_k P_j = 0$ for $k \neq j$ and defining

$$QY = \int_0^\infty \mathcal{J}(r) Y \, dr,$$

we have that

$$\lim_{\substack{\mathrm{Re}\,\lambda \to \infty}} \lambda \mathcal{R}(\lambda) Y = W(0) Y = Y$$

and hence

$$Y = \sum_{k=1}^\infty P_k Y + QY,$$

where

$$QP_k = 0 = P_k Q; \quad Q^2 = Q.$$

The mode shape vectors do not span the space $\mathcal{H}$, nor are they orthogonal.

Evanescent States

What is novel here is the existence of evanescent states that decay faster than any modal response, a phenomenon that becomes apparent only in continuum theory and lost in the discretized models of CFD.

First the Definition

The evanescent states are the states in the range space of the operator Q. We note that the response to evanescent states is given by

$$W(t) QY = \int_0^\infty e^{-rt} \mathcal{J}(r) QY \, dr$$

and are thus Laplace transforms in t, and thus are stable for all values of U. In [33] the structure is discretized and the aerodynamics is also approximated (Padé approximation) and the authors call this part the nonrational component. Here we see the exact solution and in particular it is stable for all values of the speed U.

Next we derive the time domain solution of the convolution/evolution equation using state space theory.

5.6   State Space Theory

We are ready to consider now the time domain solution of the convolution/evolution equation for given initial conditions. The solution given in [8, 22] involves a fair degree of operator theory.

Here we use a different and more self-contained technique: *state space representation* which is of interest on its own for us; but requires some knowledge of the theory of semigroups of operators.

State Space Representation: $M = 0$

Here we follow [16]. Our time domain equation is the initial value problem in $\mathcal{H}$:

$$\dot{Y}(t) = \mathcal{A}_S Y(t) + \mathcal{T} Y(t) + \int_0^t \mathcal{L}(t - \sigma)\dot{Y}(\sigma)\mathrm{d}\sigma \qquad (5.95)$$

with initial condition Y (0).

We show that this can be expressed in the "state space" form:

$$\dot{Z}(t) = \mathring{A}Z(t),$$
$$Y(t) = PZ(t) \qquad (5.96)$$

in a Banach space and $\mathring{A}$ is the infinitesimal generator of a C_0 semigroup, and P is a projection operator.

We make strong use of the fact that $L(.)$ is in $C[0, \infty]$ (the space of continuous functions with finite limit at ∞) and is absolutely continuous therein. In particular:

$$\mathcal{L}(.) \in \chi,$$

where χ is the Banach space $C[0, \infty]$ of continuous 4 by 4 matrix-valued functions. It is also absolutely continuous with derivatives therein.

Remarks: In view of these properties we could integrate by parts and get:

$$\int_0^t \mathcal{L}(t - \sigma)\dot{Y}(\sigma)\mathrm{d}\sigma = \int_0^t \dot{\mathcal{L}}(t - \sigma)Y(\sigma)\mathrm{d}\sigma - \mathcal{L}(t)Y(0)$$

but this makes it awkward, bringing in the initial value in the state equation which we want to avoid!

We begin with the convolution part which is defined by the function $\mathcal{L}(.)$. Consider the "system input–output relation" where u (.) is the input and $v(.)$ is the output

$$v(t) = \int_0^t \mathcal{L}(t - \sigma)u(\sigma)d\sigma \qquad t > 0, \tag{5.97}$$

where the input $u(.)$ takes its values in a separable Hilbert space and $\int_0^T ||u(t)||dt < \infty$ over every finite T. For representation theory, it is convenient to embed this class of inputs in a larger class where $u(.)$ is defined on $(-\infty, \infty)$ but vanishes on a negative half-line:

$$u(t) = 0 \qquad \text{for} \ t < -L \qquad \text{for some} \ L, \quad 0 < L < \infty.$$

In that case we can rewrite the relation as

$$v(t) = \int_{-\infty}^t \mathcal{L}(t - \sigma)u(\sigma)d\sigma.$$

Then of course we can extend this definition to any t, not necessarily positive. But we are only interested in the $t > 0$ with which we started. The implication is that we embed this in a controllable system [16] but this is irrelevant in our current context.

Theorem 5.26. *Let X denote the Banach space $C[[0, \infty]; H]$. The input–output relation*

$$v(t) = \int_{-\infty}^t \mathcal{L}(t - \sigma)u(\sigma)d\sigma \qquad t > 0$$

can be represented as

$$v(t) = Cx(t),$$
$$\dot{x}(t) = Ax(t) + Lu(t) \tag{5.98}$$

for $t > 0$, where $x(t)$ is in X for each $t > 0$, and A is the infinitesimal generator of a C_0 semigroup over X, and L is a linear bounded operator on H into X, and C is a linear bounded operator on X into H.

Proof. Define L by

$$Lu = x;$$
$$x(t) = \mathcal{L}(t)u, \qquad t \geq 0$$

$\square$

and in as much as

$$|| x(t) || \leq || \mathcal{L}(t) || \, ||u||,$$

it follows that x is in X.

Next define C on X by

$$Cx = x(0)$$

and C is linear bounded because

$$|| x(0) || \leq ||x||.$$

Let $S(.)$ denote the shift semigroup over X defined by:

$$S(t)x = y; \qquad y(s) = x(s + t), \qquad s \geq 0.$$

It is a C_0 semigroup. Let A denote the infinitesimal generator which is then defined by

$$\mathcal{D}(A) = [x(.) \text{ in } X \text{ such that } x'(.) \text{ is in } X],$$

$$Ax = y; \qquad y(t) = x'(t)\, t \geq 0.$$

Next let Σ denote the closed linear subspace generated by elements in X of the form:
$$S(t)Lu\ u \text{ in } H \text{ and } t \geq 0.$$

Then Σ is an invariant subspace for the semigroup: $S(t)\Sigma$ is contained in Σ for every t. We may take our state space to be Σ and let $S_c(.)$ denote the semigroup restricted to Σ, and let A_c denote the generator, being then a restriction of A.
Then we have the next lemma.

Lemma 5.27. *The nonhomogeneous Cauchy problem:*

$$\dot{x}(t) = A_c x(t) + Lu(t) \qquad t > 0$$

with $x(0)$ in the domain of A_c, has a unique solution in Σ such that:

$$x(t) \varepsilon \mathcal{D}(A_c)$$

$$||x(t) - x(0)|| \to 0 \qquad as\ t \to 0+$$

given by

$$x(t) = S_c(t)x(0) + \int_0^t S_c(t - \sigma)Lu(\sigma)d\sigma.$$

Proof. The proof is standard; see, for example, [16]. □

Next let us calculate $v(.)$. We have

$$Cx(t) = CS_c(t)x(0) + \int_0^t CS_c(t - \sigma)Lu(\sigma)d\sigma$$

$$= x(0, t) + \int_0^t L(t - \sigma)u(\sigma)d\sigma$$

$$= v(t)$$

by taking $x(0)$ to be zero, where we have used the notation

$$x(t, s) \ s \geq 0$$

to denote $x(t)$.

Remarks: As is well known (see [16]) the "method of variation of parameters" solution holds in the generalized or weak sense even if $x(0)$ is not in the domain of the generator, with the continuity at the origin, but of course the solution need not be in the domain of the generator.

Next let us calculate the resolvent of A_c.

Resolvent of A_c

Let $R(\lambda, A_c)$ denote the resolvent of A_c. Then for $\text{Re}\lambda > 0$ we have

$$R(\lambda, A_c)x = \int_0^\infty e^{-\lambda t} S_c(t)x \, dt$$

and

$$(\lambda I - A_c)x = \lambda LY, Y \text{ in } \mathcal{H}$$

has the unique solution for $\text{Re}\lambda > 0$ given by

$$x = \lambda R(\lambda, A_c)LY$$

and

$$Cx = C \int_0^\infty e^{-\lambda t} S_c(t)LY \, dt = \lambda \hat{\mathcal{L}}(\lambda)Y.$$

Lemma 5.28. *Given any Y in $\mathcal{H}$, we can find $x(\lambda)$ in Σ such that*

$$(\lambda I - A_c)x(\lambda) = \lambda LY$$

and

$$Cx = \lambda \hat{\mathcal{L}}(\lambda)Y$$

for λ in the region of analyticity of $\lambda \hat{\mathcal{L}}(\lambda)$.

Proof. For $\text{Re}\lambda > 0$

$$x(\lambda) = \lambda \int_0^\infty e^{-\lambda t} S_c(t)LY \, dt$$

is an element of Σ.

And the corresponding function

$$x(\lambda, s) = \lambda \int_0^\infty e^{-\lambda t} L(t+s) Y \, dt \qquad s \geq 0$$

$$= \lambda e^{\lambda s} \int_s^\infty e^{-\lambda \tau} L(\tau) Y \, d\tau$$

$$= \lambda e^{\lambda s} \hat{L}(\lambda) Y - \lambda e^{\lambda s} \int_0^s e^{-\lambda \tau} L(\tau) Y \, d\tau \qquad (5.99)$$

$$0 \leq s \leq \infty,$$

$$x(\lambda, \infty) = L(\infty) Y, \qquad (5.100)$$

$$x(\lambda, 0) = Cx(\lambda, .) = \lambda \hat{L}(\lambda) Y. \qquad (5.101)$$

The main point here is that thus defined for $\mathrm{Re}\lambda > 0$ $x(\lambda, s)$ can be continued analytically as an analytic function of λ in the region of analyticity of $\lambda \hat{L}(\lambda) Y$ which includes the whole plane except $\lambda \leq 0$. Hence $x(\lambda, .)$ can be continued analytically satisfying

$$(\lambda I - A_c) x(\lambda, .) = \lambda L Y \qquad (5.102)$$

as required. $\square$

Let us use $\dot{X}$ to denote the state space, the product space $\Sigma \times \mathcal{H}$ which is then a Banach space and denote the elements therein as

$$Z = \begin{pmatrix} x \\ Y \end{pmatrix} x \varepsilon \Sigma, \qquad Y \varepsilon \mathcal{H},$$

And the projection operator $\mathcal{P}$ on $\dot{X}$ into $\mathcal{H}$ by

$$\mathcal{P}Z = Y.$$

Then we can recast the convolution/evolution equation (5.72) as

$$\dot{Y}(t) = AY(t) + Cx(t),$$

$$\dot{X}(t) = A_c x(t) + L\dot{Y}(t). \qquad (5.103)$$

And further with

$$Z(t) = \begin{pmatrix} x(t) \\ Y(t) \end{pmatrix}$$

and defining the linear bounded operator B on $\dot{X}$ into itself by

$$BZ = \begin{pmatrix} LY \\ 0 \end{pmatrix},$$

we have:

$$\dot{Z}(t) = \begin{pmatrix} A_c & 0 \\ C & \mathcal{A} \end{pmatrix} Z(t) + B\dot{Z}(t).$$

Or

$$(I - B)\dot{Z}(t) = \begin{pmatrix} A_c & 0 \\ C & \mathcal{A} \end{pmatrix} Z(t).$$

Because B is nilpotent we have

$$(I - B) = (I + B)^{-1}$$

and hence finally we can express the equation as

$$\dot{Z}(t) = (I + B) \begin{pmatrix} A_c & 0 \\ C & \mathcal{A} \end{pmatrix} Z(t)$$

$$= \dot{\mathcal{A}} Z(t), \tag{5.104}$$

where

$$\dot{\mathcal{A}} = (I + B) \begin{pmatrix} A_c & 0 \\ C & \mathcal{A} \end{pmatrix} = \begin{pmatrix} A_c + LC & L\mathcal{A} \\ C & \mathcal{A} \end{pmatrix}.$$

It is immediate that thus defined $\dot{\mathcal{A}}$ is closed linear with dense domain in $\dot{X}$

(Domain of A_c) $\times$ (Domain of $\dot{\mathcal{A}}$).

Thus we have the new state space form

$$\dot{Z}(t) = \dot{\mathcal{A}} Z(t), \tag{5.105}$$

where $Z(.)$ is the state, the state space is $\dot{X}$, and $\dot{\mathcal{A}}$ is the system stability operator. To prove existence and uniqueness of the convolution/evolution equation (5.95) we only need to show that $\dot{\mathcal{A}}$ generates a C_0 semigroup.

We begin with the spectrum of $\dot{\mathcal{A}}$. In turn, we begin with the point spectrum (eigenvalues).

The point spectrum of $\dot{\mathcal{A}}$ consists of λ such that

$$\lambda x = (A_c + LC)x + L\mathcal{A}Y,$$

$$\lambda Y = Cx + \mathcal{A}Y.$$

Suppose
$$Y = 0.$$

Then
$$Cx = 0$$

and hence
$$\lambda x = A_c x.$$

But then
$$x(s) = e^{\lambda s} C x = 0.$$

Hence Y cannot be zero.
Suppose
$$x = 0.$$

Then we must have
$$L\mathcal{A}Y = 0$$

and the second equation yields
$$Y = 0$$

and hence neither x nor Y can be zero.
    Next, the first equation can be rewritten as:

$$\lambda x - A_c x = \lambda L Y$$

and can have a solution even if λ is not in the resolvent set of A_c!, and the second as

$$\lambda Y - \mathcal{A}Y = Cx,$$

where
$$Cx = \lambda \hat{\mathcal{L}}(\lambda)Y,$$

so that
$$\lambda Y - \mathcal{A}Y = Cx = \lambda \hat{\mathcal{L}}(\lambda)Y.$$

Or
$$\lambda Y - \mathcal{A}Y - \lambda \hat{\mathcal{L}}(\lambda)Y = 0$$

and thus λ is an aeroelastic mode.
    We can say more.

Theorem 5.29. *The point spectrum of the system stability operator $\acute{A}$ is the set of all aeroelastic modes.*

Proof. Let λ be an aeroelastic mode, defined by

$$(\lambda I - \mathcal{A} - \lambda \hat{\mathcal{L}}(\lambda))Y = 0, \quad Y \neq 0.$$

Or
$$(\lambda I - A)Y = \lambda\hat{\mathcal{L}}(\lambda)Y.$$

Recall that $\lambda\hat{\mathcal{L}}(\lambda)$ is defined and analytic in the whole plane except the negative real axis.

Let us next consider the equation in Σ,

$$\lambda x - A_c x = \lambda L Y;$$

this does not require that λ be in the resolvent set of A_c. Then

$$Cx = \lambda\hat{\mathcal{L}}(\lambda)Y.$$

By Lemma 5.28, this equation has a solution in Σ. And

$$(\lambda I - A)Y = \lambda\hat{\mathcal{L}}(\lambda)Y.$$

Hence let

$$Z = \begin{pmatrix} x \\ Y \end{pmatrix}.$$

Then we verify that λ is in the point spectrum of $\dot{A}$:

$$\lambda x = A_c x + \lambda L Y$$
$$= A_c x + L(AY + Cx),$$
$$\lambda Y = AY + Cx. \qquad\qquad \square$$

Let us next consider the resolvent.

Resolvent of $\dot{A}$

Theorem 5.30. *Let Ω denote the region in the complex plane*

$$\Omega = \{\text{resolvent set of } A\} \bigcap$$

$$\{\text{Complement of the set of aeroelastic modes}\} \bigcap \{\text{Re}\lambda > 0\}.$$

Then

$$\Omega \text{ is contained in the resolvent set of } \dot{A}.$$

Moreover, for λ in Ω

$$(I - \mathcal{R}(\lambda, \mathcal{A})\lambda\hat{\mathcal{L}}(\lambda)) \text{ has a bounded inverse.}$$

Define

$$\mathcal{R}(\lambda) = \left(I - \mathcal{R}(\lambda, \mathcal{A})\lambda\hat{\mathcal{L}}(\lambda)\right)^{-1} \mathcal{R}(\lambda, \mathcal{A}).$$

Then the resolvent of $\dot{\mathcal{A}}$ denoted $R(\lambda, \dot{\mathcal{A}})$ is given by

$$Z = \begin{pmatrix} x \\ Y \end{pmatrix},$$

$$R(\lambda, \dot{\mathcal{A}})Z = \begin{pmatrix} R(\lambda, A_C) & (\lambda L\Psi - LY + x) \\ & \Psi \end{pmatrix},$$

where

$$\Psi = \mathcal{R}(\lambda)(Y + CR(\lambda, A_c)x - \hat{\mathcal{L}}(\lambda)Y).$$

Proof. Note first that Ω includes a right-half-plane. For λ in Ω, we note that $\mathcal{R}(\lambda, \mathcal{A})$ is compact, and hence so is

$$\lambda\mathcal{R}(\lambda, \mathcal{A})\hat{\mathcal{L}}(\lambda).$$

Hence either

$$\left(I - \lambda\mathcal{R}(\lambda, \mathcal{A})\hat{\mathcal{L}}(\lambda)\right) \qquad Y = 0, Y \neq 0,$$

or

$$\left(I - \lambda\mathcal{R}(\lambda, \mathcal{A})\hat{\mathcal{L}}(\lambda)\right)$$

has a bounded inverse. But

$$(I - \lambda\mathcal{R}(\lambda, \mathcal{A})\hat{\mathcal{L}}(\lambda))Y = 0$$

would imply that

$$(\lambda I - \mathcal{A} - \lambda\hat{\mathcal{L}}(\lambda))Y = 0,$$

or λ is in the point spectrum of $\dot{\mathcal{A}}$. Hence

$$(I - \lambda\mathcal{R}(\lambda, \mathcal{A})\hat{\mathcal{L}}(\lambda))$$

has a bounded inverse for λ in Ω. Hence so does

$$\left(\lambda I - \mathcal{A} - \lambda\hat{\mathcal{L}}(\lambda)\right)$$

and let us denote it $\mathcal{R}(\lambda)$. Then

$$\mathcal{R}(\lambda) = \left(I - \lambda\mathcal{R}(\lambda, A)\hat{\mathcal{L}}(\lambda)\right)^{-1}\mathcal{R}(\lambda, A)$$

and is analytic in Ω. To find the resolvent of $\dot{A}$ we need to solve:

$$(\lambda I - \dot{A})Z = Z_g$$

for given

$$Z_g = \begin{pmatrix} x_g \\ Y_g \end{pmatrix}.$$

Let

$$Z = \begin{pmatrix} x \\ Y \end{pmatrix}.$$

Then

$$\lambda x - A_c x - L(AY + Cx) = x_g,$$
$$\lambda Y - AY - Cx = Y_g.$$

Because Re $\lambda > 0$, λ is in the resolvent set of A_c. Hence

$$x = R(\lambda, A_c)(\lambda LY - LY_g + x_g),$$

$$CR(\lambda, A_c)L = \hat{\mathcal{L}}(\lambda).$$

Hence

$$Cx = CR(\lambda, A_c)x_g + \lambda\hat{\mathcal{L}}(\lambda)Y - \hat{\mathcal{L}}(\lambda)Y_g,$$

$$\lambda Y - AY - \lambda\hat{\mathcal{L}}(\lambda)Y = Y_g + CR(\lambda, A_c)x_g - \hat{\mathcal{L}}(\lambda)Y_g.$$

Hence

$$Y = \mathcal{R}(\lambda)\left(Y_g + CR(\lambda, A_c)x_g - \hat{\mathcal{L}}(\lambda)Y_g\right),$$

$$x = R(\lambda, A_c)(\lambda LY - LY_g + x_g),$$

which then defines $\mathcal{R}(\lambda, \dot{A})Z$ for λ in Ω. $\square$

Remarks: The resolvent above not being compact implies that the resolvent set is properly contained in the complement of the point spectrum.

Note also that

$$PR(\lambda, \dot{A}) \begin{pmatrix} 0 \\ Y \end{pmatrix} = \mathcal{R}(\lambda)(I - \hat{\mathcal{L}}(\lambda))Y.$$

Finally let us consider the semigroup generated by $\dot{A}$.

Semigroup Generated by $\dot{A}$

Rather than estimating the resolvent, we simply construct the semigroup explicitly.

First let us note that

$$||\mathcal{R}(\lambda, A)\lambda\hat{\mathcal{L}}(\lambda)|| \to 0 \quad \text{as Re } \lambda \to \infty$$

and hence

$$||\mathcal{R}(\lambda, A)\lambda\hat{\mathcal{L}}(\lambda)|| < 1$$

in a right-half-plane contained in Ω. Hence we have the Neumann expansion

$$\left(I - \mathcal{R}(\lambda, A)\lambda\hat{\mathcal{L}}(\lambda)\right)^{-1} = I + \sum_{k=1}^{\infty} \left(\mathcal{R}(\lambda, A)\lambda\hat{\mathcal{L}}(\lambda)\right)^k,$$

where

$$\mathcal{R}(\lambda, A)\lambda\hat{\mathcal{L}}(\lambda)Y = \int_0^{\infty} e^{-\lambda t} J(t)Y \, dt$$

and

$$J(t)Y = \int_0^t \mathcal{S}(t - \sigma)\mu(\sigma)\sigma,$$

where $\mu(.)$ is the inverse Laplace transform of $\lambda\hat{\mathcal{L}}(\lambda)$.

Hence it follows that

$$\left(I - \mathcal{R}(\lambda, A)\lambda\hat{\mathcal{L}}(\lambda)\right)^{-1} = I + \hat{r}(\lambda),$$

$$\hat{r}(\lambda) = \int_0^{\infty} e^{-\lambda t} r(t) dt,$$

where

$$\hat{r}(\lambda) = \left(I - \mathcal{R}(\lambda, A)\lambda\hat{\mathcal{L}}(\lambda)\right)^{-1} - I$$

$$= \left(I - \mathcal{R}(\lambda, A)\lambda\hat{\mathcal{L}}(\lambda)\right)^{-1}\left(I - \left(\left(I - \mathcal{R}(\lambda, A)\lambda\hat{\mathcal{L}}(\lambda)\right)\right)\right)$$

$$= \mathcal{R}(\lambda)\lambda\hat{\mathcal{L}}(\lambda)$$

and hence

$$\mathcal{R}(\lambda)Y = \int_0^\infty e^{-\lambda t} W(t)Y \, dt$$

and

$$W(t)Y = \mathcal{S}(t)Y + \int_0^t \mathcal{S}(t - \sigma)r(\sigma)Y \, d\sigma,$$

$$\mathcal{R}(\lambda)Y = \mathcal{R}(\lambda, A)Y + \mathcal{R}(\lambda, A)\hat{r}(\lambda)Y.$$

The function $W(.)$ was obtained by a contour integration technique in [60, 74]. Here we have obtained it in a more constructive way that also displays explicitly the nature of the function. In particular

$$W(0) = I$$

and $W(.)$ is absolutely continuous on the domain of $\mathcal{A}$. In fact

$$\dot{W}(t)Y = AW(t)Y + r(t)Y.$$

Armed with this function we can now proceed to construct the semigroup, by simply taking the inverse Laplace transform of the resolvent. Thus, given

$$Z = \begin{pmatrix} x \\ Y \end{pmatrix},$$

where x is in the domain of A_c and Y in the domain of $\mathcal{A}$, define for $t \geq 0$:

$$Y(t) = W(t)Y + \int_0^t W(t - \sigma)(CS_c(\sigma)x - \mathcal{L}(\sigma)Y) d\sigma$$

and

$$Z(t) = \mathcal{S}_z(t)Z = \begin{pmatrix} S_c(t)x + \int_0^t S_c(t - \sigma)L\dot{Y}(\sigma)d\sigma \\ Y(t) \end{pmatrix},$$

where we note that the derivative is well defined for x in the domain of A_c and Y in the domain of $\mathcal{A}$.

Thus defined, the Laplace transform of the derivative

$$\int_0^\infty e^{-\lambda t} \dot{S}_z(t) Z \, dt = \lambda \mathcal{R}(\lambda, \dot{A}) Z - Z = \dot{A} \mathcal{R}(\lambda, \dot{A}) Z,$$

which by our construction

$$= \dot{A} \int_0^\infty e^{-\lambda t} S_z(t) Z \, dt = \int_0^\infty e^{-\lambda t} \dot{A} S_z(t) Z \, dt,$$

or

$$\dot{S}(t) Z = \dot{A} S(t) Z,$$

which is enough to prove that $\dot{A}$ generates the C_0 semigroup $S_z(\,.\,)$.

And hence we have proved the state space representation

$$\dot{Z}(t) = \dot{A} Z(t),$$

$$Y(t) = \mathcal{P} Z(t). \qquad\qquad\qquad \square$$

Time Domain Airfoil Dynamics: The Sonic Case

Next we consider the case $M = 1$ which differs radically from the case $M = 0$.

Here we follow [71]. We begin by evaluating the coefficients w_{ij} defined as before.

$$w_{11} = 4 \left(\frac{e^{-k}}{\sqrt{k\pi}} + \mathrm{erf}\sqrt{k} \right),$$

$$w_{12} = \frac{\mathrm{erf}\sqrt{k}}{k^2} - \frac{e^{-k}}{\sqrt{k^3}},$$

$$w_{21} = \frac{1}{k^2} \left(\mathrm{erf}\sqrt{k} + 2e^{-k} \frac{\sqrt{k}}{\pi} - 1 \right),$$

$$w_{22} = \frac{2}{3} \left[\left(2 + \frac{3}{k^3} \right) \mathrm{erf}\sqrt{k} + 2e^{-k} \sqrt{\frac{k}{\pi}} \left(\frac{1}{k} - \frac{2}{k^2} - \frac{3}{k^3} \right) \right].$$

Note that in contrast with the case $M = 0$, there are no polynomials in k. In fact $w_{ij}(k)$ does not converge to a finite limit as k goes to zero.

The role of the Theodorsen function is now played by the erf function

$$\text{erf} z = \frac{2}{\sqrt{\pi}} \int_0^z e^{-t^2} dt$$

and erf $\sqrt{z}$ has a logarithmic singularity along the negative real axis and hence so does w_{ij}. In fact:

$$\frac{\text{erf}\sqrt{s}}{\sqrt{s}} = \frac{2}{\sqrt{\pi}} \sum_{k=0}^{\infty} (-1)^k s^k / (2k+1)! \sim\sim \text{ entire function,}$$

$$\text{erf}\sqrt{s} = \int_0^{\infty} e^{-st}(\delta(t) - f(t)) dt,$$

where

$$f(t) = 0 \qquad 0 < t < 1$$
$$= \frac{1}{\pi} 1 \Big/ \left(t\sqrt{(t-1)}\right), \qquad t > 1.$$

The Laplace domain equations are

$$\lambda^2 m\, \hat{h}(\lambda, s) + \lambda^2 S\hat{\theta}(\lambda, s) + EI\hat{h}(\lambda, s)'''$$

$$= \hat{L}(\lambda, s)$$

$$= -\rho b U_\infty^2 \left[\frac{k}{b} w_{11}\hat{h}(\lambda, s) + (kw_{12} + (1-ka)w_{11})\hat{\theta}(\lambda, s) \right]$$

$$= -\rho b U_\infty^2 \left[\frac{k}{b} 4\left(\frac{e^{-k}}{\sqrt{k\pi}} + \text{erf}\sqrt{k} \right) \hat{h}(\lambda, s) \right.$$

$$\left. + \left(\left(\frac{\text{erf}\sqrt{k}}{k} - \frac{e^{-k}}{\sqrt{k}} \right) + (1-ka)4\left(\frac{e^{-k}}{\sqrt{k\pi}} + \text{erf}\sqrt{k} \right) \right) \hat{\theta}(\lambda, s) \right],$$

$$\lambda^2 I_\theta \hat{\theta}(\lambda, s) + S\lambda^2 \hat{h}(\lambda, s) - GJ\hat{\theta}(\lambda, s)'' = \hat{M}(\lambda, s)$$

$$= -\rho b^2 U_\infty^2 \left[\frac{k}{b}(w_{21} - aw_{11})\hat{h}(\lambda, s) \right.$$

$$\left. + (kw_{22} + (1-ak)w_{21} - akw_{12} - a(1-ak)w_{11})\hat{\theta}(\lambda, s) \right]$$

$$= -\rho b^2 U_\infty^2 \left[\frac{k}{b} \left(\frac{1}{k^2} \left(erf \sqrt{k} + 2e^{-k} \frac{\sqrt{k}}{\pi} - 1 \right) \right. \right.$$

$$- a4 \left(\frac{e^{-k}}{\sqrt{k\pi}} + erf \sqrt{k} \right) \right) \hat{h}(\lambda, s) + \left[k\frac{2}{3} \left[\left(2 + \frac{3}{k^3} \right) erf\sqrt{k} \right. \right.$$

$$+ 2e^{-k} \sqrt{\frac{k}{\pi}} \left(\frac{1}{k} - \frac{2}{k^2} - \frac{3}{k^3} \right) \right] + (1 - ak)\frac{1}{k^2} \left(erf\sqrt{k} + 2e^{-k} \frac{\sqrt{k}}{\pi} - 1 \right)$$

$$\left. \left. -a \frac{erf\sqrt{k}}{k} - \frac{e^{-k}}{\sqrt{k^3}} - 4a(1 - ak) \left(\frac{e^{-k}}{\sqrt{k\pi}} + erf\sqrt{k} \right) \right) \hat{\theta}(\lambda, s) \right].$$

Let us next isolate the non-circulatory terms by taking the nonzero limits as Re λ goes to infinity.

$$M\lambda^2 \hat{x}(\lambda, s) + A\hat{x}(\lambda, s) = \mathcal{F}(\lambda)\hat{x}(\lambda, s),$$

where

$$\mathcal{F}(\lambda) = \begin{pmatrix} F_{11}(\lambda) & F_{12}(\lambda) \\ F_{21}(\lambda) & F_{22}(\lambda) \end{pmatrix},$$

where

$$F_{11}(\lambda) = -4\rho U_\infty^2 k \left(\frac{e^{-k}}{\sqrt{k\pi}} + erf\sqrt{k} \right),$$

$$F_{12}(\lambda) = -\rho b U_\infty^2 \left(\left(\frac{erf\sqrt{k}}{k} - \frac{e^{-k}}{\sqrt{k}} \right) + 4(1 - ka) \left(\frac{e^{-k}}{\sqrt{k\pi}} + erf\sqrt{k} \right) \right),$$

$$F_{21}(\lambda) = -\rho b U_\infty^2 \left[\frac{k}{b} \left(\frac{1}{k^2} \left(erf\sqrt{k} + 2e^{-k} \frac{\sqrt{k}}{\pi} - 1 \right) - 4a \left(\frac{e^{-k}}{\sqrt{k\pi}} + erf\sqrt{k} \right) \right) \right],$$

$$F_{22}(\lambda) = -\rho b^2 U_\infty^2 \left[k\frac{2}{3} \left[\left(2 + \frac{3}{k^3} \right) erf\sqrt{k} + 2e^{-k} \sqrt{\frac{k}{\pi}} \left(\frac{1}{k} - \frac{2}{k^2} - \frac{3}{k^3} \right) \right. \right.$$

$$+ (1 - ak)\frac{1}{k^2} \left(erf\sqrt{k} + 2e^{-k} \frac{\sqrt{k}}{\pi} - 1 \right) - a\frac{erf\sqrt{k}}{k} - \frac{e^{-k}}{\sqrt{k^3}}$$

$$\left. \left. -4a(1 - ak) \left(\frac{e^{-k}}{\sqrt{k\pi}} + erf\sqrt{k} \right) \right] \right].$$

Hence

$$\lim_{\lambda \to \infty} \frac{\mathcal{F}(\lambda)}{\lambda^2} = 0,$$

$$\lim_{\lambda \to \infty} \frac{\mathcal{F}(\lambda)}{k} = \begin{pmatrix} -4\rho U_\infty^2 & -4\rho b U_\infty^2 a \\ 0 & -\rho b^2 U_\infty^2 \left(\frac{4}{3} + 4a^2 \right) \end{pmatrix},$$

$$\lim_{\lambda \to \infty} (\mathcal{F}(\lambda) - k) \begin{pmatrix} -4\rho U_\infty^2 & -4\rho b U_\infty^2 a \\ 0 & -\rho b^2 U_\infty^2 \left(\frac{4}{3} + 4a^2 \right) \end{pmatrix} = \begin{pmatrix} 0 & 4\rho b U_\infty^2 \\ 0 & 4\rho a b^2 U_\infty^2 \end{pmatrix}.$$

Having evaluated the non-circulatory terms we can formulate the equations in the Laplace domain as

$$\lambda^2 \mathcal{M} \hat{x}(\lambda) + \lambda \mathcal{D} \hat{x}(\lambda) + \mathcal{K} x(\lambda) + A \hat{x}(\lambda) = \hat{\mathcal{F}}(\lambda) \hat{x}(\lambda, .),$$

where

$$\mathcal{D} = \begin{pmatrix} 4\rho b U_\infty & -4\rho b^2 U_\infty a \\ 0 & -\rho b^2 U_\infty^2 \left(\frac{4}{3} + 4a^2 \right) \end{pmatrix},$$

$$\mathcal{K} = \begin{pmatrix} 0 & -4\rho b U_\infty^2 \\ 0 & -4\rho a b^2 U_\infty^2 \end{pmatrix},$$

$$\hat{\mathcal{F}}(\lambda) = \mathcal{F}(\lambda) - k \begin{pmatrix} -4\rho U_\infty^2 & -4\rho b U_\infty^2 a \\ 0 & -\rho b^2 U_\infty^2 \left(\frac{4}{3} + 4a^2 \right) \end{pmatrix} - \begin{pmatrix} 0 & 4\rho b U_\infty^2 \\ 0 & 4\rho a b^2 U_\infty^2 \end{pmatrix}$$

$$= \int_0^\infty e^{-\lambda t} \mathcal{L}_1(t) dt \ \mathrm{Re}\lambda > \sigma_a.$$

Aeroelastic Modes

As before

$$D(\lambda) = P e^{\ell A(\lambda) t} Q,$$

$$d(1, \lambda, U) = \det D(\lambda).$$

The aeroelastic modes are the zeros $\{\lambda_k\}$:

$$d(1, \lambda_k, U_k) = 0.$$

They are again confined to a finite strip. The corresponding mode shape vector is given by

$$Y_k = \begin{pmatrix} x_{1,k} \\ x_{2,k} \end{pmatrix},$$

$$x_{2,k} = \lambda_k x_{1,k},$$

$$D(\lambda_k) z_k = 0, \quad z_k \neq 0.$$

Assume $D(\lambda_k)$ is of rank 2

$$y_k(s) = e^{sA(\lambda_k)} Q z_k,$$

$$x_{1,k}(s) = T_{26} y_k(s) \qquad 0 < s < \ell.$$

We can now go on, as in the case $M = 0$, to define the generalized resolvent $\mathcal{R}(\lambda)$ and the modal representation.

The big difference now, unlike $M = 0$, is that

$$\hat{Y}(\lambda) = \mathcal{R}(\lambda) Y(0).$$

We can proceed as in the general case $M > 0$, treated below.

The unique feature here is that zero is not a mode; indeed $D(\lambda)$ is not defined at zero. There is no continuity at zero.

The General Case: $0 < M < 1; \alpha = 0$

We now get back to the general case $0 < M < 1$ having treated separately the cases $M = 0$ and $M = 1$ where we exploited the luxury of explicit solution of the Possio equation. We continue with the notation as in Sect. 5.3 and the normalization there, so we may take $b = 1$, with the normalized frequency $k = \lambda b / U$.

The Possio equation for $0 < M < 1$ in the abstract form is given by:

$$\frac{2}{M} \hat{w}_a(\lambda, .) = \hat{A}(\lambda, .) + \frac{\sqrt{1 - M^2}}{M} \mathbb{P}\mathbb{H}(\mathbb{B}(k) - B(\infty)) \hat{A}(\lambda, .) \qquad (5.106)$$

and we simply assume that it has a unique solution, and the solution is then given by

$$\hat{A}(\lambda,\,.) = \left(I + \frac{\sqrt{1-M^2}}{M}\mathbb{P}\mathbb{H}(\mathbb{B}(k) - B(\infty))\right)^{-1}\frac{2}{M}\hat{w}_a(\lambda,\,.). \qquad (5.107)$$

(We should recall here that we do have in Sect. 5.4 a constructive solution for small κ, which happens to be the region of major interest in practice.)

We define the coefficients $w_{ij}(\lambda)$ as in Sect. 5.4 (5.3), which are now recognized as functions of λ, and this is all we need from the solution to the Possio equation for all M.

Note that for every M:

$$\overline{w}_{ij}(\lambda) = w_{ij}(\overline{\lambda}),$$

implying that the aeroelastic modes come in conjugate pairs. The main difference for $M > 0$ from $M = 0$ is that from (5.72) we have that

$$\lim \mathrm{Re}\lambda \to \infty$$

$$w_{ij} = \frac{2}{M}\int_{-1}^{1}\xi d\xi = 0, \quad i \neq j,$$

$$w_{11} = \frac{4}{M},$$

$$w_{22} = \frac{2}{M}\int_{-1}^{1}\xi^2 d\xi = \frac{4}{3M}.$$

The implication is that in the breakdown $w_{ii}(\lambda)$ = polynomial in λ + non-polynomial term the polynomial term is a constant (of zero degree), unlike the case for $M = 0$, where it is of degree 1. The polynomial terms are called the non-circulatory terms in [6]. To see the implication of this let us consider the Laplace transform version of the structure dynamics:

$$\lambda^2 m\,\hat{h}(\lambda,s) + \lambda^2 S\hat{\theta}(\lambda,s) + EI\hat{h}(\lambda,s)'''' = \hat{L}(\lambda,s)$$

$$= -\rho b U_\infty^2\left[\frac{k}{b}w_{11}\hat{h}(\lambda,s) + (kw_{12} + (1-ka)w_{11})\hat{\theta}(\lambda,s)\right].$$

Or, collecting the non-circulatory terms into the structure state terms:

$$\lambda^2 m\,\hat{h}(\lambda,s) + \lambda^2 S\hat{\theta}(\lambda,s) + \lambda b\rho U_\infty\left(\frac{4}{M}\right)(\hat{h}(\lambda,s) - ab\hat{\theta}(\lambda,s))$$

$$+ \frac{4}{M}\rho b U_\infty^2\hat{\theta}(\lambda,s) + EI\hat{h}(\lambda,s)'''' = -\rho b U_\infty^2[\frac{k}{b}\breve{w}_{11}\hat{h}(\lambda,s)$$

$$+ (kw_{12} + (1-ka)\breve{w}_{11})\hat{\theta}(\lambda,s)],$$

where

$$\breve{w}_{11}(\lambda) = w_{11}(\lambda) - \frac{4}{M} \quad \text{and} \quad \to 0 \text{ as } \operatorname{Re} \breve{} \to \infty.$$

Similarly

$$\lambda^2 I_\theta \hat{\theta}(\lambda, s) + S\lambda^2 \hat{h}(\lambda, s) - GJ\hat{\theta}(\lambda, s)''$$

$$= \hat{M}(\lambda, s) = -\rho b^2 U_\infty^2 \left[\frac{k}{b}(w_{21} - aw_{11})\hat{h}(\lambda, s) \right.$$

$$\left. + (kw_{22} + (1 - ak)w_{21} - akw_{12} - a(1 - ak)w_{11})\hat{\theta}(\lambda, s) \right].$$

Hence collecting the non-circulatory terms into the structure side:

$$\lambda^2 I_\theta \hat{\theta}(\lambda, s) + S\lambda^2 \hat{h}(\lambda, s) + \lambda\rho b U_\infty \left(-\frac{4ab}{M} \right) \hat{h}(\lambda, s) + \lambda\rho b U_\infty \left(-\frac{4ab}{M} \right) \hat{h}(\lambda, s)$$

$$+ \lambda\rho b U_\infty \left(a^2 b^2 \frac{4}{M} + \frac{4}{3M}b^2 \right) \hat{\theta}(\lambda, s) - \frac{4}{M}\rho b U_\infty^2 ab\hat{\theta}(\lambda, s) - GJ\hat{\theta}(\lambda, s)''$$

$$= -\rho b^2 U_\infty^2 \left[\frac{k}{b}(w_{21} - a\breve{w}_{11})\hat{h}(\lambda, s) + (k\breve{w}_{22} + (1 - ak)w_{21} - akw_{12} \right.$$

$$\left. - a(1 - ak)\breve{w}_{11}) \hat{\theta}(\lambda, s) \right],$$

where

$$\breve{w}_{22}(\lambda) = w_{22}(\lambda) - \frac{4}{3M}$$

and goes to zero as $\operatorname{Re}\lambda \to \infty$.

The main thing is now to note that the structure inertia matrix is not changed unlike in the case where $M = 0$. But damping and stiffness terms are added which are reflected in the abstract version as bounded operators. Thus collecting terms containing λ^2 and λ we can now write the structure with the aerodynamic loading Laplace domain equation:

$$\lambda^2 \mathcal{M}\hat{x}(\lambda) + \lambda \mathcal{D}\hat{x}(\lambda) + \mathcal{K}\hat{x}(\lambda) + A\hat{x}(\lambda) = \mathcal{F}(\lambda)\hat{x}(\lambda, .), \qquad (5.108)$$

where

$$\mathcal{M} = \begin{pmatrix} m & s \\ s & I_\theta \end{pmatrix},$$

$$\mathcal{D} = \begin{pmatrix} b\rho U_\infty \left(\dfrac{4}{M}\right) & -ab\,b\rho U_\infty \left(\dfrac{4}{M}\right) \\ -ab\,b\rho U_\infty \left(\dfrac{4}{M}\right) & bU_\infty \left(a^2 b^2 \dfrac{4}{M} + \dfrac{4}{3M}b^2\right) \end{pmatrix},$$

$$\mathcal{K} = \begin{pmatrix} 0 & \dfrac{4}{M}\rho b U_\infty^2 \theta \\ 0 & -\dfrac{4}{M}\rho b U_\infty^2 \, ab\theta \end{pmatrix}, \tag{5.109}$$

where $\mathcal{MDK}$ are constant 2×2 matrices and $\hat{\mathcal{F}}(\lambda)$ is the Laplace transform:

$$\hat{\mathcal{F}}(\lambda)\begin{pmatrix} h \\ \theta \end{pmatrix} = \begin{pmatrix} -\rho b U_\infty^2 \left[\dfrac{k}{b}\breve{w}_{11}h + (kw_{12} + (1 - ka)\breve{w}_{11})\theta\right] \\ -\rho b^2 U_\infty^2 \left[\dfrac{k}{b}(w_{21} - a\breve{w}_{11})h \right. \\ \left. + (k\breve{w}_{22} + (1 - ak)w_{21} - akw_{12} - a(1 - ak)\breve{w}_{11})\,\theta]\right. \end{pmatrix} \tag{5.110}$$

and A is the structure differential operator as defined in Chap. 2.

Next we take inverse Laplace transforms

$$\mathcal{M}\ddot{x}(t) + \mathcal{D}\dot{x}(t) + \mathcal{K}x(t) + Ax(t) = \int_0^t \mathcal{F}(\sigma)x(t - \sigma)d\sigma, \tag{5.111}$$

where $\mathcal{F}(.)$ is the inverse Laplace transform:

$$\hat{\mathcal{F}}(\lambda) = \int_0^\infty e^{-\lambda t}\mathcal{F}(t)dt \qquad \mathrm{Re}\lambda > 0. \tag{5.112}$$

Next with

$$Y(t) = \begin{pmatrix} x(t) \\ \dot{x}(t) \end{pmatrix},$$

we obtain the convolution/evolution equation in $\mathcal{H}$:

$$\dot{Y}(t) = \mathcal{A}Y(t) + \int_0^t \mathcal{L}(t - \sigma)Y(\sigma)d\sigma,$$

where

$$\mathcal{L}(t) = \begin{pmatrix} 0 & 0 \\ \mathcal{M}^{-1}\mathcal{F}(t) & 0 \end{pmatrix} \qquad 4 \times 4 \text{ matrix function}, \tag{5.113}$$

$$\mathcal{A} = \begin{pmatrix} 0 & I \\ -\mathcal{M}^{-l}(A+\mathcal{K}) & -\mathcal{M}^{-1}\mathcal{D} \end{pmatrix} \tag{5.114}$$

with the energy space $\mathcal{H}$ as defined in Chap. 2.

Here as before, A generates a C_0 semigroup with compact resolvent; and the analysis is similar to that for $M = 1$. $M = 0$ is a special case, whereas $M = 1$ is typical of $M > 0$.

Taking Laplace transforms in

$$\dot{Y}(t) = \mathcal{A}Y(t) + \int_0^t \mathcal{L}(t-\sigma)Y(\sigma)d\sigma,$$

we have:

$$(\lambda I - \mathcal{A} - \hat{\mathcal{L}}(\lambda))\hat{Y}(\lambda) = Y(0).$$

Thus we define the aeroelastic modes as the zeros:

$$\left(\lambda I - \mathcal{A} - \hat{\mathcal{L}}(\lambda)\right) Y = 0$$

with the generalized resolvent defined again by

$$\mathcal{R}(\lambda) = \left(\lambda I - \mathcal{A}\hat{\mathcal{L}} - (\lambda)\right)^{-1}.$$

Note that the change from the $M = 0$ case is that (as in the case $M = 1$) $\lambda\hat{\mathcal{L}}(\lambda)$ is replaced by $\hat{\mathcal{L}}(\lambda)$ (the two $\hat{\mathcal{L}}(\lambda)$ are not the same functions, of course!) which makes the problem less complex in that now

$$\left(\lambda I - \mathcal{A} - \hat{\mathcal{L}}(\lambda)\right) \hat{Y}(\lambda) = Y.$$

Or,

$$\hat{Y}(\lambda) = \mathcal{R}(\lambda)Y.$$

The aeroelastic mode $\{\lambda_k\}$ are of course obtained as the zeros of the determinant

$$d(M,\lambda,U) = \det D(M,\lambda,U),$$

$$D(M,\lambda,U) = Pe^{A(\lambda)}Q$$

with $A(\lambda)$ just as defined in Sect. 5.4 except that the w_{ij} are no longer expressible explicitly.

We assume that $\hat{A}_i(\lambda, M)$ is the solution of the Possio equation (5.2) specialized to

$$\hat{A}_i(\lambda, .) = \left(I + \frac{\sqrt{1 - M^2}}{M} \mathbb{PH}(\mathbb{B}(k) - B(\infty))\right)^{-1} \frac{2}{M} f_i,$$

where

$$f_i(s) = 1, \quad -b < s < b$$
$$= s, \quad -b < s < b$$

and note that the solution is defined for all λ omitting the modes λ_k and the negative half-line. Moreover the solution is analytic in this open set and hence so are the w_{ij} and so is $A(\lambda)$.

Note that the roots again come in conjugate pairs, if complex. There can be real roots for some speeds; zero is a root for some speeds.

For the existence of the roots for U positive, we need to draw on the analyticity properties of $D(M, \lambda, U)$. These are determined by those of $\{w_{ij}(\lambda)\}$ which in turn are dependent on those of the solution to the Possio equation. However, we no longer have the luxury of an explicit solution. And thus we have to go back to (5.49), the abstract form of LDP for values other than $\mathrm{Re}k > 0$, assumed there. The Hilbert transform operator $\mathcal{H}$ is what causes trouble. The right-hand side of (5.49) is

$$\mathbb{PHPB}(k)\mathbb{P}A + T\mathbb{PH}(I - \mathbb{P})\mathbb{B}(k)\mathbb{P}A, k = \frac{\lambda b}{U},$$

where the first term is an entire function of λ and the second term

$$WA = T\mathbb{PB}(I - \mathbb{P})\mathbb{B}(k)\mathbb{P}A,$$

we now show is analytic except for the logarithmic singularity along the negative half-line, of the same kind as we have seen for $M = 0$. Thus we have

$$T\mathbb{PH}(I - \mathbb{P})\mathbb{B}(k)\mathbb{P}A = T\mathbb{PH}(I - \mathbb{P})\left[\frac{-kR(k)A(.)}{\sqrt{1 - M^2}} - \int_{-\alpha_2}^{\alpha_1} kR(ks)A(.)a(M, s)ds\right],$$

where the second term is kind of an "average" of the first term and inherits the analyticity properties of the first term. Denoting the first term by $g(.)$ we have:

$$g(x) = \frac{-k}{\pi}\sqrt{\frac{b - x}{b + x}} \int_0^\infty \frac{e^{-k\sigma}}{(\sigma + b - x)} \frac{\sqrt{2b + \sigma}}{\sqrt{\sigma}} d\sigma \int_{-b}^b e^{-k(b - \eta)} A(\eta)d\eta,$$

$$|x| < b$$

and although this function is defined for $\mathrm{Re}\, k < 0$, the integral is not defined for $\mathrm{Re}\, k < 0$.

However, it can be expressed in terms of the Bessel K functions:

$$\int_0^\infty \frac{e^{-k\sigma}}{(\sigma + s)} \frac{\sqrt{2b + \sigma}}{\sqrt{\sigma}} d\sigma = F(k), \quad \operatorname{Re} k > 0, \text{ take } b = 1,$$

$$\frac{dF}{dz} = sF(z) - (K_0(z) + K_1(z)) \qquad \operatorname{Re} z > 0$$

and because $(K_0(z) + K_1(z))$ is analytic except for $z \leq 0$, $F(z)$ can be extended analytically to the left-half-plane, but will have an essential singularity—branch cut—along the negative axis exactly as is the case for $M = 0$.

Thus the operator $\mathbb{PH}(\mathbb{B}(k) - B(\infty))$ in the Possio equation can be extended to be analytic in the whole plane except for the branch cut along the negative axis, and this property then extends to the solution, and hence to w_{ij}, and finally to $D(M, \lambda, U)$ and $d(M, \lambda, U)$.

Modes and Mode Shapes

Hence in particular the zeros of $d(M, \lambda, U)$—the aeroelastic modes—are countable. Further more they are limited to a finite strip (we call it the *spectral strip*) because the contribution of the circulatory terms:

$$\hat{\mathcal{L}}(\lambda) \to 0 \quad \text{as } \operatorname{Re} \lambda \to \infty \tag{5.115}$$

and hence we may imitate the arguments for $M = 0$ (cf. Sect. 5.3) where really only this argument is needed.

Again define x_k by:

$$D(\lambda_k)x_k = Pe^{A(\lambda_k)\ell}Q \qquad x_k = 0, \quad x_k \neq 0, \tag{5.116}$$

$D(\lambda)$ being a 3×3 matrix the null space of the matrix is at most 2, and we assume it to be 1, as is the case for $U = 0$. Similarly we assume that the multiplicity is also 1, so that

$$D'(\lambda_k)x_k \neq 0. \tag{5.117}$$

Lemma 5.31. *Let*

$$D(\lambda_k)^* z_k = 0 \qquad z_k \neq 0.$$

Then

$$[z_k, x_k] \neq 0. \tag{5.118}$$

Proof. The 3-space has the orthogonal decomposition:

$$\text{null space of } D(\lambda_k)^* + \text{ range space of } D(\lambda_k).$$

Hence if
$$[z_k, x_k] = 0,$$
we must have that x_k is in the range space of $D(\lambda_k)$. Or,

$$x_k = D(\lambda_k)h \quad \text{for } h \neq 0.$$

Hence
$$D(\lambda_k)^2 h = 0.$$

Taking the derivative with respect to λ, we have

$$D'(\lambda_k)x_k = 0,$$

which contradicts (5.11).   □

The corresponding Y_k satisfies

$$(\lambda_k I - \mathcal{A} - \hat{\mathcal{L}}(\lambda_k))Y_k = 0.$$

Or, with

$$Y_k = \begin{pmatrix} x_{1k} \\ x_{2k} \end{pmatrix},$$

we have

$$\lambda_k x_{1,k} = x_{2,k},$$

$$\lambda_k^2 \mathcal{M} x_{1,k} + \lambda \mathcal{D} x_{1,k} + \mathcal{K} x_{1,k} + \mathcal{A} x_{1,k} - \hat{\mathcal{F}}(\lambda_k)x_{1,k} = 0. \tag{5.119}$$

Hence
$$x_{2,k} = \lambda_k x_{1,k}. \tag{5.120}$$

Define the 2×6 matrix:
$$T_{26} = \begin{pmatrix} 1\,0\,0\,0\,0\,0 \\ 0\,0\,0\,1\,0\,0 \end{pmatrix}. \tag{5.121}$$

Then
$$x_{1,k}(s) = T_{26}e^{A(\lambda_k)s}Qx_k, \quad 0 < s < \ell_k$$

and x_k is defined by (5.10).

We may normalize x_k by:

$$[\mathcal{M}x_{1k}, x_{lk}] = 1; \left[\left(\int_0^\ell Q^\star e^{A^\star(\lambda_k)s}T_{26}{}^\star \mathcal{M}T_{26}e^{A(\lambda_k)s}Q\,ds\right)x_k, x_k\right] = 1. \tag{5.122}$$

This determines Y_k completely.

$$Y_k(s) = \begin{pmatrix} T_{26} e^{A(\lambda_k)s} Q x_k \\ \lambda_k T_{26} e^{A(\lambda_k)s} Q x_k \end{pmatrix}, \quad 0 < s < \ell.$$

In fact we can express it as a linear bounded transformation on C^3 into $\mathcal{H}$:

$$L_k x = \begin{pmatrix} f_1 \\ \lambda_k f_1 \end{pmatrix}; \quad f_1(s) = T_{26} e^{A(\lambda_k)s} Q x, \quad 0 < s < \ell,$$

$$Y_k = L_k x_k,$$

and

$$[Y_k, Y_k] = [L_k{}^\star L_k x_k, x_k] = |\lambda_k|^2.$$

The energy space norm of Y_k is not equal to 1.

The dimension of the null space of $\left(\lambda_k I - A - \hat{\mathcal{L}}(\lambda_k) \right)$ is the dimension of the null space of $P e^{A(\lambda_k)\ell} Q$ and is 1, with multiplicity 1.

We note that in as much as the conjugates

$$\bar{D}(\lambda) = D(\bar{\lambda}); \quad \bar{A}(\lambda) = A(\bar{\lambda}),$$

the conjugate $\bar{Y}_k$ is the eigenvector corresponding to the conjugate root $\bar{\lambda}_k$.

$$\mathcal{R}(\lambda) Y_k = \frac{Y_k}{\lambda - \lambda_k}; \quad \mathcal{R}(\lambda) \bar{Y}_k = \frac{\bar{Y}_k}{\lambda - \bar{\lambda}_k}$$

corresponding to

$$D(\lambda)^{-1} x_k = \frac{x_k}{\lambda - \lambda_k}.$$

The case where zero is an aeroelastic mode is treated separately below, as it is involved in defining the divergence speed.

We note that zero is an aeroelastic mode only for a sequence of speeds. The functions (for $M \neq 1$) are defined as the limit from above (Im λ nonzero) or from the right (Re$\lambda > 0$).

And

$$d(0,0+,0+) = 1.$$

Generalized Resolvent

The generalized resolvent is defined for λ excepting the aeroelastic modes and the negative half-line:

$$\mathcal{R}(\lambda) = \left(\lambda I - A - \hat{\mathcal{L}}(\lambda)\right)^{-1}$$

just as in the case $M = 0$ in Sect. 5.5.

$$\mathcal{R}(\lambda)Y_g = (\lambda I - A - \hat{\mathcal{L}}(\lambda))^{-1}Y_g \ : \ Y_g = \begin{pmatrix} x_{1,g} \\ x_{2,g} \end{pmatrix}.$$

Let

$$\begin{pmatrix} h_g \\ \theta_g \end{pmatrix} = \mathcal{M}x_{2,g} + (\lambda\mathcal{D} + \mathcal{K})x_{1,g}.$$

Then (as in Sect. 5.5)

$$y_g(s) = \text{col}\left[0,0,0, \frac{h_g(s)}{EI}, 0, \frac{\theta_g(s)}{GJ}\right],$$

$$y(\lambda, s) = e^{sA(\lambda)}Qz(\lambda) + \int_0^s e^{(s-\sigma)A(\lambda)}y_g(\sigma)d\sigma,$$

$$P(e^{\ell A(\lambda)}Qz(\lambda) + v_g(\lambda)) = 0; \quad v_g(\lambda) = \int_0^\ell e^{(\ell-\sigma)A(\lambda)}y_g(\sigma)d\sigma.$$

Or,

$$D(\lambda)z(\lambda) = -Pv_g(\lambda); \quad D(\lambda) = Pe^{\ell A(\lambda)}Q.$$

Hence

$$z(\lambda) = -D(\lambda)^{-1}Pv_g(\lambda),$$

which is defined where w_{ij} are defined excepting the modes $\{\lambda_k\}$ and

$$y(\lambda, s) = -e^{sA(\lambda)}QD(\lambda)^{-1}P\int_0^\ell e^{(\ell-\sigma)A(\lambda)}y_g(\sigma)d\sigma$$

$$+ \int_0^s e^{(s-\sigma)A(\lambda)}y_g(\sigma)d\sigma, \tag{5.123}$$

and finally:

$$\mathcal{R}(\lambda)Y_g = \begin{pmatrix} x_1(\lambda) \\ x_2(\lambda) \end{pmatrix},$$

$$x_1(\lambda) = T_{26}y(\lambda, .) = \begin{pmatrix} h(\lambda, .) \\ \theta(\lambda, .) \end{pmatrix},$$

$$x_2(\lambda) = \lambda x_1(\lambda) - x_{1,g}.$$

Green's Function Representation

We proceed first to deduce the Green's function for the generalized resolvent. For
this purpose it is convenient to use the notation:

$$C_6(\lambda)Y = \text{col}\left[0,0,0,\frac{h(.)}{EI},0,\frac{\theta(.)}{GJ}\right],$$

where

$$\begin{pmatrix}h(.)\\ \theta(.)\end{pmatrix} = \mathcal{M}x_2 + (\lambda\mathcal{D}+\mathcal{K})x_1; \; Y = \begin{pmatrix}x_l\\ x_2\end{pmatrix}.$$

Then $\mathcal{R}(\lambda)Y$ is the function:

$$\begin{pmatrix}x_1(\lambda,s)\\ x_2(\lambda,s)\end{pmatrix} = \int_0^\ell R(\lambda,s,\sigma)Y(\sigma)d\sigma, \quad 0<s<\ell, \tag{5.124}$$

where the kernel is deduced from

$$x_1(\lambda,s) = T_{26}e^{sA(\lambda)}QD(\lambda)^{-1}P\int_0^\ell e^{(\ell-\sigma)A(\lambda)}C_6(\lambda)Y(s)ds$$

$$+\int_0^s e^{(s-\sigma)A(\lambda)}C_6(\lambda)Y(\sigma)d\sigma x_2(\lambda,s) = \lambda x_1(\lambda,s) - x_1(s).$$

The main thing to note is that the kernel is square integrable:

$$\int_0^\ell \int_0^\ell ||R(\lambda,s,\sigma)||^2 ds d\sigma < \infty.$$

In particular this implies that the eigenvalues are square integrable:

$$\sum \frac{1}{|\lambda-\lambda_k|^2} < \infty,$$

where the summation is over all roots, the index k being chosen such that the
imaginary part

$$|\omega_{k+1}| \geq |\omega_k|.$$

Our convention for the modes is:

$$\lambda_k = \sigma_k + i\omega_k, \quad \omega_k > 0$$

and the conjugate is of course always a mode as well, if complex. And in particular

$$\sum \frac{1}{|\lambda_k|^2} < \infty \text{ nonzero } \lambda_k. \qquad (5.125)$$

Laplace Inversion Formula

Equation (5.9) shows that the generalized resolvent has simple poles at the aeroelastic modes confined to a finite strip. We can use this to determine the inverse Laplace transform by the inversion formula

$$W(t)Y_0 = \lim_{L \to \infty} \frac{1}{2\pi i} \int_{-L}^{L} e^{(\sigma+i\omega)t} \mathcal{R}(\sigma + i\omega) Y_0 d\omega, \qquad (5.126)$$

where σ is to the right of the strip containing the point spectrum,

$$W(t)Y_0 = \begin{pmatrix} x_1(t) \\ x_2(t) \end{pmatrix},$$

$$Y_0 = \begin{pmatrix} x_{1,0} \\ x_{2,0} \end{pmatrix},$$

$$W(0)Y_0 = Y_0 \text{ in } \mathcal{D}(\mathcal{A}),$$

and

$$x_2(t) = \frac{d}{dt} x_1(t),$$

$$\begin{pmatrix} h_0\,(\lambda) \\ \theta_0\,(\lambda) \end{pmatrix} = \mathcal{M}x_{2,0} + (\lambda \mathcal{D} + \mathcal{K})x_{1,0},$$

$$y_0(\lambda, s) = \mathrm{col} \left[0, 0, 0, \frac{h_0(\lambda, s)}{EI}, 0, \frac{\theta_0(\lambda, s)}{GJ} \right],$$

$$x_1(t, s) = \lim_{L \to \infty} \frac{1}{2\pi i} \int_{-L}^{L} e^{\lambda t} T_{26}$$

$$\times \left(-e^{sA(\lambda)} Q D(\lambda)^{-1} P \int_0^{\ell} e^{(\ell-\xi)A(\lambda)} y_0(\lambda, \xi) d\xi \right.$$

$$\left. + \int_0^s e^{(s-\xi)A(\lambda)} y_0(\lambda, \xi) d\xi \right) i d\omega|_{\lambda=\sigma+i\omega} \qquad 0 < s < \ell,$$

where we have seen that $D(\lambda)^{-1}$ has simple poles at $\lambda = \lambda_k$ with residue

$$D_k^{-1} = \frac{1}{2\pi i} \oint D(\lambda)^{-1} d\lambda \qquad \lambda = \lambda_k + re^{i\theta}.$$

(The convergence of the infinite series is assumed here for the moment; the proof is given below.)

Hence we deform the contour as before following [33,44] where the functions are meromorphic but with a branch cut along the negative axis.

Thus we have

$$x_1(t,s) = \sum_{k=1}^{\infty} T_{26} \left(-e^{sA(\lambda_k)} QD_k^{-1} P \int_0^{\ell} e^{(\ell-\sigma)A(\lambda_k)} y_0(\sigma) d\sigma \right) e^{\lambda_k t} + x_{1,R}(t,s),$$

$$x_2(t,s) = \dot{x}_1(t,s), \qquad Y_0 \text{ in } \mathcal{D}(\mathcal{A}), \tag{5.127}$$

where the second term $x_{1,R}(t, .)$ is the "evanescent" term we have already treated in Sect. 5.5.

The evanescent term can be expressed:

$$x_{1,R}(t, .) = \frac{U}{b} \int_0^{\infty} e^{-RU/bt} R_J(-R) Y_g dR, \tag{5.128}$$

where $R_J(.)$ is the jump across the line of singularity: $-\infty < \lambda \le 0$ and for all $M \ge 0$ including $M = 0$.

Note the slight difference in notation from Sect. 5.5. Thus we define

$$R_J(-r)Y = \frac{1}{2\pi i} \lim_{\omega \to 0} [R(-r + i\omega) - R(-r - i\omega)]Y, \qquad \omega \ne 0,$$

where the subscript J is for "jump" and

$$Y = \begin{pmatrix} x_l \\ x_2 \end{pmatrix}$$

and $R(\lambda)Y$ is the function defined by

$$(R(\lambda)Y)(s) = T_{26} \left[\left(-e^{sA(\lambda)} QD(\lambda)^{-1} P \right) \int_0^{\ell} e^{(\ell-\sigma)A(\lambda)} y(\sigma) d\sigma \right.$$

$$\left. + \int_0^s e^{(s-\sigma)A(\lambda)} y(\sigma) d\sigma \right], \qquad 0 < s < \ell,$$

$$y(s) = \mathrm{col} \left[0, 0, 0, \frac{h(\lambda, s)}{EI}, 0, \frac{\theta(\lambda, s)}{GJ} \right],$$

$$\begin{pmatrix} h(\lambda) \\ \theta(\lambda) \end{pmatrix} = \mathcal{M} x_2 + (\lambda \mathcal{D} + \mathcal{K}) x_1$$

defined and analytic for $\lambda \neq \lambda_k$, and omitting $\lambda \leq 0$.

$$(R(\lambda)Y)(s) = T_{26} \left[\left(-e^{sA(\lambda)} QD(\lambda)^{-1} P \right) \int_0^\ell e^{(\ell-\sigma)A(\lambda)} y(\sigma) d\sigma \right.$$
$$\left. + \int_0^s e^{(s-\sigma)A(\lambda)} y(\sigma) d\sigma \right], \qquad 0 < s < \ell$$

is defined and analytic for $\lambda \neq \lambda_k$; and omitting $\lambda < 0$ which is a logarithmic branch cut.

Moreover the jump $[R(-r+i\omega) - R(-r-i\omega)]Y$, $\omega \neq 0$ converges strongly as we approach the line of singularity as ω goes to zero for each $r > 0$. And the limit $R_J(-r)Y$ goes to zero as r goes to zero.

Hence finally:

$$W(t)Y_0 = \begin{pmatrix} x_1(t) \\ x_2(t) \end{pmatrix}; \qquad Y_0 = \begin{pmatrix} x_{1,0} \\ x_{2,0} \end{pmatrix},$$

$$x_1(t,s) = \sum_{k=1}^\infty T_{26} \left(-e^{sA(\lambda_k)} QD_k^{-1} P \int_0^\ell e^{(\ell-\sigma)A(\lambda_k)} y_g(\sigma) d\sigma \right) e^{\lambda_k t}$$

$$+ \int_0^\infty e^{-rt} (R_J(-r)Y_0)(s) dr,$$

$$x_2(t,s) = \dot{x}_1(t,s), \qquad t > 0, \; 0 < s < \ell. \tag{5.129}$$

Again: let us take the Laplace transform

$$\hat{x}_1(\lambda, \, .) = \int_0^\infty e^{-\lambda t} x_1(t, \, .) dt$$

$$= \sum T_{26} \left(-e^{sA(\lambda_k)} QD_k^{-1} P \int_0^\ell e^{(\ell-\sigma)A(\lambda_k)} y_g(\sigma) d\sigma \right) \frac{1}{\lambda - \lambda_k}$$

$$\times \frac{1}{\pi} \int_0^\infty \frac{(R_J(-r)Y_g)(s)}{\lambda + r} dr, \qquad 0 < s < \ell,$$

where the sum is over all roots here and below.

$$\int_0^\infty e^{-\lambda t} x_2(t, \, .) dt = \hat{x}_2(\lambda) = \lambda \hat{x}_1(\lambda) - x_{1,g}.$$

Hence we have the representation $\mathcal{R}(\lambda)Y$ is the function

$$= \begin{pmatrix} \sum \frac{1}{\lambda - \lambda_k} T_{26} \left(-e^{sA(\lambda_k)} QD_k^{-1} P \int_0^\ell e^{(\ell-\sigma)A(\lambda_k)} y_g(\sigma)d\sigma \right) \\ +\frac{1}{\pi} \int_0^\infty \frac{(R_J(-r)Y)(s)}{\lambda + r} dr \\ \sum T_{26} \left(-e^{sA(\lambda_k)} QD_k^{-1} P \int_0^\ell e^{(\ell-\sigma)A(\lambda_k)} y_g(\sigma)d\sigma \right) \frac{\lambda}{\lambda - \lambda_k} \\ +\frac{1}{\pi} \int_0^\infty \frac{\lambda(R_J(-r)Y)(s)}{\lambda + r} dr - x_{1,g}(s) \end{pmatrix},$$

$$0 < s < \ell. \tag{5.130}$$

Let us define the linear bounded operator on $\mathcal{H}$ into $\mathcal{H}$ by the contour integral

$$P_k Y = \int_{\Omega(\lambda_k)} \mathcal{R}(\lambda)d\lambda; \quad \Omega(\lambda_k) = \{\lambda = \lambda_k + re^{i\theta}, \ 0 \le \theta \le 2\pi\}$$

$$= \int_0^{2\pi} \frac{1}{2\pi} \mathcal{R}\left(\lambda_k + re^{i\theta}\right) Y d\theta \quad \text{for } 0 < r,$$

small enough to include no other poles. $P_k Y$ is the function

$$\begin{pmatrix} T_{26} \left(-e^{sA(\lambda_k)} QD_k^{-1} P \int_0^\ell e^{(\ell-\sigma)A(\lambda_k)} y_g(\sigma)d\sigma \right) \\ \lambda_k T_{26} \left(-e^{sA(\lambda_k)} QD_k^{-1} P \int_0^\ell e^{(\ell-\sigma)A(\lambda_k)} y_g(\sigma)d\sigma \right) \end{pmatrix}, \quad 0 < s < \ell. \tag{5.131}$$

And more generally:

$$\int_{\Omega(\lambda_k)} f(\lambda)\mathcal{R}(\lambda)d\lambda = f(\lambda_k) P_k.$$

For $Y = \begin{pmatrix} x_l \\ x_2 \end{pmatrix}$,

$$\mathcal{R}(\lambda)Y = \sum \frac{P_k Y}{\lambda - \lambda_k} + \begin{pmatrix} \frac{l}{\pi} \int_0^\infty \frac{R_J(-r)Y}{\lambda + r} dr \\ +\frac{1}{\pi} \int_0^\infty \frac{\lambda R_J(-r)Y}{\lambda + r} dr - x_1(s) \end{pmatrix}.$$

Now

$$Y = \left(\lambda I - A - \hat{\mathcal{L}}(\lambda) \right) \mathcal{R}(\lambda)Y \sum \frac{\left(\lambda I - A - \hat{\mathcal{L}}(\lambda) \right) P_k Y}{\lambda - \lambda_k}$$

$$+ \left(\lambda I - \mathcal{A} - \hat{\mathcal{L}}(\lambda) \right) \begin{pmatrix} \dfrac{1}{\pi} \displaystyle\int_0^\infty \dfrac{R_J(-r)Y}{\lambda + r} dr \\[2ex] \dfrac{1}{\pi} \displaystyle\int_0^\infty \dfrac{\lambda R_J(-r)Y}{\lambda + r} dr - x_1(s) \end{pmatrix}.$$

Then the contour integral $\oint (1/2\pi i) Y d\lambda, \lambda = \lambda_k + r e^{i\theta}$ yields

$$0 = \left(\lambda_k I - \mathcal{A} - \hat{\mathcal{L}}(\lambda_k) \right) P_k Y.$$

Hence $P_k Y$ is the slant (not orthogonal because $P_k^* \neq P_k$) projection on the null space of $\left(\lambda_k I - \mathcal{A} - \hat{\mathcal{L}}(\lambda_k) \right)$ which is of dimension 1 and is spanned by Y_k. Hence the dimension of the range of P_k is 1. Or

$$P_k Y = \alpha_k Y_k = \frac{[P_k Y, Y_k]}{[Y_k, Y_k]} Y_k,$$

$$P_k Y_k = Y_k,$$

$$\mathcal{R}(\lambda) Y = \sum_k \frac{P_k Y}{\lambda - \lambda_k} + \begin{pmatrix} \dfrac{1}{\pi} \displaystyle\int_0^\infty \dfrac{R_J(-r)}{\lambda + r} Y dr \\[2ex] \dfrac{1}{\pi} \displaystyle\int_0^\infty -r R_J(-r) Y dr \end{pmatrix}.$$

Note that this runs over all roots: the conjugates are treated as separate roots.

Next let Q denote the linear bounded operator

$$QY = \begin{pmatrix} \frac{1}{\pi} \int_0^\infty R_J(-r) Y dr \\[2ex] \frac{1}{\pi} \int_0^\infty -r R_J(-r) Y dr \end{pmatrix}. \tag{5.132}$$

The range of Q are again the evanescent states as in the incompressible case, stable for all U. From (6.9)

$$\mathcal{R}(\lambda) P_k Y = \frac{P_k Y}{\lambda - \lambda_k}.$$

Now

$$\lim_{\mathrm{Re}\lambda \to \infty} \lambda \mathcal{R}(\lambda) Y = Y.$$

Using which in particular we have the representation:

$$Y = \sum P_k Y + QY \tag{5.133}$$

consistent with (5.11).

Lemma 5.32.

$$P_k P_j = P_j P_k = 0, \qquad j \neq k$$
$$= P_k \qquad j = k,$$
$$Q^2 = Q,$$
$$P_k Q = 0 = Q P_k.$$

The P_k as well as Q are idempotents, but not projections.

Proof. Here we essentially follow the theory of pseudoresolvents in [10, p. 208 et seq].

Thus we note that for $k \neq j$,

$$P_k P_j x = \frac{1}{(2\pi i)^2} \int_{\Omega(\lambda_k)} \int_{\Omega(\lambda_j)} \mathcal{R}(\lambda) \mathcal{R}(\mu) d\lambda d\mu,$$

where

$$\mathcal{R}(\lambda) - \mathcal{R}(\mu) = \left[(\mu - \lambda)I - \left(\hat{\mathcal{L}}(\lambda) - \hat{\mathcal{L}}(\mu) \right) \right] \mathcal{R}(\lambda) \mathcal{R}(\mu)$$

and hence

$$\mathcal{R}(\lambda) \mathcal{R}(\mu) = \frac{\mathcal{R}(\lambda) - \mathcal{R}(\mu)}{\mu - \lambda} + \left(\frac{\hat{\mathcal{L}}(\lambda) - \hat{\mathcal{L}}(\mu)}{\mu - \lambda} \right) \mathcal{R}(\lambda) \mathcal{R}(\mu) \qquad (5.134)$$

and

$$\frac{1}{(2\pi i)^2} \int_{\Omega(\lambda_k)} \int_{\Omega(\lambda_j)} \frac{\mathcal{R}(\lambda)}{\mu - \lambda} d\lambda d\mu = 0, \quad k \neq j = P_k, \ k = j,$$

$$\frac{1}{(2\pi i)^2} \int_{\Omega(\lambda_k)} \int_{\Omega(\lambda_j)} \frac{\mathcal{R}(\mu)}{\mu - \lambda} d\lambda d\mu = 0, \quad k \neq j = -P_k, \ k = j,$$

$$\frac{1}{(2\pi i)^2} \int_{\Omega(\lambda_k)} \int_{\Omega(\lambda_j)} \left(\frac{\hat{\mathcal{L}}(\lambda) - \hat{\mathcal{L}}(\mu)}{\mu - \lambda} \right) \mathcal{R}(\lambda) \mathcal{R}(\mu) d\lambda d\mu$$

$$= \frac{\hat{\mathcal{L}}(\lambda_k) - \hat{\mathcal{L}}(\lambda_j)}{\lambda_j - \lambda_k} P_k P_j \quad k \neq j.$$

Or

$$P_k P_j = \frac{\hat{\mathcal{L}}(\lambda_k) - \hat{\mathcal{L}}(\lambda_j)}{\lambda_j - \lambda_k} P_k P_j \qquad k \neq j.$$

Hence

$$P_k P_j = 0 \qquad k \neq j.$$

Also by a similar integration procedure:

$$P_k Q x = 0 = Q P_k.$$

□

Next, using (5.22)

$$P_k x = P_k \left(\sum_j P_j x + Q x \right) = P_k P_k x,$$

$$Q x = Q \left(\sum_j P_j x + Q x \right) = Q Q x.$$

Hence the P_k and Q are idempotents but not necessarily orthogonal projections because not self-adjoint.

To proceed further we need to go on to the operator adjoints.

Adjoint Theory

Let us consider now the properties of the adjoint operator: (Note that the adjoint now is with respect to the energy inner product.) We calculate

$$(\lambda I - A - \hat{\mathcal{L}}(\lambda))^\star Z = 0$$

by

$$[Z, (\lambda I - A - \hat{\mathcal{L}}(\lambda)) Y] = 0 \quad \text{for every } Y.$$

Or

$$[A z_1, \lambda x_1 - x_2]$$
$$+ \left[z_2, \lambda^2 \mathcal{M} x_1 + \lambda \mathcal{D} x_1 + \mathcal{K} x_1 + A x_1 + \lambda \mathcal{D} x_1 + \mathcal{K} x_1 - \hat{\mathcal{F}}(\lambda) x_1 \right] = 0$$

with

$$x_1 = \begin{pmatrix} h \\ \theta \end{pmatrix};$$

the second term

$$= \left[h_2, \lambda^2 mh + \lambda^2 S\theta + EIh'''' - \hat{L}(\lambda, .) \right]$$

$$+ \left[\theta_2, \lambda^2 I_\theta \theta + S\lambda^2 h - GJ\theta'' - \hat{M}(\lambda, .) \right],$$

where

$$\hat{L}(\lambda, s) = -\rho b U_\infty^2 \left[\frac{k}{b} w_{11} h(s) + (kw_{12} + (1 - ka)w_{11})\theta(s) \right] \hat{M}(\lambda, s)$$

$$= -\rho b^2 U_\infty^2 \left[\frac{k}{b}(w_{21} - aw_{11})h(s) \right.$$

$$\left. + (kw_{22} + (1 - ak)w_{21} - akw_{12} - a(1 - ak)w_{11})\,\theta(s) \right].$$

As in the case of Y_k we can calculate

$$z_k = \begin{pmatrix} z_1 k \\ z_2 k \end{pmatrix},$$

$$z_{2k}(s) = T_{26} e^{\tilde{A}(\lambda_k)s} Q z_k \qquad 0 < s < \ell, \tag{5.135}$$

$$\tilde{\mathcal{D}}(\lambda_k) z_k = 0, \qquad z_k \neq 0, \tag{5.136}$$

$$\tilde{\mathcal{D}}(\lambda) = P e^{\tilde{A}(\lambda)\ell},$$

$$\tilde{A}(\alpha) = \begin{pmatrix} 0 & 1 & 0 & 0 & 0 & 0 \\ 0 & 0 & 1 & 0 & 0 & 0 \\ 0 & 0 & 0 & 1 & 0 & 0 \\ \overline{w_1} & 0 & 0 & 0 & \overline{w_3} & 0 \\ 0 & 0 & 0 & 0 & 0 & 1 \\ \overline{w_2} & 0 & 0 & 0 & \overline{w_4} & 0 \end{pmatrix},$$

$$\bar{\lambda}_k^2 \mathcal{M} z_{2k} + A z_{2k} = \left(-\mathcal{K}^\star + \hat{\mathcal{F}}(\lambda_k)^\star - \bar{\lambda}_k \mathcal{D}^\star \right) z_{2k}, \tag{5.137}$$

$$\bar{\lambda}_k z_{1k} = \left(-I - A^{-1} \mathcal{K}^\star + A^{-1} \hat{\mathcal{F}}(\lambda_k)^\star \right) z_{2k}. \tag{5.138}$$

This time z_{1k} is determined from z_{2k}. The zeros of det $\tilde{D}(\lambda)$ are the conjugates of those of det $\mathcal{D}(\lambda)$.

Recall that we normalized Y_k so that

$$||Y_k||^2 = [x_{1k}, Ax_{1k}] + |\lambda_k|^2$$

and asymptotically in k,

$$\frac{[x_{1k}, Ax_{1k}]}{|\lambda_k|^2} \to 1$$

and

$$\frac{||Y_k||^2}{|\lambda_k|^2} \to 2.$$

We would want a similar normalization for Z_k. We now normalize so that

$$[\mathcal{M}z_{2k}, z_{2k}] = |\lambda_k|^2.$$

Then

$$\left[\left(\int_0^\ell Q^\star e^{\tilde{A}(\lambda_k)^\star s} (T_{26})^\star \mathcal{M} T_{26} e^{\tilde{A}(\lambda_k)s} Q \, ds \right) z_k, z_k \right] = |\lambda_k|^2,$$

which is then how z_k has to be normalized.

Then we calculate

$$[z_{1k}, Az_{1k}] = \frac{1}{|\lambda_k|^2} \left[\left(I + A^{-1}\mathcal{K}^\star - A^{-1}\hat{\mathcal{F}}(\lambda_k)^\star \right) z_{2k}, \left(A + \mathcal{K}^\star - \hat{\mathcal{F}}(\lambda_k)^\star \right) z_{2k} \right].$$

Now

$$\bar{\lambda}_k^2 \mathcal{M} z_{2k} + A z_{2k} = \left(-\mathcal{K}^\star + \hat{\mathcal{F}}(\lambda_k)^\star - \bar{\lambda}_k \mathcal{D}^\star \right) z_{2k}$$

using which we have:

$$[z_{1k}, Az_{1k}] = -\lambda_k^2 \left[z_{2k}, \left(\mathcal{M}z_{2k} + \frac{1}{\bar{\lambda}_k} \mathcal{D}^\star z_{2k} \right) \right] \bigg/ [\mathcal{M}z_{2k}, z_{2k}]$$

$$- \lambda_k^2 \left[\left(A^{-1}\mathcal{K}^\star - A^{-1}\hat{\mathcal{F}}(\lambda_k)^\star \right) z_{2k}, \left(\mathcal{M}z_{2k} + \frac{1}{\bar{\lambda}_k} \mathcal{D}^\star z_{2k} \right) \right]$$

$$\bigg/ [\mathcal{M}z_{2k}, z_{2k}].$$

Hence asymptotically in k

$$\frac{[z_{1k}, Az_{1k}]}{|\lambda_k|^2} \bigg| \to 1, \quad \text{noting that} \quad \frac{-\lambda_k^2}{|\lambda_k|^2} \to 1.$$

Hence it follows in particular that

$$[Z_k, Z_k]\big/|\lambda_k|^2 \to 2.$$

Second Normalization

We renormalize Z_k without changing norms. We have seen that $[Y_k, Z_k] \neq 0$, therefore we define

$$\tilde{Z}_k = \mu_k Z_k, \quad \text{where } |\mu_k| = 1, \quad \text{and}$$

$$[Y_k, \, \tilde{Z}_k]\Big/ \sqrt{[\tilde{Z}_k, \tilde{Z}_k][Y_k, Y_k]} = -1. \tag{5.139}$$

We simply keep using Z_k again in this renormalized version. Again we can write

$$z_k = \tilde{L}_k z_k,$$

where $\tilde{L}_k$ is a linear bounded transformation on C^3 into $\mathcal{H}$ given by

$$\tilde{L}_k z = \begin{pmatrix} g_1 \\ g_2 \end{pmatrix}; \qquad g_2(s) = T_{26} e^{\tilde{A}(\lambda_k)s} Q z, \qquad 0 < s < \ell,$$

$$\tilde{\lambda}_k g_1 = -g_2 - \left(A^{-1} K^\star - A^{-1} \hat{\mathcal{F}}(\lambda_k)^\star \right) g_2,$$

$$[\tilde{L}_k^* \mathcal{M} \tilde{L}_k g_2, g_2] = |\lambda_k|^2.$$

Note that

$$[P_k^\star Z, \, P_j Y] = 0 \qquad j \neq k.$$

Hence

$$[Y_k, Z_j] = 0 \quad \text{for} \quad k \neq j$$

and similarly:

$$[QY, Z_j] = 0 \quad \text{for every } j.$$

The sequences $\{Y_k\}, \{Z_k\}$ are biorthogonal. Note the set of eigenvalues is the same. Then we have the representations:

$$Y = \sum \alpha_k Y_k + QY \tag{5.140}$$

$$= \sum \beta_k Z_k + Q^* Y. \tag{5.141}$$

Problem: Given any two sequences: Can we always biorthogonalize them?

We note that the $\{P_k Y\}$ do not span the space. In particular

$$P_k Q Y = 0 = Q P_k Y.$$

So then the question is: Given $P_k Y = \alpha_k Y_k$, how do we determine α_k?

We use the representation:

$$[Y, Z_j] = [P_j Y, Z_j]$$
$$= \alpha_j [Y_j, Z_j].$$

If

$$[Y_j, Z_j] = 0 \quad \text{for any } j, \text{ then for every } Y,$$

$$[Y_j, Y] = \sum_{k=1}^{\infty} [Y_j, \beta_k z_k] + [Y_j, Q^* Y] = 0,$$

which is a contradiction. Hence

$$\alpha_k = \frac{[Y, Z_k]}{[Y_k, Z_k]}.$$

Next

$$|[Y_k, Z_k]| \leq \|Y\| \sqrt{[Z_k, Z_k]}.$$

Next

$$[Y_k, Z_k] = (-1)\sqrt{[Y_k, Y_k]}\sqrt{[Z_k, Z_k]}.$$

Hence

$$| \alpha_k | \|Y_k\| \leq \frac{\|Y\|}{\sqrt{[Y_k, Y_k]}},$$

which asymptotically

$$\leq \frac{\|Y\|}{|\lambda_k|}.$$

Because

$$\sum_{k=1}^{\infty} \frac{1}{|\lambda_k|^2} < \infty,$$

it follows that

$$\sum_{k=1}^{\infty} |\alpha_k|^2 \|Y_k\|^2 < \infty. \tag{5.142}$$

And similarly that

$$\sum_{k=1}^{\infty} |\beta_k|^2 ||Z_k||^2 < \infty. \tag{5.143}$$

This is called the Riesz property of the biorthogonal sequence proved in a more abstract way in [22].

Note that

$$[Y, \ Y] = \sum \alpha_k \bar{\beta}_k [Y_k, Z_k]. \tag{5.144}$$

This is enough to prove the convergence of all the infinite sums considered above.

Next we can calculate the Z_k in a different way. Thus

$$\mathcal{A}_k P_k Y \ = \lambda_k P_k Y.$$

Hence for any Z

$$[\mathcal{A}_k P_k Y, Z] = \lambda_k [P_k Y, Z] = [Y, \bar{\lambda}_k P_k^* Z].$$

Hence

$$P_k^* Z \epsilon \mathcal{D}(\mathcal{A}_k^*)$$

and

$$\left(\bar{\lambda}_k I - \mathcal{A}_k^* \right) P_k^* Z = 0.$$

We define:

$$P_k^* Z_k = Z_k.$$

To calculate P_k^* we use

$$[P_k Y, Z], \ Z = \begin{pmatrix} z_l \\ z_2 \end{pmatrix}; \quad Y = \begin{pmatrix} x_l \\ x_2 \end{pmatrix}; \quad C_{62} x = \mathrm{col} \left[0, 0, 0, \frac{h}{EI}, 0, \frac{\theta}{GJ} \right]$$

$$= \int_0^{\ell} \left[T_{26} e^{-sA(\lambda_k)} (QD_k^{-1} P) C_6 Y(s), (Az_1)(s) \right] \mathrm{d}s$$

$$+ \int_0^{\ell} \left[\lambda_k T_{26} e^{-sA(\lambda_k)} (QD_k^{-1} P) C_6 Y(s), \mathcal{M} z_2(s) \right] \mathrm{d}s$$

$$= \int_0^{\ell} \left[x_1(s), (C_6)^* (QD_k^{-1} P)^* e^{sA(\lambda_k)^*} (T_{26})^* (Az_1)(s) \right] \mathrm{d}s$$

$$+ \int_0^{\ell} \left[x_2(s), \bar{\lambda}_k (C_6)^* (QD_k^{-1} P)^* e^{sA(\lambda_k)^*} (T_{26})^* \mathcal{M} z_2(s) \right] \mathrm{d}s$$

$$= \int_0^\ell \left[A x_1(s), A^{-1}(C_6)^* (QD_k^{-1} P)^* e^{sA(\lambda_k)^*} (T_{26})^* (Az_1)(s) \right] ds$$

$$+ \int_0^\ell \left[M x_2(s), \bar{\lambda}_k M^{-1}(C_6)^* (QD_k^{-1} P)^* e^{sA(\lambda_k)^*} (T_{26})^* M z_2(s) \right] ds.$$

Hence

$$P_k^* Z = \begin{pmatrix} A^{-1}(C_6)^* (QD_k^{-1} P)^* e^{sA(\lambda_k)^*} (T_{26})^* (Az_1)(s) \\ \bar{\lambda}_k M^{-1}(C_6)^* (QD_k^{-1} P)^* e^{sA(\lambda_k)^*} (T_{26})^* M z_2(s) \end{pmatrix},$$

$$(QD_k^{-1} P)^* = P^* (D_k^{-1})^* Q^*, \dots, a\ 6 \times 6 \text{ matrix},$$

$$(D_k^{-1})^* = \frac{1}{2\pi i} \oint D^*(\lambda)^{-1} d\lambda \qquad \lambda = \bar{\lambda}_k + r\, e^{i\theta}.$$

Finally let

$$Q(t)Y = \begin{pmatrix} \dfrac{1}{\pi} \displaystyle\int_0^\infty e^{-rt} R_J(-r) Y\, dr \\ \dfrac{1}{\pi} \displaystyle\int_0^\infty (-r) e^{-rt} R_J(-r) Y\, dr \end{pmatrix} \tag{5.145}$$

so that in particular $Q = Q(0)$.

Furthermore, $[QY, Z_k] = 0$ for every k. Hence the range of Q—the space of evanescent states—is orthogonal to the modal space spanned by the sequence $\{Z_k\}$. And similarly

$$[Q^* Z, Y_k] = 0.$$

Then finally the solution to the aeroelastic equation can be expressed:

$$Y(t) = \sum e^{\lambda_k t} P_k Y(0) + Q(t) Y(0), \tag{5.146}$$

where

$$P_k Y = \frac{[Y, Z_k]}{[Y_k, Z_k]} Y_k.$$

Here the most intriguing part is the term that cannot be expressed in terms of modes! Now we can return to the properties of the nonrational component: the evanescent states:

$$Q(t)Y, t \geq 0.$$

Lemma 5.33. *For each U:*

$$S(t)Y = W(t)(Y - QY), \qquad t \geq 0$$

$$= \sum_k e^{\lambda_k t} P_k Y$$

$$= W(t)Y - Q(t)Y$$

defines a C_0 semigroup, actually a group, over the subspace spanned by the modes $\{Y_k\}$.

$$S(t)QY = QS(t)Y = 0$$

$$Q(t)Y \text{ is stable for all } U.$$

Proof. Is immediate from the representation (5.29).   □

Green's Function: Time Domain

From (5.35) or directly from (5.16)we have the Green's function representation for the time domain solution:

$$W(t)Y(0) = Y(t)$$

is the function:

$$Y(t, y) = \int_0^\ell W(t, y, s)Y(0, s)ds, \qquad (5.147)$$

where

$$\int_0^\infty e^{-\lambda t} W(t, y, s)dt = R(\lambda, y, s). \qquad (5.148)$$

Next we extend the state space theory for $M > 0$.

State Space Theory: $M \neq 0$

In view of the fact that the convolution/evolution is different from that for $M = 0$:

$$\dot{Y}(t) = AY(t) + \int_0^t \mathcal{L}(t - \sigma)Y(\sigma)d\sigma$$

(no derivative in the integral) we need to examine the changes needed in the state space representation (5.2)

$$\dot{Z}(t) = \mathcal{A}Z(t) \quad Y(t) = \mathcal{P}Z(t).$$

Thus we begin with the analogue of Theorem 5.22. The main question is the nature of the function $\mathcal{L}(.)$. We begin with:

$$\hat{A}_i(0,.) = \frac{2}{\sqrt{1-M^2}} T f_i,$$

$$\hat{A}_1(0,x) = \frac{2}{\sqrt{1-M^2}} \sqrt{\frac{1-x}{1+x}},$$

$$\hat{A}_2(0,x) = \frac{2}{\sqrt{1-M^2}} \sqrt{1-x^2}.$$

Hence

$$w_{11}(0) = \frac{2\pi}{\sqrt{1-M^2}},$$

$$\check{w}_{11}(0) = \frac{2\pi}{\sqrt{1-M^2}} - \frac{4}{M},$$

$$w_{22}(0) = 0,$$

$$\check{w}_{22}(0) = -\frac{4}{3M}.$$

Which is enough to show that $\mathcal{L}(.)$ here enjoys the same sort of properties as in the case for zero M. Hence Theorem 5.22 holds in this case as well and so does the definition of $\sum$ and that in turn of $\dot{X}$, of C and A_c, and Z denoting the elements therein, as there. But now with

$$Z(t) = \begin{pmatrix} x(t) \\ Y(t) \end{pmatrix},$$

we have:

$$\dot{x}(t) = A_c x(t) + LY(t),$$

$$\dot{Y}(t) = \mathcal{A}Y(t) + Cx(t)$$

and the representation is thus much simpler:

$$\dot{Z}(t) = \dot{\mathcal{A}} Z(t); \qquad \dot{\mathcal{A}} = \begin{pmatrix} A_c & L \\ C & \mathcal{A} \end{pmatrix},$$

$$\mathcal{P}Z(t) = Y(t). \tag{5.149}$$

Let us examine the changes needed with this definition. The domain of $\dot{\mathcal{A}}$ remains the same: i.e., Domain of A_c = Domain of $\mathcal{A}$.

Spectrum of $\dot{A}$

The spectrum of $\dot{A}$ consists of λ such that

$$\lambda x = A_c x + LY,$$
$$\lambda Y = Cx + \mathcal{A}Y.$$

Or

$$\lambda x - A_c x = LY.$$

Theorem 5.34. *The point spectrum of $\dot{A}$ consists of the aeroelastic modes.*

Proof. For Re $\lambda > 0, \lambda$ is in the resolvent set of A_c so that given any Y in $\mathcal{H}$, we have:

$$x(\lambda) = R(\lambda, A_c)LY$$
$$= \int_0^\infty e^{-\lambda t} S_c(t)LY\,dt.$$

Hence the function

$$x(\lambda, s) = \int_0^\infty e^{-\lambda t} L(s+t)Y\,dt, \qquad 0 \le s \le \infty$$

and hence

$$Cx = x(\lambda, 0) = \int_0^\infty e^{-\lambda t} L(t)Y\,dt = \hat{L}(\lambda)Y.$$

Hence we have the representation:

$$x(\lambda, s) = e^{\lambda s}\hat{L}(\lambda)Y - \int_0^s e^{\lambda(s-\sigma)} L(\sigma)Y\,d\sigma, \qquad (5.150)$$

where

$$x(\lambda, \infty) = \frac{L(\infty)Y}{\lambda}.$$

But this defines an analytic function of λ in the region of analyticity of $\hat{L}(\lambda)Y/\lambda$ which is the whole plane omitting the branch cut along $\lambda < 0$. In other words we have proved that given Y, we can find $x(\lambda)$ in Σ such that

$$x(\lambda) - A_c x(\lambda) = LY.$$

Next we are given

$$\lambda x - A_c x = LY$$

even if λ is not in the resolvent set of A_c, and thus not necessarily unique. Hence

$$\lambda Y - \mathcal{A}Y = \hat{\mathcal{L}}(\lambda)Y$$

if λ is in the spectrum of $\dot{\mathcal{A}}$. But this means that λ is an aeroelastic mode.

Next suppose λ is an aeroelastic mode. Then

$$\lambda Y - \mathcal{A}Y = \hat{\mathcal{L}}(\lambda)Y.$$

But given Y we have shown that we can find $x(\lambda)$ such that

$$x(\lambda) - A_c x(\lambda) = LY.$$

Hence λ is in the spectrum of $\dot{\mathcal{A}}$. □

Resolvent of $\dot{\mathcal{A}}$

To find the resolvent we need to solve:

$$\lambda Z - \ddot{A}z = \begin{pmatrix} x_g \\ Y_g \end{pmatrix},$$

$$\lambda x - A_c x - LY = x_g,$$

$$\lambda Y - \mathcal{A}Y - Cx = Y_g. \tag{5.151}$$

We show the following analogous to Theorem 5.24.

Theorem 5.35. *Let Ω denote the region in the complex plane:* $\Omega =$ {resolvent set of $\mathcal{A}$} $\cap$ {complement of the point spectrum of $\dot{\mathcal{A}}$} $\cap$ {Re $\lambda > 0$}. *Then Ω is contained in the resolvent set of $\dot{\mathcal{A}}$. Moreover, it has the representation*

$$R(\lambda, \dot{\mathcal{A}})Z = \begin{pmatrix} R(\lambda, A_c)(x + L\Phi) \\ \Phi \end{pmatrix},$$

$$\Phi = \mathcal{R}(\lambda)(CR(\lambda, A_c)x + Y),$$

which is simpler than the case for $M = 0$.

Proof. For λ in Ω we can solve (5.40) as:

$$\begin{pmatrix} x \\ Y \end{pmatrix} = \begin{pmatrix} R(\lambda, A_c)(x_g + LY) \\ R(\lambda, \mathcal{A})(Cx + Y_g) \end{pmatrix}.$$

But

$$Cx = CR(\lambda, A_c)x_g + \hat{\mathcal{L}}(\lambda)Y$$

$$= CR(\lambda, A_c)x_g + \hat{\mathcal{L}}(\lambda)Y,$$

$$\lambda Y - \mathcal{A}Y - CR(\lambda, A_c)x_g - \hat{\mathcal{L}}(\lambda)Y = Y_g,$$

$$\lambda Y - \mathcal{A}Y - \hat{\mathcal{L}}(\lambda)Y = Y_g + CR(\lambda, A_c)x_g.$$

But λ is not an aeroelastic mode and hence

$$Y = \mathcal{R}(\lambda)(Y_g + CR(\lambda, A_c)x_g),$$

where $\mathcal{R}(\lambda)$ is the generalized resolvent and in turn

$$LY = L\mathcal{R}(\lambda)(CR(\lambda, A_c)x_g + Y_g).$$

Hence

$$\begin{pmatrix} x \\ Y \end{pmatrix} = \begin{pmatrix} R(\lambda, A_c)(x_g + L\Phi) \\ \Phi \end{pmatrix},$$

$$\Phi = \mathcal{R}(\lambda)(CR(\lambda, A_c)x_g + Y_g)$$

as required.   □

Semigroup Generated by $\dot{\mathcal{A}}$

We show next that $\dot{\mathcal{A}}$ generates a C_0 semigroup. The main step here is again to evaluate the inverse Laplace transform of the generalized resolvent. We note that

$$||R(\lambda, \mathcal{A})\hat{\mathcal{L}}(\lambda)|| \to 0 \quad \text{as Re } \lambda \to \infty.$$

Hence we have that

$$||R(\lambda, \mathcal{A})\hat{\mathcal{L}}(\lambda)|| < 1$$

in a right-half-plane and furthermore, also the Neumann expansion therein

$$\left(I - R(\lambda, \mathcal{A})\hat{\mathcal{L}}(\lambda)\right)^{-1} = \sum_{k=0}^{\infty} \left(R(\lambda, \mathcal{A})\hat{\mathcal{L}}(\lambda)\right)^k,$$

where

$$R(\lambda, A)\hat{\mathcal{L}}(\lambda)Y = \int_0^\infty e^{-\lambda t} J_1(t)Y\,dt,$$

$$J_1(t) = \int_0^t S(t-\sigma)\mathcal{L}(\sigma)d\sigma,$$

whose Laplace transform is defined for $\mathrm{Re}\lambda >$ the growth bound of the semigroup S.

Define

$$J_n(t)Y = \int_0^t J_1(t-\sigma)J_{n-1}(\sigma)d\sigma, \quad n \geq 2,$$

whose Laplace transform is:

$$\left(R(\lambda, A)\hat{\mathcal{L}}(\lambda)\right)^n.$$

Let

$$\hat{r}(\lambda) = \left(I - R(\lambda, A)\hat{\mathcal{L}}(\lambda)\right)^{-1} - I,$$

which is defined for $\mathrm{Re}\,\lambda >$ the growth bound of the semigroup $S(\,.\,)$.

Then

$$\hat{r}(\lambda) = \left(I - R(\lambda, A)\hat{\mathcal{L}}(\lambda)\right)^{-1} R(\lambda, A)\hat{\mathcal{L}}(\lambda) = R(\lambda)\hat{\mathcal{L}}(\lambda).$$

Let

$$r(t)Y = \sum_{n=1}^\infty J_n(t)Y, \quad t \geq 0.$$

Then the Laplace transform

$$\int_0^\infty e^{-\lambda t} r(t)dt = \sum_{k=1}^\infty \left(R(\lambda, A)\hat{\mathcal{L}}(\lambda)\right)^k = \left(I - R(\lambda, A)\hat{\mathcal{L}}(\lambda)\right)^{-1} - I$$

$$= \hat{r}(\lambda).$$

Next let

$$W(t)Y = S(t)Y + \int_0^t S(t-\sigma)r(\sigma)Y\,d\sigma \quad t > 0. \tag{5.152}$$

Then the Laplace transform

$$\int_0^\infty e^{-\lambda t} W(t) Y \, dt = R(\lambda, \mathcal{A}) \, (I + \hat{r}(\lambda)) \, Y$$

$$= R(\lambda, \mathcal{A}) \left(I - R(\lambda, \mathcal{A}) \hat{\mathcal{L}}(\lambda) \right)^{-1} Y$$

$$= R(\lambda) Y.$$

Define

$$S_z(t) Z = \begin{pmatrix} S_c(t) x + \int_0^t S_c(t - \sigma) L \Phi(\sigma) d\sigma \\ \Phi(t) \end{pmatrix}, \qquad Z = \begin{pmatrix} x \\ Y \end{pmatrix},$$

$$\Phi(t) = W(t) Y + \int_0^t W(t - \sigma) C S_c(\sigma) \, x d\sigma.$$

The Laplace transform is the resolvent of $\dot{\mathcal{A}}$, proving that it is the semigroup generated by $\dot{\mathcal{A}}$. Thus we have the state space representation:

$$\dot{Z}(t) = \dot{\mathcal{A}} Z(t),$$

$$Y(t) = \mathcal{P} Z(t). \tag{5.153}$$

This establishes in particular the existence and uniqueness of solution to the aeroelastic equations. This result is essential for control theory treated in Chap. 8.

We can now go on to the main objective of the theory: flutter analysis next.

5.7 Flutter Analysis

Divergence Speed

We begin with the divergence speed (4.42) we have already seen in connection with the steady-state or "static" solution in Chap. 4. Here we examine it from the dynamic side as corresponding to zero frequency.

Thus here we consider U such that $d(M, 0, U) = 0$. Setting $\lambda = 0$ is equivalent to setting all the time derivative terms to be zero.

Theorem 5.36.

$$d(M, 0, U) = \cosh \ell \sqrt{w_4},$$

where

$$w_4 = -\frac{\pi\rho b^2(1+2a)}{GJ\sqrt{1-M^2}}U^2.$$

Proof. We calculate $w_{ij}(0)$. Using:

$$\hat{A}(0,.) = \frac{2}{\sqrt{1-M^2}}T\hat{w}_a(0,.),$$

$$\hat{A}(0,.) = \frac{2}{\sqrt{1-M^2}}\sqrt{\frac{1-x}{1+x}} \quad |x| < 1.$$

Hence

$$w_{11}(0) = \frac{2\pi}{\sqrt{1-M^2}},$$

$$w_{12}(0) = \frac{\pi}{\sqrt{1-M^2}},$$

$$w_{21}(0) = \frac{-\pi}{\sqrt{1-M^2}},$$

$$w_{22}(0) = 0.$$

From which we calculate the entries in the matrix A(0):

$$w_1 = 0,$$

$$w_3 = 0,$$

$$w_2 = bU^2\rho\frac{2\pi}{\sqrt{1-M^2}},$$

$$w_4 = -\frac{\pi\rho b^2(1+2a)}{GJ\sqrt{1-M^2}}U^2,$$

which in turn yields

$$d(M,0,U) = \cosh\ell\sqrt{-\frac{\pi\rho b^2(1+2a)}{GJ\sqrt{1-M^2}}U^2}. \qquad \square$$

Hence $d(M,0,U) = 0$ yields $\cosh\ell\sqrt{w_4} = 0$.
    Or

$$\cos\left(\ell\sqrt{\frac{\pi\rho b^2(1+2a)}{GJ\sqrt{1-M^2}}U^2}\right) = 0.$$

Hence

$$U = (2n + 1)\left(1 - M^2\right)^{1/4} \frac{1}{2b\ell} \frac{\sqrt{\pi GJ}}{\sqrt{\rho(1 + 2a)}},$$

n positive integer and hence the divergence speed:

$$U_d = \left(1 - M^2\right)^{1/4} \frac{1}{2b\ell} \frac{\sqrt{\pi GJ}}{\sqrt{\rho(1 + 2a)}}, \qquad (5.154)$$

which of course checks our previous result based on the time-invariant solution (5.42).

Remarks: A question of interest here (see below) is what happens as we increase speed at the divergence speed. We want to calculate the nearest aeroelastic mode as we increase the speed. We show that the derivative with respect to U is positive at this point, thus as we increase speed the system becomes unstable, with a positive aeroelastic mode.

Thus we want to calculate $\Delta\lambda$ as a function of $U_d + \Delta U$ such that

$$d(M, \Delta\lambda, U_d + \Delta U) = 0.$$

Because

$$d(M, 0, U) = \cosh \ell \sqrt{w_4},$$

we see that

$$\frac{\partial d(M, 0, U_d)}{\partial U} = -\ell \sqrt{\frac{\pi\rho b^2(1 + 2a)}{GJ\sqrt{1 - M^2}}}.$$

However we run into the difficulty that $\partial d(M, 0, U_d)/\partial\lambda$ is not defined at $\lambda = 0$– singularity of the function at $\lambda = 0$! We hence use the perturbation formula: for small λ

$$d(M, \lambda, U_d) = \det \times Pe^{A(\lambda)}Q$$

$$= \det\left(Pe^{A(0)}Q + P\int_0^\ell e^{A(0)(\ell-s)}(A(\lambda) - A(0))e^{A(0)s}Qds\right),$$

where we may use the approximation for small λ for the $w_{ij}(\lambda)$ given in [21], following the expansion in Sect. 5.4.

   Here, however, we follow a less computation-intense type of approximation. Thus note that $d(M, \lambda, U)$ is a function of the variables, $w_1, \ldots w_4$ and we may denote it $d(w_1, \ldots w_4)$ We note that

$$d(0, w_2, 0, w_4) = \cosh\left[\ell\sqrt{w_4}\right]$$

and for small k:

$$d(M, \lambda, U) = \frac{\partial d(0, w_2(0), 0, w_4(0))}{\partial w_1} w_1 + \frac{\partial d(0, w_2(0), 0, w_4(0))}{\partial w_3} w_3$$
$$+ \cosh\left[\ell \sqrt{w_4}\right].$$
(5.155)

This is in fact a relation we use many times. And as a further approximation we may simply use

$$d(M, \lambda, U) = \cosh\left[\ell \sqrt{w_4}\right]$$

$$= \cosh$$

$$\times \left[\ell \sqrt{\frac{1}{GJ}\left[\lambda^2 I_\theta + \rho b^2 U^2 (w_{21} + k w_{22} - a(1 - ak)w_{11}] - ak(w_{21} + w_{12}))\right]}\right]$$

$$k = \frac{\lambda b}{U}.$$

Hence we need to find λ such that for

$$U = \Delta U + U_d, \Delta U > 0,$$

$$\lambda^2 I_\theta + \rho b^2 U^2 (w_{21} + k w_{22} - a(1 - ak)w_{11} - ak(w_{21} + w_{12})) = -GJ\frac{4\pi^2}{\ell^2},$$

where we know that

$$\rho b^2 U_d^2 (w_{21}(0) - a w_{11}(0)) = -GJ\frac{4\pi^2}{\ell^2}.$$

Subtracting the bottom from the top we have

$$\lambda^2 I_\theta + \rho b^2 U^2 \left(w_{21} - w_{21}(0) + k w_{22} - a^2 k w_{11} - a(w_{11} - w_{11}(0))\right.$$
$$+ (w_{21}(0) - a w_{11}(0)) \rho b^2 (U^2 - U_d^2) - ak(w_{21} + w_{12}) = 0.$$

We can approximate the w_{ij} here by their limit values at $\lambda = 0$, and obtain the quadratic equation in λ:

$$\lambda^2 \sqrt{1 - M^2} I_\theta - \rho b^2 U a^2 2\pi b \lambda - \pi(1 + 2a)\rho b^2 (U^2 - U_d^2) = 0.$$

But for $U > U_d$ we see that the term independent of λ is negative, and the roots are

$$\frac{1}{\sqrt{1 - M^2} I_\theta} \left(\rho b^2 U a^2 \pi b \pm\right.$$

$$\sqrt{\left(\left(\rho b^2 U a^2 \pi b\right)^2 + \sqrt{1 - M^2} I_\theta \pi(1 + 2a)\rho b^2 \left(U^2 - U_d^2\right)\right)}\right),$$

and we note that there is a positive root which also depends on M and I_θ. This verifies our claim. The main point is that we wind up with a quadratic equation in λ, where the term independent of λ is negative; this is what continues to hold even with higher-order approximations.

The divergence speed for nonzero α and the transonic dip has been covered in Chap. 4. So far we have considered the CF case. Following similar lines it is easy to see that for the FF or CC case the speeds are determined by

$$\sinh\ell \sqrt{w_4} = 0$$

or the divergence speed is now given by

$$U_d = \left(1 - M^2\right)^{1/4} \frac{1}{4b\ell} \frac{\sqrt{\pi GJ}}{\sqrt{\rho(1 + 2a)}},$$

one-half the value for the CF case.

Root Locus and Stability Curve

Our first objective is to make precise the notion of flutter speed. For this we begin with a closer examination of the aeroelastic modes. The function $d(M, \lambda, U)$ fixed M, is analytic in both λ and U (even though we are interested only in positive values of U), and for fixed U, with a branch cut along the negative axis, including zero, in λ. For each U there is a countable number of roots of $d(M, \lambda, U) = 0$ which we denote by $\lambda(U)$. We consider only nonzero roots, and assume that there are no multiple roots. If there are, we need to work with each branch in a similar way. We have in fact seen that there are two near the divergence speed.

We can apply the implicit function theorem by which we see that $\lambda(M, U)$ is an analytic function of U, omitting isolated singularities where

$$\frac{\partial d(M, \lambda, U)}{\partial \lambda} = 0; \quad d(M, \lambda, U) = 0$$

and in particular we see that:

$$\frac{d}{d\lambda}\lambda(M, U) = -\frac{\partial d(M, \lambda, U)}{\partial U} \bigg/ \frac{\partial d(M, \lambda, U)}{\partial \lambda}\bigg|\lambda = \lambda(M, U).$$

Also

$$\frac{d^2}{d\lambda^2}\lambda(M,U) = \frac{\dfrac{\partial^2 d(.)}{\partial U^2} + 2\dfrac{\partial^2 d(.)}{\partial\lambda\partial U} - \left(\dfrac{d\lambda}{dU}\right)^2 \dfrac{\partial^2 d(.)}{\partial\lambda^2}}{\underbrace{\dfrac{\partial d(M,\lambda,U)}{\partial\lambda}}}\Big|\lambda = \lambda(M,U),$$

where of course the denominator cannot be zero. At $U = 0$, the aeroelastic modes
are the structure modes, and we assume that the damping can be neglected, actually
that the coupling S can be neglected. As before, let us order these modes in terms of
increasing magnitudes, within each class—bending or torsion—so we can talk for
instance about the "first few modes": $\lambda_k(0)$ where $\lambda_k(0) = i\omega_k(0)$. For nonzero U
we have difficulty in classifying them. So we use *root locus*. Thus we consider the
function $\lambda_k(U)$ starting with the value at zero, the structure mode $i\omega_k(0)$. Hence we
can talk about a "bending" mode, or "torsion" mode. Of course the mode and the
mode shape will change as U increases. We call this function a root locus starting
at $i\omega_k(0)$. We use the notation

$$\lambda_k(U) = \sigma_k(U) + i\omega_k(U)$$

and refer to $\sigma_k(U)$ as the "damping" term–stable mode if the real part is strictly
negative.

We note that if complex, the modes occur in complex conjugate pairs; so we may
only consider the ones with the positive imaginary part for most purposes.

By a stability curve we mean the function $\sigma_k(U), U > 0$ so that we have
instability if it is nonnegative, an unstable mode. Our first result is that the damping
decreases at $U = 0$, whatever the mode.

Theorem 5.37.

$$\frac{\partial\sigma_k(0)}{\partial U} = -\frac{2b\rho}{mM} \text{ at every bending mode} \tag{5.156}$$

$$= -\frac{2b^3\left(a^2 + \frac{1}{3}\right)\rho}{MI_\theta}\text{at every torsion mode,} \tag{5.157}$$

$$\frac{\partial\omega_k(0)}{\partial U} = 0. \tag{5.158}$$

Remarks: Note that the slope does not depend on the mode number. This result does
not appear to be in the aeroelastic lore.

Proof. We again use the local perturbation formula

$$Pe^{\ell(A+\Delta A)}Q = Pe^{\ell A}Q + P\int_0^\ell e^{(\ell-s)A}\Delta Ae^{sA}ds Q,$$

where A is the matrix $A(\lambda)$ corresponding to $U = 0$, the structure-only case we have already considered in Chap. 2

$$A(\lambda) = \begin{pmatrix} 0 & 1 & 0 & 0 & 0 & 0 \\ 0 & 0 & 1 & 0 & 0 & 0 \\ 0 & 0 & 0 & 1 & 0 & 0 \\ -\frac{m\lambda^2}{EI} & 0 & 0 & 0 & 0 & 0 \\ 0 & 0 & 0 & 0 & 0 & 1 \\ 0 & 0 & 0 & 0 & \frac{\lambda^2 I_\theta}{GJ} & 0 \end{pmatrix}$$

and ΔA is the increment to get back to $A(\lambda)$. Now, recall from Chap. 2:

$$P e^{\ell A} Q =$$

$$\begin{pmatrix} \frac{1}{2}\left(\cos[w1^{1/4}\ell] + \cosh[w1^{1/4}\ell]\right) & \frac{\sin[w1^{1/4}\ell] + \sinh[w1^{1/4}\ell]}{2w1^{1/4}} & 0 \\ \frac{1}{2}w1^{1/4}\left(-\sin[w1^{1/4}\ell] + \sinh[w1^{1/4}\ell]\right) & \frac{1}{2}\left(\cos[w1^{1/4}\ell] + \cosh[wl^{1/4}\ell]\right) & 0 \\ 0 & 0 & \cosh[\sqrt{w4\ell}] \end{pmatrix}$$

and

$$d_0 = \text{determinant } P e^{\ell A} Q$$

$$= \frac{1}{2}\left(1 + \cos[w1^{1/4}\ell]\cosh[w1^{1/4}\ell]\right)\cosh\left[\sqrt{w4\ell}\right]$$

and the perturbation term (see [21])

$$P \int_0^\ell e^{(\ell-s)A} \Delta A\, e^{sA}\, ds\, Q =$$

$$\begin{pmatrix} 0 \\ 0 \\ \frac{w3\left(2e^{-\sqrt{w4\ell}}w1^{1/4}\sqrt{w4} - 2e^{\sqrt{w4\ell}}w1^{1/4}\sqrt{w4} - e^{-w1^{1/4}\ell}\left(\sqrt{w1}+w4\right) + e^{w1^{1/4}\ell}\left(\sqrt{w1}+w4\right) - 2\left(\sqrt{w1}-w4\right)\sin[w1^{1/4}\ell]\right)}{4w1^{1/4}\left(w1-w4^2\right)} \\ 0 \\ 0 \\ \frac{w3\left(-2e^{-\sqrt{w4\ell}}\sqrt{w1} - 2e^{\sqrt{w4\ell}}\sqrt{w1} + e^{-w1^{1/4}\ell}\left(\sqrt{w1}+w4\right) + e^{w1^{1/4}\ell}\left(\sqrt{w1}+w4\right) + 2\left(\sqrt{w1}-w4\right)\cos[w1^{1/4}\ell]\right)}{\left(4\sqrt{w1}(w1-w4^2)\right)} \\ \frac{w2\left(-2e^{-\sqrt{w4\ell}}w1^{1/4}\sqrt{w4} + 2e^{\sqrt{w4\ell}}w1^{1/4}\sqrt{w4} + e^{-w1^{1/4}\ell}\left(\sqrt{w1}+w4\right) - e^{w1^{1/4}\ell}\left(\sqrt{w1}+w4\right) + 2\left(\sqrt{w1}-w4\right)\sin[w1^{1/4}\ell]\right)}{\left(4w1^{1/4}(w1-w4^2)\right)} \\ \frac{w2\left(\left(-\sqrt{w1}+w4\right)\cos[w1^{1/4}\ell] + \left(\sqrt{w1}+w4\right)\cosh[w1^{1/4}\ell] - 2w4\,\cosh[\sqrt{w4\ell}]\right)}{2(w1-w4^2)} \\ 0 \end{pmatrix}$$

and we see that we can write

$$d(M, \lambda, U) = d_0 + w_2 w_3 \det\Delta(\gamma, \mu), \tag{5.159}$$

where $\Delta(. , .)$ is a 3 by 3 matrix depending only on γ, μ.

From (5.6) we can calculate that the derivatives we need:

$$\frac{\partial d}{\partial U} = \frac{\partial d_0}{\partial \gamma} \frac{\partial \gamma}{\partial U} + \frac{\partial d_0}{\partial \mu} \frac{\partial \mu}{\partial U} + \frac{\partial}{\partial U}(w_2 w_3 \mathrm{Det}.\Delta(\gamma, \mu)),$$

$$\frac{\partial d}{\partial \lambda} = \frac{\partial d_0}{\partial \gamma} \frac{\partial \gamma}{\partial \lambda} + \frac{\partial d_0}{\partial \mu} \frac{\partial \mu}{\partial \lambda} + \frac{\partial}{\partial \lambda}(w_2 w_3 \mathrm{Det}.\Delta(\gamma, \mu)).$$

And hence at $U = 0$:

$$\frac{d\lambda(U)}{dU} = -\frac{\dfrac{\partial w_4}{\partial U}}{\dfrac{\partial w_4}{\partial \lambda}} \quad \text{at a pitching (torsion) mode}$$

$$= -\frac{\dfrac{\partial wr_1}{\partial U}}{\dfrac{\partial w_1}{\partial \lambda}} \quad \text{at a bending (plunge) mode.}$$

Using:

$$\frac{\partial w_{ij}}{\partial U} = \left(\frac{\partial w_{ij}}{\partial k}\right) \frac{-\lambda b}{U^2},$$

$$\frac{\partial w_{ij}}{\partial \lambda} = \left(\frac{\partial w_{ij}}{\partial k}\right) \frac{b}{U},$$

we have:

$$\frac{\partial w_1}{\partial U} = -\frac{1}{EI}\left(\rho b \lambda w_{11} - \left(\frac{\partial w_{11}}{\partial k}\right) k b \rho \lambda\right),$$

$$\frac{\partial w_1}{\partial \lambda} = -\frac{1}{EI}\left(\left(\frac{\partial w_{11}}{\partial k}\right) b^2 \rho \lambda + \rho b U w_{11} + 2m\lambda\right),$$

$$\frac{\dfrac{\partial w_1}{\partial U}}{\dfrac{\partial w_1}{\partial \lambda}} = \frac{\rho b \left(w_{11} - k\dfrac{\partial w_1}{\partial k}\right)}{2m + \dfrac{\rho w_{11}}{k} + \rho b^2 \dfrac{\partial w_{11}}{\partial k}},$$

which we note depends only on k.

Similarly we can calculate that

$$\frac{\frac{\partial w_4}{\partial U}}{\frac{\partial w_1}{\partial \lambda}} = \frac{2b^3 \rho \frac{c(k)}{k} - \rho b^3 \frac{\partial c(k)}{\partial k}}{2I_\theta + \frac{b^2 \rho}{k} \frac{\partial c(k)}{\partial k}},$$

where

$$c(k) = w_{21} + k w_{22} - a(1 - ak)w_{11} - ak(w_{21} + w_{12})$$

and again the ratio depends only on k. Note that $U = 0$ means that $k = \infty$.

For $M > 0$ we see from (5.1) that

$$\hat{A}_i(\lambda, .) \to \frac{2}{M} f i \quad \text{as } k \to \infty$$

and hence we see that

$$w_{11}(\infty) = \frac{4}{M},$$

$$w_{ij}(\infty) = 0, \quad i \neq j,$$

$$w_{22}(\infty) = \frac{4}{3M},$$

$$k \frac{\partial w_{ij}(k)}{\partial k} \to 0 \quad \text{as } k \to \infty.$$

Hence

$$w_2 \to 0 \quad \text{as } U \to 0(S = 0),$$

$$w_3 \to 0 \quad \text{as } U \to 0(S = 0!).$$

Hence

$$\frac{\frac{\partial w_1}{\partial U}}{\frac{\partial w_1}{\partial \lambda}} \bigg|_{U = 0} = \frac{\rho b}{2m} \frac{4}{M},$$

which yields (5.3).

Next:

$$\frac{c(k)}{k} \to \frac{4}{3M} + a^2 \frac{4}{M} \quad \text{as } U \to 0,$$

$$\frac{\partial c(k)}{\partial k} \to \frac{4}{3M} + a^2 \frac{4}{M} \quad \text{as } U \to 0.$$

Hence (5.4) and (5.5) follow. $\square$

Remarks: The appearance of $(1/M)$ is interesting, showing in particular that these are not continuous with respect to M at zero. In fact we have to deal with the case $M = 0$ separately.

The stability curve for $M = 0$
   Noting that

$$T(k) \to 1 \quad \text{as } k \to \infty,$$

we have:

$$\left. \frac{\partial \lambda(U)}{\partial U} \right|_{U = 0} = - \frac{\pi \rho b}{2(m + \pi b^2 \rho)} \quad \text{at every bending mode}$$

$$= \frac{-\pi \rho b^3 \left(a - \frac{1}{2} \right)^2}{2 \left(I_\theta + \pi \rho (a^2 + \frac{1}{8}) \right)} \quad \text{at every pitching mode.} \quad (5.160)$$

Definition of Flutter Speed

We can now make a precise definition of flutter speed.
   We begin with the *stability curve*:

$$\sigma_k(U) \text{ as a function of } U \text{ with } \lambda_k(0) = i \omega_k, \quad \omega_k > 0.$$

Theorem 5.37 shows that the slope of this curve is strictly negative at $U = 0$. Hence the speed U at which $\sigma_k(U)$ becomes zero again for the first time we call the flutter speed for this mode denoted $U_F(k)$. Thus $U_F(k)$ is defined by:

$$\sigma_k(U_F(k)) = 0; \qquad \sigma_k(U) < 0 \quad \text{for} \quad U < U_F(k);$$

$$\frac{d}{dU} \sigma_k(U_F(k)) > 0.$$

The second condition means that the system becomes more unstable as we increase the speed. Of course at this point, Im $\lambda_k(U_F)$ is not necessarily $= \omega_k$ and may have changed considerably.

Flutter Speed

By "flutter speed" we mean

$$\overset{\inf}{k} U_F(k) = U_F, \quad (5.161)$$

where the infimum is taken over all structure modes.

In the event that no structural mode flutters, we would have to define it to be infinity.

The situation changes if we modify the definition (as we do) to be instead the infimum over all aeroelastic modes (rather than merely structure modes) because this would then include zero. First we prove that the flutter speed in either definition is positive.

Corollary 5.38. *The flutter speed is positive. The infimum in (5.8) is attained.*

Proof. Suppose the infimum in (5.8) is zero. Then we can find a sequence of mode frequencies ω_n and corresponding speeds U_{Fn} such that U_{Fn} converges to zero. Suppose the infimum is attained for some n. Then the corresponding flutter speed must be positive because the slope at zero speed has to be negative by Theorem 5.37. Suppose then it is not attained. Consider the stability curve for each n:

$$\sigma_n(U) = \text{Re } \lambda_n(U) \quad \text{for } 0 < U < U_{Fn}.$$

We have for all n: by Theorem 5.37

$$\frac{\mathrm{d}}{\mathrm{d}U}\sigma_n(0) < -\beta < 0$$

and by definition

$$\frac{\mathrm{d}}{\mathrm{d}U}\sigma_n(U_{Fn}) \geq 0.$$

The function $(\mathrm{d}/\mathrm{d}U)\sigma_n(U)$ is continuous on finite intervals. Hence for some U— denoted U_n—the slope

$$\frac{\mathrm{d}}{\mathrm{d}U}\sigma_n(U_n) = 0, \quad \text{where } U_n < U_{Fn}$$

and U_n goes to zero by hypothesis.

Let us look at the corresponding mode sequence ω_n. Suppose the sequence is bounded. Then we can find a convergent subsequence with finite limit ω_∞ which would then attain the infimum. Hence the sequence must be unbounded.

Then

$$K_n = \frac{\lambda_n b}{U_n}; \quad d(M, \lambda_n, U_n) = 0$$

and in as much as $\omega_n \to \infty$, we must have that

$$|\kappa_n| \to \infty.$$

The Mikhlin multiplier converges to a finite limit as $|\kappa_n| \to$ infinity, and

$$\frac{\partial \lambda}{\partial U} = -\frac{\partial d}{\partial U} \Big/ \frac{\partial d}{\partial \lambda}$$

and

$$\frac{\partial^2 \lambda}{\partial U^2} = \frac{\dfrac{\partial d}{\partial U}\dfrac{\partial^2 d(.)}{\partial \lambda \partial U} - \dfrac{\partial^2 d(.)}{\partial U^2}\dfrac{\partial d}{\partial \lambda}}{\left(\dfrac{\partial d}{\partial \lambda}\right)^2}\bigg|_{\lambda = \lambda(M,\,U)}$$

converge to finite values as $|\kappa_n| \to \infty$.

Now for each n we have, with primes denoting derivative with respect to U:

$$0 = \sigma_n'(U_n) = \sigma_n'(0) + \int_0^{U_n} \sigma_n''(U)\, dU,$$

where the integral $\to 0$ and the first term goes to a nonzero value, leading to a contradiction.

Hence the infimum is attained and is positive. $\qquad\qquad\qquad\qquad\Box$

A computer program for generating stability curves in incompressible flow is given in the Appendix of this chapter with numerical calculations for the Goland model. We note that the first torsion mode flutters.

Theorem 5.39. *Zero is an aeroelastic mode for* $0 \le M < 1$.

Proof. For $M = 0$:

We note that from Sect. 5.4

$$\lambda \hat{\mathcal{L}}(\lambda) = 0 \quad \text{at } \lambda = 0.$$

Hence to prove that zero is an aeroelastic mode we need to find Y such that

$$\mathcal{A}Y = 0, \quad Y \ne 0.$$

Or, we need to find a nonzero solution of (see Sect. 5.4)

$$Y = \begin{pmatrix} x_1 \\ x_2 \end{pmatrix} x_2 = 0,$$

$$\mathcal{A}x_1 + \mathcal{K}x_1 = 0.$$

Let $x_1 = \begin{pmatrix} h \\ \theta \end{pmatrix}$,

$$-GJ\theta''(Y) - \pi\rho b^2 U^2 \theta(Y) = 0,$$

$$EI h''''(y) = 0 \text{ plus end conditions.} \tag{5.162}$$

But this is precisely the linear steady-state aeroelastic equation treated in Chap. 3. For a nonzero solution we have an eigenvalue problem, which has a solution for only a sequence of far field speeds: under CF end conditions

$$U_\infty = U_n = (2n+1)\frac{1}{2b\ell}\frac{\sqrt{\pi GJ}}{\sqrt{\rho(1+2a)}},$$

with $\theta(y)$ determined by (5.9.) and $h(y) = 0$ and also, of course $d(0,0,U_n) = 0$.
For $0 < M < 1$ we have

$$-\mathcal{A}Y - \mathcal{L}(0)Y = 0$$

and from the evaluation of the $w_{ij}(0)$ we see that

$$Y = \begin{pmatrix} x_1 \\ 0 \end{pmatrix}; \qquad x_1 = \begin{pmatrix} h \\ \theta \end{pmatrix},$$

$$EIh''''(y) + \frac{4}{M}\rho b U^2 \theta(y) = 0,$$

$$-GJ\theta''(y) = \pi\rho b^2 U^2 \frac{1}{\sqrt{1-M^2}}\,\theta(y), \qquad (5.163)$$

plus end conditions. Thus for nonzero solution we have an eigenvalue problem with solution (under CF end conditions):

$$U_\infty = U_n = (2n+1)(1-M^2)^{1/4}\frac{1}{2b\ell}\frac{\sqrt{\pi GJ}}{\sqrt{\rho(1+2a)}}$$

and $\theta(y), h(y)$ are nonzero, and of course $d(M, 0, U_n) = 0$.   □

Remarks: Note that it is required that $M < 1$. For $M = 1$, the formula for U_n yields zero. Indeed we have seen that for $M = 1$ zero is not an aeroelastic mode.

So what is new is that we have evaluated the mode shape corresponding to the aeroelastic mode equal to zero. And we see that U_n is the sequence of flutter speeds corresponding to the flutter mode zero, which however is not a structure mode.

Moreover if we consider this as corresponding to $k = 0$, the stability curve starting at $\sigma_0(U_n)$, we see that the slope is nonnegative, by the Remark under Theorem 5.37.

In particular this shows that there are zeros other than those given by the root loci of the structure modes.

Which definition do we use? Note that we do not know whether $U_F(k)$ is $< U_d$.

This depends on whether "a" is positive or negative—whether the cg is "above or below" the elastic axis; see [6]—but of course we have offered no mathematical proof. We use $k = 0$ to indicate that the mode frequency is zero.

Note that for the mode zero we were essentially calculating the mode shape (x_1 in our notation above).

Note that by our definition, if we include $k = 0$ in (5.9) we have

$$0 < U_{\mathrm{F}} \leq U_d < \infty. \tag{5.164}$$

The appendix presents a computer program for calculating the flutter speed in incompressible flow for the Goland model. Here the first torsion mode flutters.

At the present time we have no idea, no formula for determining which mode will flutter and which won't, without numerical calculation. Fortunately we are interested in practice only in the "first few" modes, which makes it doable by computation.

Now we return to proving the convergence of the many series representations for the solution of the linear equation. The basic idea here is that as the mode number increases in the limit they all are like the structure modes $\{\lambda_k(0)\}$. Also we know that "physically" the modes cannot grow indefinitely (there are no microwaves) but the model shows that they are of diminishing importance as the mode number increases.

Dependence on Far Field Speed

The roots $\{\lambda_k\}$ depend on U_∞ which we shorten to simply U and use the notation $\lambda_k(U)$ when we need to emphasize the dependence on U.
Thus

$$\mathcal{A}(0)Y_k = i\omega_k Y_k; \quad \mathcal{A}(0)^* Z_k = -i\omega_k Z_k; \quad Z_k = -Y_k; \quad [Y_k, Z_k] = -2\omega_{k^2}.$$

Normalization

$$[\mathcal{M}x_{1k}, x_{1k}] = 1; Y_k = \begin{pmatrix} x_1 k \\ i\omega_k x_1 k \end{pmatrix}.$$

Similarly we now use $\mathcal{R}(\lambda, U)$ to emphasize the dependence on U.
We can express

$$\mathcal{A} + \hat{\mathcal{L}}(\lambda)$$

$$\mathcal{A}(U) = \mathcal{A}(0) + T(U) + \hat{\mathcal{L}}(\lambda, U),$$

where $T(U)$ depends on U but not on λ:

$$T(U) = \begin{pmatrix} 0 & 0 \\ -\mathcal{M}^{-1} \mathcal{K}(U) & -\mathcal{M}^{-l} \mathcal{D}(U) \end{pmatrix}$$

and defines a compact linear bounded operator on $\mathcal{H}$ into $\mathcal{H}$ for each U.

$$\hat{\mathcal{L}}(\lambda, U) = \begin{pmatrix} 0 & 0 \\ -\mathcal{M}^{-1} \ \hat{\mathcal{F}}(\lambda, U) & 0 \end{pmatrix},$$

which is also a bounded linear operator as $T(U)$ but goes to zero in operator norm as $|\lambda|$ goes to infinity in the spectral strip. Our main result is the asymptotic equivalence of the aeroelastic modes with the modes of the structure.

Note that it holds in particular if the convolution term $\hat{\mathcal{L}}(\lambda)$ is absent.

Theorem 5.40. *For each U*

$$\frac{\lambda_n(U)}{\lambda_n(0)} \to 1 \quad \text{as } n \to \infty.$$

Proof. The Mikhlin multiplier:

$$\frac{1}{2} \frac{\sqrt{M^2 \kappa^2 + 2M\kappa i\omega + (1 - M^2)\omega^2}}{\kappa + i\omega}$$

behaves badly as $\kappa \to 0$, however, it goes to a finite limit as $|\kappa| \to \infty$. This is what we exploit.

As $|\kappa| \to \infty$ in any strip that houses the roots, the multiplier goes to $M/2$ and

$$\Psi(\kappa, \ M) \to \frac{2}{M},$$

$$w_{11} \to \frac{2}{M},$$

$$w_{ij} \to 0 \quad i \neq j,$$

$$w_{22} \to$$

and similarly for the derivatives with respect to κ.

We use the same technique as in proving Lemma 5.41.

The (nonzero) roots $\{\lambda_n(U)\}$ are defined along the root locus and we have:

$$\lambda_n(U) = \lambda_n(0) + \int_0^U \frac{\partial \lambda_n(u)}{\partial u} du. \tag{5.165}$$

As n increases so does $|\kappa_n(U)|$, and

$$\left| \int_0^U \frac{\partial \lambda_n(u)}{\partial u} du \right| \text{ goes to a finite limit as } n \to \infty.$$

This follows from the calculations above. Dividing by $\lambda_n(0)$ in (5.8), we obtain

$$\frac{\lambda_n(U)}{\lambda_n(0)} = 1 + \frac{\int_0^U \frac{\partial \lambda_n(u)}{\partial u}\,du}{\lambda_n(0)}, \tag{5.166}$$

where the second term goes to zero with n. Hence the ratio

$$\frac{\lambda_n(U)}{\lambda_n(0)} \to 1 \quad \text{as } n \to \infty.$$

$\square$

In particular we see that

$$\text{Re } \lambda_n(U) \to 0 \quad \text{as } n \to \infty \tag{5.167}$$

and that

$$\lambda_n(U) = O(n) \tag{5.168}$$

because this is true of $\lambda_n(0)$.

What we have proved is that the sequences $\{\lambda_k(U)\}$ as U varies are "asymptotically equivalent" to $\{\lambda_k(0)\}$; see [2] for similar property in control problems.

Dependence on Mach Number

The next big question is the dependence of the flutter speed on M, $U_F(M)$.

Actually we are interested in the flutter speed at a given altitude. The speed of sound a_∞ depends on the altitude. Thus the system is stable if

$$U < U_F\left(\frac{U}{a_\infty}\right).$$

Dependence of Flutter Speed on Mach Number

The Federal Aviation Administration (FAA) in the United States mandates a 15% margin. The question of whether the Flutter speed decreases or increases with M has been of interest from the early days of aeroelasticity. The first attempt at this was by the pioneer Garrick [38] where he determines an empirical formula drawing on the Possio integral equation for nonzero M for M up to 0.6 or so. We may express this as the Garrick formula:

$$U_F(M) = (1 - M^2)^{1/4} U_F(0).$$

Note that according to this, the flutter speed decreases as M increases, and goes to zero as M goes to 1. This is of course for zero angle of attack. It is remarkable that there is a marked change for nonzero angle of attack as seen for $\kappa = 0$ in Chap. 4.

Because we are interested in the dependence on M, we now write $\lambda(M,\ U)$ in place of $\lambda(U)$; and similarly $\lambda_k(M, U), U_F(M, k)$ for the kth mode.

It has been observed [19] that a mode which flutters at $M = 0$ does not flutter at $M = 1$, and vice versa. This would indicate that for some value of M there are multiple roots for the same U. This is true near the flutter speed for $k = 0$, as we have already noted. We return to this question in Chap. 6.

Numerical calculations indicate that typically for current commercial aircraft (heavy wings) the value of k for flutter is small, around 0.1 or so. In as much as we know that

$$d(M, 0,\ U) = \cosh[\ell \sqrt{w_4}\,],$$

we may use the perturbation formula to obtain:

$$
\begin{aligned}
d(M, \lambda, U) = \frac{1}{5760 w4^9}\Big(&-2\big(-2880(4w_2^3 w_3^3 \\
&- 4w_1 w_2^2 w_3^2 w_4 + w_4^9 + 2w_2 w_3 w_4^4(w_1 - w_4^2)) \\
&+ 1440 w_2 w_3 w_4(-4w_2^2 w_3^2 + w_4^6 - 2w_2 w_3 w_4(-2w_1 + w_4^2))\ell^2 \\
&+ 120 w_4^2(-w_2 w_3 + w_1 w_4)(12w_2^2 w_3^2 + 2w_4^6 + w_2 w_3 w_4(2w_1 - 5w_4^2))\ell^4 \\
&- 4w_2 w_3 w_4^3(w_2 w_3 - w_1 w_4)(24w_2 w_3 - 4w_1 w_4 - 5w_4^3)\ell^6 \\
&+ w_4^4(w_2 w_3 - w_1 w_4)^2(w_2 w_3 - w_4^3)\ell^8\big)\cosh[\sqrt{w_4}\ell] \\
&+ w_2 w_3(-1440(12w_2^2 w_3^2 - 12w_1 w_2 w_3 w_4 + 5w_2 w_3 w_4^3 + 4w_1 w_4^4 - 8w_4^6) \\
&+ 2880 w_4(-3w_2^2 w_3^2 - 2w_1 w_4^4 + 3w_2 w_3 w_4(w_1 + w_4^2))\ell^2 \\
&+ 60 w_4^2(-w_2 w_3 + w_1 w_4)(-5w_2 w_3 + 4(3w_1 w_4 + w_4^3))\ell^4 \\
&+ 8w_4^3(-w_2 w_3 + w_1 w_4)(-21w_2 w_3 + 26w_1 w_4)\ell^6 + 6w_4^4(w_2 w_3 - w_1 w_4)^2\ell^8 \\
&- 60(4w_1 w_4^4(24 + w_1\ell^4) + w_2^2 w_3^2(96 + w_4\ell^2(48 + 5w_4\ell^2)) \\
&- 3w_2 w_3 w_4(40w_4^2 + w_1(32 + w_4\ell^2(16 + 3w_4\ell^2))))\cosh[2\sqrt{w_4}\ell] \\
&+ \sqrt{w_4}\ell((2880(-4w_2^2 w_3^2 + 2w_1 w_4^4 + 3w_4^6 + 4w_2 w_3 w_4(w_1 - 2w_4^2)) \\
&- 480 w_4(17w_2^2 w_3^2 + 2w_1 w_4^2(w_1 + 3w_4^2) - w_2 w_3 w_4(19w_1 + 3w_4^2))\ell^2 \\
&- 48 w_4^2(-w_2 w_3 + w_1 w_4)(-21w_2 w_3 + w_1 w_4 + 10w_4^3)\ell^4 \\
&+ 4w_4^3(-w_2 w_3 + w_1 w_4)(w_2 w_3 + 4wl w_4)\ell^6 + w_4^4(w_2 w_3 - w_1 w_4)^2\ell^8) \\
&\times \sinh[\sqrt{w_4}\ell] + 24(w_2 w_3 - w_1 w_4)(-w_4^2(120w_4 + 20w_1\ell^2 + w_1 w_4\ell^4) \\
&+ w_2 w_3(240 + w_4\ell^2(50 + w_4\ell^2)))\sinh[2\sqrt{w_4}\ell]))\Big).
\end{aligned}
$$

$$\hspace{12cm}(5.169)$$

In addition we may use the approximation for w_{ij} following [21] using the Neumann expansion in Sect. 5.4. Thus we have:

$$
w_{11} = \frac{2Pi}{\sqrt{1-M^2}}\left[2 - \frac{k}{\sqrt{1-M^2}}\mathrm{Bessel}K(0,k) \right.
$$

$$
+ \mathrm{Bessel}K(1,k))(\mathrm{Bessel}I(0,k)
$$

$$
- \mathrm{Bessel}I(1,k)) - k\int_0^{\alpha_1} a(s)(\mathrm{Bessel}K(0,ks) + \mathrm{Bessel}K(1,ks))
$$

$$
\times (\mathrm{Bessel}I(0,ks) - \mathrm{Bessel}I(1,ks))ds
$$

$$
+ \int_0^{\alpha_2} a(-s)(\mathrm{Bessel}K(0,ks)
$$

$$
\left. - \mathrm{Bessel}K(1,ks))(\mathrm{Bessel}I(0,ks) + \mathrm{Bessel}I(1,ks))ds \right],
$$

$$
w_{12} = \frac{Pi}{\sqrt{1-M^2}}\left[1 + \frac{k}{\sqrt{1-M^2}} \right.
$$

$$
\times \left(\frac{1}{k} - \frac{2\mathrm{Bessel}I(\alpha_1,k)}{k}(\mathrm{Bessel}K(0,\ k) + \mathrm{Bessel}K(1,k)) \right)
$$

$$
+ k\int_0^1 a(s)\left(\frac{1}{ks} - \frac{2\mathrm{Bessel}I(1,ks)}{ks}(\mathrm{Bessel}K(0,ks) \right.
$$

$$
\left. + \mathrm{Bessel}K(1,ks)) \right)ds + k\int_0^{\alpha_2} a(-s)
$$

$$
\left. \times \left(-\frac{1}{ks} + \frac{2\mathrm{Bessel}I(1,ks)}{ks}(\mathrm{Bessel}K(1,ks) - \mathrm{Bessel}K(0,ks)) \right)ds \right],
$$

$$
w_{21} = \frac{2Pi}{\sqrt{1-M^2}}\left[\frac{-1}{2} + \frac{k}{\sqrt{1-M^2}} \right.
$$

$$
\left(\frac{1}{k}\left(\frac{1}{k} - \frac{1}{2} \right) - \frac{\mathrm{Bessel}K(1,k)}{k}\mathrm{Bessel}I(0,k) - \mathrm{Bessel}I(1,k)) \right)
$$

$$
+ k\int_0^{\alpha_1} a(s)\left(\frac{1}{ks}\left(\frac{1}{ks} - \frac{1}{2} \right) \right.
$$

$$
\left. - \frac{\mathrm{Bessel}K(1,ks)}{ks}\mathrm{Bessel}I(0,ks) - \mathrm{Bessel}I(1,ks)) \right)ds
$$

$$
+ k\int_0^{\alpha_2} a(-s)\left(\frac{1}{ks}\left(\frac{1}{ks} + \frac{1}{2} \right) - \frac{\mathrm{Bessel}K(1,ks)}{ks} \right.
$$

$$
\left. \times (\mathrm{Bessel}I(0,ks) + \mathrm{Bessel}I(1,ks)) \right)ds,
$$

$$w_{22} = \frac{1}{\sqrt{1-M^2}}\left[\frac{1}{\sqrt{1-M^2}}\left(\frac{Pi}{k} - \frac{1}{k}2Pi\,\text{Bessel}I(1,k)\text{Bessel}K(1,k)\right)\right.$$

$$+ \int_0^{\alpha_1} a(M,s)\frac{1}{ks^2}(Pi - 2Pi\,\text{Bessel}I(1,ks)\text{Bessel}K(1,ks))ds$$

$$\left. + \int_0^{\alpha_2} a(M,-s)\frac{1}{ks^2}(Pi - 2Pi\,\text{Bessel}I(1,ks)\text{Bessel}K(1,ks))ds.\right]$$

Here we have a still further approximation:

$$d(M,\lambda,U) = w_1\left(-\frac{1}{12}\ell^4\cosh[\sqrt{w_4}\ell]\right) + w_3\left(\frac{1}{5760w_4^9}w_2(11520w_4^6\right.$$

$$- 2(5760\,w_4^6 + 1440\,w_4^7\ell^2 - 240w_4^8\ell^4)\cosh[\sqrt{w_4}\ell]$$

$$\left. + 8640\,w_4^{13/2}\ell\,\sinh[\sqrt{w_4}\ell])\right) + \cosh[\ell\sqrt{w_4}].$$

We may use this to calculate

$$\frac{\partial U_F(0,\kappa)}{\partial M} = -\frac{\dfrac{\partial d}{\partial M}}{\dfrac{\partial d}{\partial U}},$$

at $\lambda = 0$, so that $U_F(0,k) = U_d$, and

$$w_4 = -\frac{\pi\rho b^2(1+2a)}{GJ\sqrt{1-M^2}}U^2,$$

and hence

$$\frac{\partial U_F(0,\kappa)}{\partial M} = -\frac{\dfrac{\partial w_4}{\partial M}}{\dfrac{\partial w_4}{\partial U}} = -\frac{MU_d}{2(1-M^2)}$$

as obtained in [21].

This is consistent with the Garrick formula above but of course it is limited to small κ. It is also consistent with divergence as a flutter speed for zero frequency, except that zero is not a structure mode as we have noted.

5.8 Nonlinear Structure Models

Finally we extend the theory to the nonlinear structure models we introduced in Chap. 2.

Linearization: Beran–Straganac Model

We are fortunate that for this model in spite of its extreme complexity, the static solution is the zero solution for the structure and constant air flow. Hence the linearization yields exactly the same equations as in the case of the linear Goland model.

Linearization: Dowell–Hodges Model

Here as we have noted the presence of the gravity terms complicates matters. Of course they have to be considered only along with nonzero far field velocity, as we have noted before.

The static solution which is nonzero, is given in Chap. 4, Sect. 4.5. Here we go on to linearize the solution to the aeroelastic equations about this solution. This requires substantial effort because the static structure solution is no longer zero. An important question is the role played by the gravity terms.

Thus we seek the solution to the structure equations in the power series form in the structure state variables. Using the parameter λ, we define

$$x(\lambda, t, s) = x(0, s) + \lambda x(t, s), \qquad 0 < s < \ell,$$

$$x = \begin{pmatrix} h \\ v \\ \theta \end{pmatrix} \tag{5.170}$$

and determine the linear equation characterizing $x(t, .)$ by retaining only linear terms in λ. Correspondingly $\phi(\lambda, t, x, z)$ satisfies the field equation for every λ:

$$\frac{\partial^2 \phi(\lambda, .)}{\partial t^2} + \frac{\partial}{\partial t} \|\nabla \phi(\lambda, .)\|^2$$

$$= a_\infty^2 \left(1 + \frac{\gamma - 1}{a_\infty^2} \left(\frac{U_\infty^2}{2} - \frac{\|\nabla \phi(\lambda, .)\|^2}{2} - \frac{\partial \phi(\lambda, .)}{\partial t} \right) \right) \Delta \phi(\lambda, .)$$

$$- \nabla \phi(\lambda, .) \cdot \nabla \left(\frac{\|\nabla \phi(\lambda, .)\|^2}{2} \right) \tag{5.171}$$

and

$$\psi(\lambda, t, x, z) = \frac{\partial \phi}{\partial t} + \frac{|\nabla \phi|^2}{2},$$

$$\delta p(\lambda, .) = -\rho \delta \psi(\lambda, ..)$$

with the boundary conditions detailed in Chap. 3 plus the flow tangency condition and the Kutta Joukowsky condition.

And

$$\phi_0(x, z) = \text{the static potential,}$$

$$L(\lambda, t, y) = \int_{-b}^{b} -\rho_\infty \delta \psi(\lambda, x, y) dx,$$

$$M(\lambda, t, y) = \int_{-b}^{b} -\rho_\infty (x - ab) \delta \psi(\lambda, x, y) dx$$

and $x(\lambda, t, s)$ satisfies the equations:

$$m\ddot{h}(\lambda, t, y) + EI_1 h''''(\lambda, t, y) + (EI_2 - EI_1)(\theta(\lambda, t, y) v(\lambda, t, y)'')''$$

$$= mg \sin\varphi + L(\lambda, t, y), \tag{5.172}$$

$$m\ddot{v}(\lambda, t, y) + EI_2 v''''(\lambda, t, y) + (EI_2 - EI_1)(\theta(\lambda, t, y) h(\lambda, t, y)'')''$$

$$= mg \cos\varphi, \tag{5.173}$$

$$I\ddot{\theta}(\lambda, t, y) - GJ\theta''(\lambda, t, y) + (EI_2 - EI_1) h(\lambda, t, y)'' v(\lambda, t, y)''$$

$$= M(\lambda, t, y) \qquad 0 < t; \quad 0 < y < \ell. \tag{5.174}$$

Taking the derivative with respect to λ at zero (or equivalently equating coefficients of the first power of λ), we see that $x(t, .)$ satisfies the linear equation:

$$m\ddot{h}(t, y) + EI_1 h''''(t, y) + (EI_2 - EI_1)(\theta(t, y) v_o(y)''$$

$$+ \theta_o(y) v(t, y)'')'' = L_1(t, y),$$

$$m\ddot{v}(t, y) + EI_2 v''''(t, y) + (EI_2 - EI_1)(\theta(t, y) h_o(y)''$$

$$+ \theta_o(y) h(t, y)'')'' = 0,$$

$$I\ddot{\theta}(t, y) - GJ\theta''(t, y)$$

$$+ (EI_2 - EI_1)(h(t, y)'' v_o(y)'' + h_o(y)'' v(t, y)'') = M_1(t, y),$$

where

$$L_1(t, y) = \int_{-b}^{b} \delta p_1(t, x, y)dx,$$

$$M_1(t, y) = \int_{-b}^{b} (x - ab)\delta p_1(t, x, y)dx,$$

$$\delta p_1 = \frac{d}{d\lambda}\delta p(\lambda, .)\big|\lambda = 0 \quad 0 < t; \quad 0 < y < \ell$$

and

$$h_o(y) = h(0, y),$$

$$v_o(y) = v(0, y),$$

$$\theta_o(y) = \theta(0, y). \tag{5.175}$$

Note that these functions depend on the gravity terms.

Consider first the case where the air speed is zero, $U_\infty = 0$, but we retain the gravity terms, so that we have that $x(0, s)$ is the solution of:

$$EI_1 h''''(s) + (EI_2 - EI_1)(\theta(s)v(s)'')'' = mg \cos\varphi,$$

$$EI_2 v''''(s) + (EI_2 - EI_1)(\theta(s)h(s)'')'' = mg \sin\varphi,$$

$$GJ\theta''(s) + (EI_2 - EI_1)v(s)''h''(s) = 0.$$

Plus CF end conditions.

A power series solution is given Chap. 4, Sect. 4.5. From which we can see in particular that the functions are real-valued.

The linear structure dynamics equations then become:

$$m\ddot{h}(t, y) + EI_1 h''''(t, y) + (EI_2 - EI_1)(\theta(t, y)v_0(y)'' + \theta_0(y)v(t, y)'')'' = 0,$$

$$m\ddot{v}(t, y) + EI_2 v''''(t, y) + (EI_2 - EI_1)(\theta(t, y)h_0(y)'' + \theta_0(y)h(t, y)'')'' = 0,$$

$$I\ddot{\theta}(t, y) - GJ\theta''(t, y) + (EI_2 - EI_1) \cdot (h(t, y)''v_0(y)'' + h_0(y)''v(t, y)'') = 0.$$

This can be given a Hilbert space formulation as in the case of the Goland model in Chap. 2.

Let $H = L_2[0, \ell]^3$ with elements

$$x = \begin{pmatrix} h \\ v \\ \theta \end{pmatrix}.$$

Let A denote the operator with domain:

$$\mathcal{D}(A) = \{x | h'''', v'''', \theta'' \epsilon L_2[0, \ell],$$

with CF or FF end conditions as in Chap. 2

$$Ax = \begin{pmatrix} EI_1 h'''' \\ EI_2 v'''' \\ -GJ\theta'' \end{pmatrix}.$$

Define the operator B with domain: $\mathcal{D}(B) = \mathcal{D}(A)$.

Then because the coefficient functions $h_0(y)$, $v_0(y)$, and $\theta_0(y)$ have continuous fourth derivatives in $[0, \ell]$, it follows that

$$Bx = (EI_2 - EI_1) \begin{pmatrix} (\theta v_0'' + \theta_0 v'')'' \\ (\theta h_0'' + \theta_0 h'')'' \\ h'' v_0'' + h_0'' v'' \end{pmatrix}$$

is in H. Because $x_0(.\,)$ is real-valued, it is readily verified that

$$B = B^* = B^{**}$$

and is closed on the domain of A. Hence

$$(A + B)^* = (A + B)$$

and is closed on the domain of A. Hence the equations can be written in abstract form:

$$\mathcal{M}\ddot{x}(t) + Ax(t) + Bx(t) = 0,$$

where

$$\mathcal{M} = \mathrm{diag}[m, m, I_\theta].$$

We note that, as in Chap. 2, A is self-adjoint and nonnegative definite, and $Ax = 0$ implies $x = 0$.

Hence we may, as in Chap. 2, introduce the energy norm space

$$\mathcal{H} = \mathcal{D}(\sqrt{A}) \times H$$

with inner product:

$$[Y,\ Z] = [\sqrt{A}x_1, \sqrt{A}z_1] + [\mathcal{M}x_2, z_2]$$

and rewrite the structure equation as

$$\dot{Y}(t) = \mathcal{A}Y(t) + \mathcal{B}Y(t),$$

where

$$\mathcal{A} = \begin{pmatrix} 0 & I \\ -\mathcal{M}^{-l}(A+B) & 0 \end{pmatrix},$$

where we note that $\mathcal{D}(A)$ is contained in $\mathcal{D}(B)$, and

$$\mathcal{D}(\mathcal{A}) = \left\{ \begin{pmatrix} x_l \\ x_2 \end{pmatrix}, x_2 \epsilon H,\ x_1 \epsilon \mathcal{D}(A) \right\}.$$

Let us calculate first the resolvent of $\mathcal{A}$. Thus we consider the equation:

$$\lambda Y - \mathcal{A}Y = Z; \qquad X = \begin{pmatrix} x_l \\ x_2 \end{pmatrix} \epsilon \mathcal{D}; \qquad Z = \begin{pmatrix} z_l \\ z_2 \end{pmatrix}.$$

Or

$$\lambda^2 \mathcal{M}x_1 + Ax_1 + Bx_1 = \mathcal{M}(z_2 + \lambda z_1) \qquad x_2 = \lambda x_1 - z_1. \qquad (5.176)$$

Lemma 5.41.

$$B(\lambda^2 \mathcal{M} + A)^{-1}$$

is bounded (but not necessarily compact) and the norm

$$\|B(\lambda^2 \mathcal{M} + A)^{-1}\| = 0\left(\frac{1}{\lambda^2}\right) \quad \text{for } \lambda > 0.$$

Proof. Only the last part is new. Here we note B is closed and that the range of $(\lambda^2 \mathcal{M} + A)^{-1}$ is contained in the domain of A which is contained in the domain of B. And we can write

$$B\left(\lambda^2 \mathcal{M} + A\right)^{-1} = BA^{-1}\left(\lambda^2 \mathcal{M}A^{-1} + I\right)^{-1},$$

$$\left|\left(\lambda^2 \mathcal{M}A^{-1} + I\right)^{-1} x\right|^2 = \sum_{k=1}^{\infty} \left(\frac{\lambda_k}{\lambda^2 + \lambda_k}\right)^2 \left|[x, \phi_k]\right|^2,$$

which is enough to show via the uniform boundedness principle that

$$\|\left(\lambda^2 \mathcal{M}A^{-1} + I\right)^{-1}\| = 0\left(\frac{1}{\lambda^2}\right)$$

and in as much as BA^{-1} is bounded, this is enough to show that

$$\| B(\lambda^2 \mathcal{M} + A)^{-1} \| = 0 \left(\frac{1}{\lambda^2} \right).$$

Hence we can go back to (5.7).

Lemma 5.42. *The resolvent* $\mathcal{R}(\lambda, A)$ *is compact for all* λ *in the resolvent set that includes a half-plane*

$$X = \mathcal{R}(\lambda, A)Z,$$

$$x_1 = \left(\lambda^2 \mathcal{M} + A \right)^{-1} \left(I + B \left(\lambda^2 \mathcal{M} + A \right)^{-1} \right) \mathcal{M}(z_2 + \lambda z_1),$$

$$x_2 = \lambda x_1 - z_1 \tag{5.177}$$

for all λ large enough.

Proof.

The compactness follows from (5.8).   □

We next proceed to calculate the modes following the procedure outlined in Chap. 2, Sect. 2.4, for the Goland beam. Here we simplify the analysis by taking advantage of the fact that $mg(EI_2 - EI_1)$ is small.

Let

$$Y = \mathrm{Col}[h, h', h'', h''', v, v', v'', v''', \theta, \theta'].$$

Then (5.7) can be expressed:

$$Y'(\lambda) = (\dot{A}(\lambda) + \dot{B})Y(\lambda),$$

where

$$\dot{A}(\lambda) = \begin{pmatrix}
0 & 1 & 0 & 0 & 0 & 0 & 0 & 0 & 0 & 0 \\
0 & 0 & 1 & 0 & 0 & 0 & 0 & 0 & 0 & 0 \\
0 & 0 & 0 & 1 & 0 & 0 & 0 & 0 & 0 & 0 \\
w_1 & 0 & 0 & 0 & 0 & 0 & 0 & 0 & 0 & 0 \\
0 & 0 & 0 & 0 & 0 & 1 & 0 & 0 & 0 & 0 \\
0 & 0 & 0 & 0 & 0 & 0 & 1 & 0 & 0 & 0 \\
0 & 0 & 0 & 0 & 0 & 0 & 0 & 1 & 0 & 0 \\
0 & 0 & 0 & 0 & w_2 & 0 & 0 & 0 & 0 & 0 \\
0 & 0 & 0 & 0 & 0 & 0 & 0 & 0 & 1 & 0 \\
0 & 0 & 0 & 0 & 0 & 0 & 0 & 0 & 0 & w_3
\end{pmatrix},$$

$$w_1 = -\lambda^2 \frac{m}{EI_1},$$

$$w_2 = -\lambda^2 \frac{m}{EI_2},$$

$$w_3 = \lambda^2 \frac{I_\theta}{GJ},$$

where $\dot{B}Y(s) = (EI_2 - EI_1)\mathrm{Col}[0,0,0,(\theta v_0'' + \theta_0 v'')'',0, 0, 0, (\theta h_0'' + \theta_0 h'')'',0,(h''v_0'' + h_0''v'')]$, which we solve as

$$Y(s) = e^{\dot{A}(\lambda)s}Y(0) + \int_0^s e^{\dot{A}(\lambda)(s-\sigma)}\dot{B}Y(\sigma)d\sigma,$$

which is a Volterra integral equation that we solve by retaining only the linear term in $\dot{B}$ to obtain:

$$Y(s) = e^{\dot{A}(\lambda)s}Y(0) + \int_0^s e^{\dot{A}(\lambda)(s-\sigma)}\dot{B}e^{\dot{A}(\lambda)\sigma}Y(0)d\sigma,$$

so that

$$Y(\ell) = \left(e^{\dot{A}(\lambda)\ell} + \int_0^\ell e^{\dot{A}(\lambda)(\ell-\sigma)}\dot{B}\,e^{\dot{A}(\lambda)\sigma}d\sigma\right)Y(0).$$

The eigenvalues for the CF end conditions then are the zeros of

$$d(\lambda) = \det P\left(e^{\dot{A}(\lambda)\ell} + \int_0^\ell e^{\dot{A}(\lambda)(\ell-\sigma)}\dot{B}\,e^{\dot{A}(\lambda)\sigma}d\sigma\right)Q,$$

the determinant of a 5×5 matrix, where $P = Q^*$, where now P is 5×10, given by:

$$P = \begin{pmatrix} 0 & 0 & 0 & 0 & 0 \\ 0 & 0 & 0 & 0 & 0 \\ 1 & 0 & 0 & 0 & 0 \\ 0 & 1 & 0 & 0 & 0 \\ 0 & 0 & 1 & 0 & 0 \\ 0 & 0 & 0 & 1 & 0 \\ 0 & 0 & 0 & 0 & 0 \\ 0 & 0 & 0 & 0 & 0 \\ 0 & 0 & 0 & 0 & 0 \\ 0 & 0 & 0 & 0 & 1 \end{pmatrix}.$$

The main point to note here is that $d(\lambda)$ is an entire function of λ and has a countable number of zeros with no limit point in the finite plane.

Let us see what we can say next about the perturbed eigenvalues.
First we see that

$$d_0(\lambda) = \det \mathrm{P} e^{\acute{A}(\lambda)\ell} Q,$$

$$\lambda = i\omega_k = (1 + \cosh\gamma_k \, \cos\gamma_k)(1 + \cosh\beta_k \cos\beta_k)\cos\mu\omega_k^\ell,$$

where the mode frequencies are:

$$\omega_k = \frac{1}{\ell}\left(\frac{E I_1}{m}\right)^{1/4} \gamma_k$$

or

$$\omega_k = \frac{1}{\ell}\left(\frac{E I_2}{m}\right)^{1/4} \beta_k$$

or

$$\omega_k = (2k+1)\frac{\pi}{2\ell}\sqrt{\frac{GJ}{I_\theta}}.$$

Let the perturbation be denoted:

$$d_1(\lambda) = \mathrm{Det}\, P \int_0^\ell e^{\acute{A}(\lambda)(\ell-\sigma)} \acute{B} \, e^{\acute{A}(\lambda)\sigma} d\sigma \, Q.$$

We note that this is a function of λ^2. And hence $d_1(i\omega_k)$ is real-valued, although the derivative is imaginary. Hence a one Newton step shows that the root continues to be imaginary. Hence it follows that the perturbed eigenvalues are also imaginary. Usually for perturbation by a controller, for example, we assume that the eigenfunctions are approximately the same and only the eigenvalues change. If we do that here we would find no change in the eigenvalues because $[B\phi_k, \phi_k] = 0$.

Linearizing the Euler Full Potential Equation

Next let us linearize the potential equation about the static solution. Let $\phi_0(.,.)$ denote the static solution, and $\phi(t, ., .)$ the linearized potential. Then $\phi(t, .)$ satisfies the linear equation with nonconstant coefficients:

$$\frac{\partial^2 \phi}{\partial t^2} + 2\nabla\phi_0 \cdot \frac{\partial \nabla\phi}{\partial t} = a_\infty^2 \left(1 + \frac{\gamma - 1}{a_\infty^2} \left(\frac{U_\infty^2}{2} - \frac{|\nabla\phi_0|^2}{2}\right)\right) \Delta\phi$$

$$- (\gamma - 1)\Delta\phi_0 \left(\nabla\phi_0 \cdot \nabla\phi + \frac{\partial \phi}{\partial t}\right)$$

$$- \nabla\phi_0 \cdot \nabla(\nabla\phi_0 \cdot \nabla\phi) - \nabla\left(\frac{|\nabla\phi_0|^2}{2}\right) \cdot \nabla\phi$$

with the boundary condition

$$\frac{\partial \phi(t, x, 0\pm)}{\partial z} = -\left[\dot{h}(t, \ y) + (x - ab)\dot{\theta}(t, y) - \frac{\partial \phi_0}{\partial x}\theta(t, y)\right]$$

$$- \theta_0(y)(\partial \phi(t, x, 0\pm))/\partial x, \quad |x| < b.$$

This is of course quite a formidable task—as stated—because the coefficients are no longer constant. So we seek only an approximate solution, as in the case above for the structure. Thus we break up the linearized equation with

$$\phi_{00}(x, \ z) = xU$$

and correspondingly

$$\phi(t, \ x, \ z) = \phi_{00}(t, \ x, \ z) + \phi_{01}(t, \ x, \ z)$$

with the boundary conditions

$$\frac{\partial \phi(t, x, 0\pm)}{\partial z} = -\left[\dot{h}(t, y) + (x - ab)\dot{\theta}(t, \ y) - \frac{\partial \phi_0}{\partial x}\theta(t, \ y)\right.$$

$$\left. + \theta_0(y)(\partial \phi \ (t, \ x, \ 0\pm))/\partial x\right]. \tag{5.178}$$

The boundary conditions don't involve the in-plane bending $v(t, \ . \ )$.

Now $\phi(t, \ . \ )$ satisfies the field equation:

$$\frac{\partial^2 \phi}{\partial t^2} + 2U \frac{\partial^2 \phi}{\partial x \partial t} + a_\infty^2 \left((1 - M^2)\frac{\partial^2 \phi}{\partial x^2} + \frac{\partial^2 \phi}{\partial z^2}\right)$$

$$= \varphi(t, x, z) = (\nabla\phi_{00} - \nabla\phi_0) \cdot \frac{\partial}{\partial t}\nabla\phi + (\gamma - 1)\left(\frac{U_\infty^2}{2} - \frac{|\nabla\phi_0|^2}{2}\right)\Delta\phi$$

$$-(\gamma - 1)\Delta\phi_0 \left(\nabla\phi_0 \cdot \nabla\phi + \frac{\partial\phi}{\partial t}\right) - (\nabla\phi_0 - \nabla\phi_{00})$$

$$\cdot\nabla((\nabla\phi_0 - \nabla\phi_{00}) \cdot \nabla\phi) - \nabla\left(\frac{|\nabla\phi_0|^2}{2}\right) \cdot \nabla\phi.$$

So we express the solution as the sum of two: One (denote it ϕ_F) that satisfies the nonhomogeneous field equation with zero boundary conditions and the other (denote it ϕ_B) that satisfies the homogeneous field equation:

$$\frac{\partial^2\phi}{\partial t^2} + 2U\frac{\partial^2\phi}{\partial x \partial t} + a_\infty^2\left((1 - M^2)\frac{\partial^2\phi}{\partial x^2} + \frac{\partial^2\phi}{\partial z^2}\right) = 0$$

with the boundary conditions given by (5.7).

The solution to the first is given by (as in [7])

$$\phi_F(t, x, z) = \int_0^t d\sigma \int_{R^2} L(t - \sigma, x - \xi, z - \zeta)\varphi(\sigma, \xi, \zeta)d\xi d\zeta,$$

where

$$L(t, x, z) = \frac{1}{2\pi a_\infty^2\sqrt{1 - M^2}} \frac{1}{\sqrt{\left(t - U\frac{x}{c_{12}}\right)^2 - \frac{1}{1 - M^2}\left(\frac{x^2}{c_{12}} + \frac{z^2}{c_2^2}\right)}},$$

where

$$c_1^2 = a_\infty^2(1 - M^2); \qquad c_2^2 = a_\infty^2,$$

$$\phi_B(t, x, z) = \frac{z}{\pi}\int_0^t d\sigma\,\frac{A(t - \sigma, \xi)d\xi}{((x - \xi - U\sigma)^2 + z^2)}$$

and

$A(t, .)$ is the solution of the Possio equation:

$$\int_{-b}^b P(t - \sigma, x - \xi)A(\sigma, \xi)d\sigma d\xi = -\left[\dot{h}(t, y) + (x - ab)\dot{\theta}(t, y)\right.$$

$$\left. - \frac{\partial\phi_0}{\partial x}\theta(t, y) + \theta_0(y)\left(-A(t, x) + \frac{d}{dt}\int_0^t A(t - \sigma, x - U\sigma)d\sigma\right)\right],$$

where $P(., .)$ is the linear TDP kernel as given in Sect. 5.6.

This equation is a little bit more involved than the standard TDP, because of the function $\theta_0(y)$ but we can still take advantage of our solution of the Possio equation and rewrite the equation as

$$A + \theta_0(y)\psi(M)\mathcal{L}A = \psi(M)w, \tag{5.179}$$

where

$$\mathcal{L}A = A - \frac{\mathrm{d}}{\mathrm{d}t}\int_0^t \mathcal{S}(U\sigma)A(t - \sigma, \, .\,)\mathrm{d}\sigma,$$

$$w(t, \, x) = -\left[\dot{h}(t, \, y) + (x - ab)\dot{\theta}(t, \, y) + \frac{\partial \phi_0}{\partial x}\theta(t, \, y)\right] \quad \text{for fixed } y.$$

And hence

$$A = (I + \theta_0(y)\, \psi(M)\mathcal{L})^{-1}\psi(M)w,$$

which may be approximated by

$$A = (I - \theta_0(y)\, \psi(M)\mathcal{L})\psi(M)w. \tag{5.180}$$

Hence we have:

$$\phi(t, x, z) = \phi_B(t, x, z) + \int_0^t \int_{R^2} \mathrm{d}\sigma L(t - \sigma, x - \xi, z - \zeta)\varphi(t, \xi, \zeta)\mathrm{d}\xi\mathrm{d}\zeta,$$

which is a Volterra equation in the time domain, and has the solution (up to the first term):

$$\phi(t, \, x, \, z) = \phi_B(t, \, x, \, z) + \int_0^t \mathrm{d}\sigma \int_{R^2} L(t - \sigma, \, x - \xi, \, z - \zeta).$$

$$\left[(\nabla\phi_{00} - \nabla\phi_0) \cdot \frac{\partial}{\partial t}\nabla\phi_B + (\gamma - 1)\left(\frac{U_\infty^2}{2} - \frac{|\nabla\phi_0|^2}{2}\right)\Delta\phi_B\right.$$

$$- (\gamma - 1)\Delta\phi_0\,(\nabla\phi_0.\nabla\phi_B) - (\nabla\phi_0 - \nabla\phi_{00}) \cdot \nabla((\nabla\phi_0 - \nabla\phi_{00}) \cdot \nabla\phi_B)$$

$$\left. - \nabla(|\nabla\phi_0|^2) \cdot \nabla\phi_B\right]\mathrm{d}\xi\mathrm{d}\zeta.$$

What is unique about this example is that we have modes (eigenvalues) even though the system is not linear which continue to be the modes for the linearized approximation.

We stop here because further analysis would involve explicit use of the function $\phi_0(.,\,.\,)$.

5.9   Appendix: Computer Program for Flutter Speed in Incompressible Flow: Goland Model

Dynamic Response of a Wing in Unsteady Aerodynamics

This Program Calculates the Flutter Speed of a Wing in Unsteady Aerodynamics Using Full 2D Continuum Model

```
(*Initialization & System Parameters*)
(*Initialization & Output Log File*)
Off [General::"spelll", General :: "spell"];
  (*Initialize*)
ClearGlobal[ ] :=ClearAll ["Global'*"]; Clear Derivative];
Remove ["Global'*"];);
ClearGlobal[ ];
sciNumut[num_] := ToString [MantissaExponent
[Abs[num]] [[1]]] <>
  "E"<> ToString" [MantissaExponent[Abs[num]] [[2]]];
writeNum [name_, num_, desc:-""]:= name <>" = \ t"<>
ToString[N[num]] <> "\ t" <> desc <>"\ n";


nbInfo = NotebookInformation [EvaluationNotebook[ ]];
sLogFileName =
"z. RunLog."<> StringReplaceList Extract ["FileName"/.
nbInfo, 2],
". nb" → ""];
nbFileName = Extract["FileName"/. nbInfo, 2];
nbDir = DirectoryName "ToFileName["FileName"/. nbInfo]];
nbFilePath =
  "FileName"/. nbInfo /. FrontEnd 'FileName [dir_,
    fname_,] ↔ ToFileName [dir, fname];

sWinTtl =": Initializing, creating log file. . . ";
Setoptions [EvaluationNotebook[ ], WindowTitle → Dynamic
[nbFileName <> sWinTtl]];
SetDirectory [nbDir];
sTimeStampForm = {"Year", "-", "Month", "-", "Day",
  ",", "Hour", ".", "Minute", ".", "Second",",",
    "Millisecond"},
sOutFile = DateString Join [{sLogFileName, ", "},
sTimeStampForm]];
sLogFile = sOutFile <>".xls";
sPDFFile = sOutFile <>". pdf";
sStartTime = DateString Join [{"Start time: \t"},
sTimeStampForm, {"\n}]];
```

```
(*formatted time stamp, write in log file *)
dtStartTime = DateList [ ]; (*raw time stamp for run
time
calculation *)
fTimeUsedO = TimeUsed [ ]; (*total CPU processing time
used by
Mathematica Kernel *)

strm = OpenWrite [sLogFile];
```

(*Parameters for the Structure*)
```
m = 0.7;          (* mass per unit length *)
a = -0.3;          (* location of elastic axis *)
b = 3;          (* half-chord length *)
1 = 20;          (* span *)
S = 0.447;
```
$I = 1.943;$
$GJ = 2.39 * 10^6$
$EI = 23.6 * 10^6;$
$mt = 0;$
$s_t = 0;$
$I_{yt} = 0;$
$I_t = 0;$

(*Parameters for Control*)
$ghi = 0; \quad gi = 0; \quad steu = 010; \quad numu = 100;$

(*Parameters for the Air*)
$\rho = .\ 0022973;$
$\kappa = \dfrac{\rho b^2 \pi}{m};$
2.39×10^6

(*Calculation of 1st Bending and 1st Torsion Modes of the Structure *)
```
sWinTtl ="": Calc 1st bending & 1st torsion of
    structure...";
```
$x = \dfrac{s}{mb};$
$r = \sqrt{\dfrac{I}{mb^2}};$
$n1 = \dfrac{I_{yt}}{EI}{\check{}}^2 + {\check{}}\dfrac{gh}{EI}; \qquad n2 = -mt/EI{\check{}}^2; \qquad n3 = S_t/EI{\check{}}^2;$
$n4 = -S_t/GJ{\check{}}^2; \qquad n5 = I_t/GJ{\check{}}^2 + {\check{}}\dfrac{g}{GJ};$
$tw1 = -\dfrac{m}{EI}{\check{}}^2(1 + {\check{}});$
$tw2 = -\dfrac{m}{EI}b{\check{}}^2(x - a{\check{}});$

$$\text{tw3} = \frac{m}{GJ} b^{\smile 2} (x - a^{\smile});$$

$$\text{tw4} = \frac{m}{GJ} b^{2 \smile 2} \left(r^2 + {}^{\smile} \left(\frac{1}{8} + a^2 \right) \right);$$

$$A = \begin{pmatrix} 0 & 1 & 0 & 0 & 0 & 0 \\ 0 & 0 & 1 & 0 & 0 & 0 \\ 0 & 0 & 0 & 1 & 0 & 0 \\ \text{tw1} & 0 & 0 & 0 & \text{tw2} & 0 \\ 0 & 0 & 0 & 0 & 0 & 1 \\ \text{tw3} & 0 & 0 & 0 & \text{tw4} & 0 \end{pmatrix} /. \ U-> 0;$$

$$\text{p3} = \begin{pmatrix} 0 & \text{n1} & 1 & 0 & 0 & 0 \\ \text{n2} & 0 & 0 & 1 & \text{n3} & 0 \\ \text{n4} & 0 & 0 & 0 & \text{n5} & 1 \end{pmatrix};$$

$$\text{q3} = \begin{pmatrix} 0 & 0 & 0 \\ 0 & 0 & 0 \\ 1 & 0 & 0 \\ 0 & 1 & 0 \\ 0 & 0 & 0 \\ 0 & 0 & 1 \end{pmatrix};$$

$$\omega b1 = (0.597)^2 \frac{\pi^2}{\ell^2} \sqrt{\frac{EI}{m}};$$

$$\omega t1 = \sqrt{\frac{GJ}{I\alpha}} \frac{\pi}{21};$$

$$\omega b2 = (1.49)^2 \frac{\pi^2}{\ell^2} \sqrt{\frac{EI}{m}};$$

$$\omega t2 = \sqrt{\frac{GJ}{I\alpha}} \frac{3\pi}{21};$$

```
g = Det[p3.MatrixExp[A * 1].q3]/.λ- > ωI/.gh- > 0/.gα- > 0;
ω2 = ω/.FindRoot[g == 0,{ω,ωb1,ωb1 + 0.01}];
ω3 = ω/.FindRoot[g == 0,{ω,ωb2,ωb2 + 0.1};
ω1 = ω/.FindRoot[g == 0,{ω,ωt1,ωt1 + 0.1}];
ω4 = ω/.FindRoot[g == 0,{ω,ωt2,ωt2 + 0.01}];
mode = Chop[{!1, !2, !3, !4}];
mode/2/π;
```

(*System Equations*)
```
sWinTtl = ":Calc system equations...";
ClearAll[A,p3,q3];
z = ˘b/U;
w11 = Pi z + 2 Pi T;
w12 = Pi T;
w21 = −Pi T;
w22 = Pi/2(1 + z/4 − T);
```

```
T = BesselK[1, z]/(BesselK[0, z] + BesselK[1, z]);
```

$$w1 = -1/EI(\check{}^2m + \text{æUb}\check{}w11);$$

$$w2 = -1/EI(\check{}^2S + \text{æbU}^2((1 - az)w11 + zw12));$$

$$w3 = 1/GJ(\check{}^2S + \text{æb}^2U\check{}(w21 - aw12));$$

$$w4 = 1/GJ(\check{}^2I + \text{æb}^2U^2(w21 + zw22 - a(1 - az)w11 - az(w21 + w12)));$$

$$A = \begin{pmatrix} 0 & 1 & 00 & 0 & 0 \\ 0 & 0 & 10 & 0 & 0 \\ 0 & 0 & 01 & 0 & 0 \\ w1 & 0 & 0 & 0 & w2 & 0 \\ 0 & 0 & 00 & 0 & 1 \\ w3 & 0 & 0 & 0 & w4 & 0 \end{pmatrix}$$

$$p3 = \begin{pmatrix} 0 & n1 & 1 & 0 & 0 & 0 \\ n2 & 0 & 0 & 1 & n3 & 0 \\ n4 & 0 & 0 & 0 & n5 & 1 \end{pmatrix};$$

$$q3 = \begin{pmatrix} 0 & 0 & 0 \\ 0 & 0 & 0 \\ 1 & 0 & 0 \\ 0 & 1 & 0 \\ 0 & 0 & 0 \\ 0 & 0 & 1 \end{pmatrix};$$

$$d\check{}0 = \text{Det}[p3.\text{MatrixExp}[A1].q3]/.Cz- > \frac{1}{2};$$

$$d\check{}1 = \text{Det}[p3.\text{MatrixExp}[A1].q3]/.gh- > ghi/.g- > gi/.$$

$$Cz- > \frac{\text{BesselK}[1, z]}{\text{BesselK}[0, z] + \text{BesselK}[1, z]};$$

(*Root Loci with U as the parameter*)
(*Root Loci*)
```
sWinTtl =": Calc Root Loci w/ param U.";
ClearAll[˘, gh, g, U]; gr = {}; tp = {};

st0 = "Root Loci with U as the parameter:"; (*Prepare
Plot Titles*)
sPlotTtl = {st0 <> "1st Torsion", st0 <> "1st Bending",
st0 <> "2nd Bending", st0 <> "3rd Bending", st0 <> "2nd
   Torsion";};
10 = ", Listing for";
11 = {10 <> "1st Torsion", 10 <> "1st Bending", 10 <>
"2nd Bending",
10 <> "2nd Torsion"};

Do [U = 0; ht = {}; rp = {}; ip = {}; tmp = mode[[k]]i;
WriteString [strm, "U\tSigma=Re(L) \tf=Im(L) \tRe/Im\t
   |dL|\
tmode\ n"];
```

```
(* column headers *)

Do [U = U + steu; ˘ = ˘/.FindRoot[d˘1 == 0, {˘,
tmp, tmp + 0.01'i}]; tmp = ˘; Um = U;
sWinTt1 =": Calc Root Loci w/ param U = "<> ToString[U];
WriteString [strm, StringForm ["" \ t"\ t"\ t"\ t"\ t"\
n",
NumberForm[Um, 10], NumberForm[Re[λ], 10], NumberForm
   [Im[λ], 10],
   NumberForm[Re[˘]/Im[˘], 10], sciNumOut[Abs[d˘1]],
NumberForm[mode[[k]], 10]]];
```

$$rp = \text{Append}\left[rp, \left\{Um, \frac{\text{Re}[˘]}{\text{Im}[˘]}\right\}\right];$$

$$ip = \text{Append}\left[ip, \left\{Um, \frac{\text{Im}[˘]}{2\text{ß}}\right\}\right];$$

$$ht = \text{Append}\left[ht, \left\{\frac{\text{Re}[˘]}{\text{Im}[˘]}, \frac{\text{Im}[˘]}{2\text{ß}}\right\}\right],$$

```
{j, 1, numu}
];
   sTimeUsed = "CPU time used (min): \t"<> ToString
   [(TimeUsed
[]-- fTimeUsed0)/60.0]<> "\n\n"; WriteString [strm ,
sTimeUsed];
tp = Append [tp, {rp, ip}];
p1 = ListLinePlot [rp, AxesLabel → {"U", "Re[˘]"}];
p2 = ListLinePlot [ip, AxesLabel → {"U", "Im[˘]"}];
p3 = ListLinePlot [ht, AxesLabel → {"Re[λ]", "Im[λ]"}];
Print [GraphicsGrid [{{p1, p2, p3}}, ImageSize → Full,
   Frame → True,
   FrameStyle → Directive [Blue, Dotted], PlotLabel →
sPlotTt1
[[k]]]], {k, 1, 4}
];
sEndTime = DateString[Join[{"End time: \t"},
   sTimeStampForm, {"\n"}]];
sTimeUsed =
"CPU time used (min): \t" <> ToString
[(TimeUsed[]-fTimeUsed0)/60.0] <> "\n"; WriteString[strm,
sStartTime, sEndTime, sTimeUsed];
   (*Print time stamps for performance review*)
WriteString[strm, writeNum["m", m, "mass per unit
   length"] <>
writeNum["a", a, "location of elastic axis"]<>
writeNum["b", b, "half-chord length"] <> writeNum["1",
1,
"wing span"]<>
```

```
writeNum["S", S]<> writeNum ["I_alpha", I]<> writeNum
["GJ", GJ]<>
writeNum["EI", sciNumOut[EI]] <> writeNum["mt", mt] <>
writeNum
["S_t",S_t]<>
writeNum ["I_yt",I_yt] <> writeNum ["I_t",I_t] <> writeNum
["ghi", ghi]<>
writeNum ["g-alpha-i", gi] <> writeNum ["steu", steu]
<>
writeNum ["numu", numu] <> writeNum ["rho", æ, ""]<>
writeNum
["kappa","]];
Close [strm];
NotebookPrint [EvaluationNotebook[], nbDir <> sPDFFile];
(*Save nb file with graphs as PDF*)
sWinTtl =": Done! See log file.";
```

Root Loci with U as the parameter:1ˢᵗ Torsion

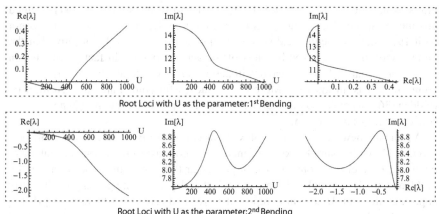

Root Loci with U as the parameter:1ˢᵗ Bending

Root Loci with U as the parameter:2ⁿᵈ Bending

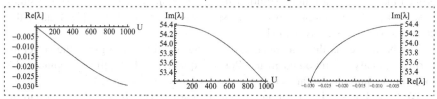

Root Loci with U as the parameter:3ʳᵈ Bending

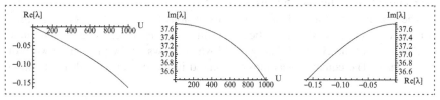

Notes and Comments

It is humbling to note that in spite of all the theory developed, we still cannot answer some of the simplest of questions. For example: which mode is going to flutter first without carrying out a computer program for the given parameters? The dependence of the flutter speed on the parameters is just a little too complicated. We do have, however, a closed-form formula for the divergence speed which can give some idea of the range of the flutter speed but not much more. This is further confounded by the fact that in the case of axial flow treated in Chap. 10 the divergence speed is simply not defined!

The Possio integral equation, which is the heart and soul of our theory, was derived by Possio in 1938 [37] for the "oscillating wing"; it was customary at that time to distinguish between the "steady" oscillatory motion and the transient "unsteady motion," a distinction that persists in the aeroelastic literature even today.

The Laplace Transform version that encompasses both was given in 2003 in [4]. It is not merely a matter of replacing $i\omega$ by λ. The Fourier transform integrals in the original Possio version are not convergent whereas the Laplace transform integrals are shown to be convergent and the Fourier transform version obtained by taking limits as in modern theory [10, 41]. The well-known book by Fung [47] claimed that the existence of a solution was proved but no reference was given. It was not until 1976 that an unequivocal statement appeared in [49] in the negative but the Fung assertion was believed by most of the aeroelasticians. In fact when I began my research many of my colleagues in aeroelasticty would ask me, "Why are you doing this? It is all known!" Moreover, a successful computational algorithm by Rodden [36] has replaced the analytical version in practice even though no proof of convergence has been given.

A radically new version that employs Fourier transforms in the space domain was given in [4] with a simple explicit function (in contrast to the two-page statement involving Hankel functions in [6]). In particular this shows the lack of analyticity in M and k (the normalized "frequency") thereby questioning many of the early approximations in M and in k reported in [6] and references therein. In [4] it is proved that there is a unique solution in L_p, $1 < p < 2$ for each M for small enough $|k|$. A time domain abstract version was given in 2007 in [5]. Finally the nonlinear version was given in [14].

The basic idea is that the calculation of the pressure jump given the structure normal velocity is embodied in the Possio equation and effectively replaces the Euler field equation insofar as the structure dynamics is concerned.

We have thus an "input–output" problem characterized by the Possio equation, the recurrent theme throughout this work.

The Possio equation and its various extensions including the nonlinear time domain version are used systematically in this work, even for the incompressible case in place of the classical theory in [6] which draws on the pioneering work of Theodorsen. The equation was mostly ignored in the literature, with no mention in the recent standard references [17] or [5], for example. It is ironic that Possio was killed in the last Allied air raid on Turin, Italy, in World War II. Indeed even his name would seem to be largely forgotten in his country of birth because of the perception that he was a collaborator!

Chapter 6
Nonlinear Aeroelasticity Theory in 2D Aerodynamics: Flutter Instability as an LCO

6.1 Introduction

In this chapter we return to the full nonlinear aeroelastic problem as stated in Chap. 3, with the structure models both linear and nonlinear, described in Chap. 2 and the isentropic aerodynamics as treated in Chap. 3 with the flow tangency and the Kutta–Joukowsky boundary conditions. Recall that we use continuum models without immediately approximating them by finite-dimensional models as in all the current aeroelastic literature.

Our major result is a characterization of Flutter as a limit cycle oscillation with the flutter speed as a Hopf bifurcation point determined by the linear equations, obtained by linearizing the nonlinear equations about the equilibrium or rest structure state and constant air flow. The point of departure is the input–output point of view and the key role is played by the Possio equation which we need to generalize to the nonlinear case, perforce in the time domain rather than the Laplace domain. We limit the theory to 2D air flow because almost no results are available for the Possio equation except in this case. For the same reason the angle of attack is taken to be zero. Furthermore we consider only M such that $0 < M < 1$. The case $M = 0$ is treated in Chap. 10.

A basic assumption is that the structure displacements are neglible compared to the air displacement in the same time. This means that the structure is essentially not moving compared to the air, which is implicit in the statement of the fluid-structure boundary conditions.

As in the time invariant case in Chap. 4, which we follow closely, we provide a constructive existence theorem where the main tool is the power series expansion in terms of the structure variables. This is unique to our approach and is consistent with our view that our interest is in the structure dynamics and how it is affected by the air flow primarily, and in the air flow per se only secondarily.

A V Balakrishnan, *Aeroelasticity: The Continuum Theory*,
DOI 10.1007/978-1-4614-3609-6_6, © Springer Science+Business Media, LLC 2012

6.2   The Aeroelastic Equations: Linear Structure Model

We begin with a statement of the full dynamics with the linear Goland structure model. Thus we have:

$$m\ddot{h}(t, y) + S\ddot{\theta}(t, y) + EIh''''(t, y) = L(t, y)$$

$$= \int_{-b}^{b} \delta p(t, x, y)dx, \quad 0 < y < \ell, \quad (6.1)$$

$$I_\theta\ddot{\theta}(t, y) + S\ddot{h}(t, y) - GJ\theta''(t, y) = M(t, y)$$

$$= \int_{-b}^{b} (x - a)\delta p(t, x, y)dx$$

$$0 < y < \ell, \quad (6.2)$$

with CF or FF end conditions.

$$\delta p(t, x, y) = -\rho_\infty \delta\psi(t, x, y). \quad (6.3)$$

For each $y, 0 < y < \ell$, the 2D aerodynamic equations in x, z:

$$q(t, x, z) = \nabla\phi(t, x, z),$$

$$\psi(t, x, z) = \frac{D\phi}{Dt} = \frac{\partial\phi}{\partial t} + \frac{1}{2}\nabla\phi \cdot \nabla\phi, \quad (6.4)$$

$$\frac{\partial^2\phi}{\partial t^2} + \frac{\partial}{\partial t}|\nabla\phi|^2 + (\gamma - 1)\frac{\partial\phi}{\partial t}\Delta\phi$$

$$= a_\infty^2 \left(1 + \frac{\gamma - 1}{2a_\infty^2}\left(U^2 - |\nabla\phi|^2\right)\right)\Delta\phi - \nabla\phi \cdot \nabla\frac{|\nabla\phi|^2}{2},$$

$$-\infty < x, z < \infty, \quad \text{excepting } z = 0, \quad (6.5)$$

where we consider only:

$$0 < M < 1; \quad M = \frac{U_\infty}{a_\infty}.$$

(Omit $M = 0$ and $M = 1$.)

The boundary conditions are:

1. Flow Tangency.

Allowing for flow discontinuity between the top and bottom of the wing and again assuming wing displacement to be small compared to that of the air.

$$\frac{\partial \phi}{\partial z}(t, x, 0+) = \frac{\partial \phi_\infty}{\partial z} + (-1)\left[\dot{h}(t, y) + (x - a)\dot{\theta}(t, y)\right.$$
$$\left. + \frac{\partial \phi}{\partial y}(t, x, 0+)\theta(t, y)\right], \quad |x| < b. \tag{6.6}$$

And

$$\frac{\partial}{\partial z}\phi(t, x, 0-) = \frac{\partial \phi_\infty}{\partial z} + (-1)\left[\dot{h}(t, y) + (x - a)\dot{\theta}(t, y)\right.$$
$$\left. + \frac{\partial}{\partial x}\phi(t, x, 0-)\theta(t, y)\right], \quad |x| < b. \tag{6.7}$$

2. Kutta–Joukowski Conditions:

The pressure jump is zero off the wing:

$$\delta\psi(t, x) = 0 \qquad |x| > b. \tag{6.8}$$

The Kutta condition at trailing edge

$$\delta\psi(t, x) = 0 \quad \text{as } x \to b - .$$

This completes the description of the problem.

The Aero-Structure Dynamics/Input–Output Problem

From a general viewpoint we may consider the fluid-structure interaction problem as an "input–output" problem where the structure velocity $w_a(t, .)$ (the "downwash") can be considered the input and the pressure jump $\delta p(t, .)$ as the output. The lift and moment are just linear functionals of $\delta p(t, .)$.

Heuristically, we can then invoke the Duhamel principle, familiar in electromagnetic theory or more generally (see [1, 44]), as the response to a boundary input in wave motion. Thus $\delta p(t, .)$ must be a physically realizable response, albeit nonlinear, extending the theory from the steady-state case in Chap. 4.

Power Series Expansion/Flow Decomposition

The far field potential is

$$\phi_\infty(t, x, y, z) = xU,$$

where U is the speed parameter, $U \geq 0$. We omit the subscript ∞.

As in Chap. 4, we consider the solution potential $\phi(\lambda, t, x, y, z)$ corresponding to $\lambda h(t, y), \lambda \theta(t, y)$ where λ is a scalar-scale parameter, and make it $0 \leq |\lambda| \leq 1$, yielding the expansion:

$$\phi(\lambda, t, x, z) = \sum_{k=0}^{\infty} \frac{\lambda^k}{k!} \phi_k(t, x, z), \qquad (6.9)$$

where

$$\phi_0(t, x, z) = \phi_\infty(t, x, z) = xU,$$

$$\phi_k(t, x, z) = \frac{\partial^k}{\partial \lambda^k} \phi(0, t, x, y, z), \quad k \geq 1,$$

where we refer to $\phi_k(.)$ as the kth-order potential.

We calculate these by taking derivatives in the field equation with respect to λ. We see that $\phi_1(t, x, z)$ satisfies the linear field equation:

$$\Xi(\phi_1) = 0, \qquad -\infty < x, z < \infty, \quad \text{omitting the wing,}$$

where

$$\Xi(\phi) = \frac{\partial^2 \phi}{\partial t^2} + 2U \frac{\partial^2 \phi}{\partial t \partial x} - \left[a_\infty^2 \left(1 - M^2\right) \frac{\partial^2 \phi}{\partial x^2} + a_\infty^2 \frac{\partial^2 \phi}{\partial z^2} \right]. \qquad (6.10)$$

It is interesting to note that the constant γ does not appear in this equation, and hence if we are interested only in the linear equation we may take $\gamma = 1$.

More generally, for ϕ_k, $k > 1$, as in Chap. 4, we expand both sides of (6.5) in a power series, obtaining:

$$\sum_{k=1}^{\infty} \frac{\lambda^k}{k!} \left(\frac{\partial^2 \phi_k}{\partial t^2} \right) + \frac{\partial}{\partial t} \left(\sum_{k=0}^{\infty} \sum_{j=0}^{\infty} \frac{\lambda^k \lambda^j}{k! \, j!} q_k \cdot q_j \right) + (\gamma - 1) \sum_{k=1}^{\infty} \sum_{j=1}^{\infty} \frac{\lambda^k \lambda^j}{k! \, j!} \frac{\partial \phi_j}{\partial t} \Delta \phi_k$$

$$= a_\infty^2 \left(\sum_{k=1}^{\infty} \frac{\lambda^k}{k!} \Delta \phi_k \right) \left(1 + \frac{\gamma - 1}{2a_\infty^2} \left(-\sum_{i=1}^{\infty} \sum_{j=1}^{\infty} \frac{\lambda^i \lambda^j q_i . q_j}{i! \, j!} - U \sum_{i=1}^{\infty} \frac{\lambda^i \partial \phi_i}{\partial x} \right) \right)$$

$$-\sum_{n=0}^{\infty}\sum_{m=0}^{\infty}\sum_{p=1}^{\infty}\frac{1}{n!m!p!}\lambda^{(n+m+p)}\left[\frac{\partial\phi_n}{\partial x}\left(\frac{\partial\phi_m}{\partial x}\frac{\partial^2\phi_p}{\partial x^2}+\frac{\partial\phi_m}{\partial z}\frac{\partial^2\phi_p}{\partial x\partial z}\right)\right.$$

$$\left.+\frac{\partial\phi_n}{\partial z}\left(\frac{\partial\phi_m}{\partial z}\frac{\partial^2\phi_p}{\partial z^2}+\frac{\partial\phi_m}{\partial x}\frac{\partial^2\phi_p}{\partial x\partial z}\right)\right], \tag{6.11}$$

where

$$\Delta=\frac{\partial^2}{\partial x^2}+\frac{\partial^2}{\partial z^2},$$

$$q_k=\nabla\phi_k,$$

$$q_i\cdot q_j=\frac{\partial\phi_i}{\partial x}\frac{\partial\phi_j}{\partial x}+\frac{\partial\phi_i}{\partial z}\frac{\partial\phi_j}{\partial z}.$$

These of course reduce to the equations in Chap. 4 upon setting the time derivatives to zero.

Theorem 6.1. *The* k*th-order potential satisfies the linear nonhomogeneous equation*

$$\Xi(\phi_k)=g_{k-1},$$

where g_k *only involves potentials of order* $\le k$, *and* $g_0=0$.

$$g_{k-1}=-\frac{\partial}{\partial t}\left(\sum_{j=1}^{k-1}C_{k,j}\,q_{k-j}\cdot q_j\right)-\sum_{j=1}^{k-1}C_{k,j}\frac{\partial\phi_j}{\partial t}\Delta\phi_{k-j}$$

$$+(1-\gamma)U\sum_{j=1}^{k-1}C_{k,j}\frac{\partial\phi_{k-j}}{\partial x}\Delta\phi_j$$

$$+\frac{1-\gamma}{2}k!\sum_{i=1}\sum_{j=1}\sum_{k=1}\frac{1}{i!}\frac{1}{j!}\frac{1}{m!}q_i\cdot q_j\Delta\phi_m,\qquad i+j+m=k$$

$$+\sum_{n=1}\sum_{m=1}\sum_{p=1}\frac{k!}{n!m!p!}\left(\frac{\partial\phi_n}{\partial x}\frac{\partial\phi_m}{\partial x}\frac{\partial^2\phi_p}{\partial x^2}\right.$$

$$\left.+\frac{\partial\phi_n}{\partial x}\frac{\partial\phi_m}{\partial z}\frac{\partial^2\phi_p}{\partial x\partial z}+\frac{\partial\phi_n}{\partial z}\frac{\partial\phi_m}{\partial x}\frac{\partial^2\phi_p}{\partial z\partial x}\right.$$

$$+ \frac{\partial \phi_n}{\partial z} \frac{\partial \phi_m}{\partial z} \frac{\partial^2 \phi_p}{\partial z^2} \bigg), \qquad n + m + p = k$$

$$- 2U \sum_{p=1}^{k-1} C_{k,p} \left(\frac{\partial \phi_{k-p}}{\partial x} \frac{\partial^2 \phi_p}{\partial x^2} + \frac{\partial \phi_{k-p}}{\partial z} \frac{\partial^2 \phi_p}{\partial z^2} \right).$$

Proof. We go back to (6.12). Note that we can write

$$\frac{\partial}{\partial t} \left(\sum_{k=0}^{\infty} \sum_{j=0}^{\infty} \frac{\lambda^k}{k!} \frac{\lambda^j}{j!} q_k \cdot q_j \right) = 2U \sum_{k=1}^{\infty} \frac{\lambda^k}{k!} \frac{\partial^2 \phi_k}{\partial x \partial t} + \frac{\partial}{\partial t} \left(\sum_{k=1}^{\infty} \sum_{j=1}^{\infty} \frac{\lambda^k}{k!} \frac{\lambda^j}{j!} q_k \cdot q_j \right).$$

And

$$- \sum_{n=0}^{\infty} \sum_{m=0}^{\infty} \sum_{p=1}^{\infty} \frac{1}{n! m! p!} \lambda^{(n+m+p)} \left[\frac{\partial \phi_n}{\partial x} \left(\frac{\partial \phi_m}{\partial x} \frac{\partial^2 \phi_p}{\partial x^2} + \frac{\partial \phi_m}{\partial z} \frac{\partial^2 \phi_p}{\partial x \partial z} \right) \right.$$

$$\left. + \frac{\partial \phi_n}{\partial z} \left(\frac{\partial \phi_m}{\partial z} \frac{\partial^2 \phi_p}{\partial z^2} + \frac{\partial \phi_m}{\partial x} \frac{\partial^2 \phi_p}{\partial x \partial z} \right) \right]$$

as

$$- U^2 \sum_{p=1}^{\infty} \lambda^p \frac{\partial^2 \phi_p}{\partial x^2} - 2U \sum_{p=1}^{\infty} \left(\frac{\partial \phi_1}{\partial x} \frac{\partial^2 \phi_p}{\partial x^2} + \frac{\partial \phi_1}{\partial z} \frac{\partial^2 \phi_p}{\partial x \partial z} \right) \lambda^{p+1}$$

$$- \sum_{n=1}^{\infty} \sum_{m=1}^{\infty} \sum_{p=1}^{\infty} \frac{1}{n! m! p!} \lambda^{(n+m+p)} \left[\frac{\partial \phi_n}{\partial x} \left(\frac{\partial \phi_m}{\partial x} \frac{\partial^2 \phi_p}{\partial x^2} + \frac{\partial \phi_m}{\partial z} \frac{\partial^2 \phi_p}{\partial x \partial z} \right) \right.$$

$$\left. + \frac{\partial \phi_n}{\partial z} \left(\frac{\partial \phi_m}{\partial z} \frac{\partial^2 \phi_p}{\partial z^2} + \frac{\partial \phi_m}{\partial x} \frac{\partial^2 \phi_p}{\partial x \partial z} \right) \right].$$

Hence (6.12) can be expressed:

$$\sum_{k=1}^{\infty} \frac{\lambda^k}{k!} \left(\frac{\partial^2 \phi_k}{\partial t^2} \right) + 2U \sum_{k=1}^{\infty} \frac{\lambda^k}{k!} \frac{\partial^2 \phi_k}{\partial x \partial t} + \frac{\partial}{\partial t} \left(\sum_{m=1}^{\infty} \sum_{j=1}^{\infty} \frac{\lambda^m}{m!} \frac{\lambda^j}{j!} q_m \cdot q_j \right)$$

$$+ (\gamma - 1) \sum_{m=1}^{\infty} \sum_{j=1}^{\infty} \frac{\lambda^m}{m!} \frac{\lambda^j}{j!} \frac{\partial \phi_j}{\partial t} \Delta \phi_m$$

$$= a_\infty^2 \sum_{k=1}^{\infty} \frac{\lambda^k}{k!} \Delta \phi_k - \sum_{k=1}^{\infty} U^2 \frac{\lambda^k}{k!} \frac{\partial^2 \phi_k}{\partial x^2} - 2U \sum_{k=1}^{\infty} \frac{\lambda^{k+1}}{k!} \left(\frac{\partial \phi_1}{\partial x} \frac{\partial^2 \phi_k}{\partial x^2} + \frac{\partial \phi_1}{\partial z} \frac{\partial^2 \phi_k}{\partial x \partial z} \right)$$

$$+ (1-\gamma) \sum_{j=1}^{\infty} \sum_{k=1}^{\infty} \frac{\lambda^j}{j!} \frac{\lambda^k}{k!} \Delta\phi_j q_\infty \cdot q_k + \frac{1-\gamma}{2} \sum_{i=1}^{\infty} \sum_{j=1}^{\infty} \sum_{k=1}^{\infty} \frac{\lambda^i}{i!} \frac{\lambda^j}{j!} \frac{\lambda^k}{k!} q_i$$

$$\times q_j \Delta\phi_k - \sum_{n=1}^{\infty} \sum_{m=1}^{\infty} \sum_{p=1}^{\infty} \frac{1}{n!m!p!} \lambda^{(n+m+p)}$$

$$\times \left(\frac{\partial\phi_n}{\partial x} \frac{\partial\phi_m}{\partial x} \frac{\partial^2\phi_p}{\partial x^2} + \frac{\partial\phi_n}{\partial x} \frac{\partial\phi_m}{\partial z} \frac{\partial^2\phi_p}{\partial x \partial z} + \frac{\partial\phi_n}{\partial z} \frac{\partial\phi_m}{\partial x} \frac{\partial^2\phi_p}{\partial z \partial x} \right)$$

$$+ 2U \sum_{k=1}^{\infty} \sum_{p=1}^{\infty} \frac{\lambda^n}{n!} \frac{\lambda^p}{p!} \left(\frac{\partial\phi_n}{\partial x} \frac{\partial^2\phi_p}{\partial x^2} + \frac{\partial\phi_n}{\partial z} \frac{\partial^2\phi_p}{\partial z^2} \right). \tag{6.12}$$

Hence collecting the coefficients of λ^k, we obtain

$$\Xi(\phi_k) = g_{k-1}, \tag{6.13}$$

where

$$g_{k-1} = -\frac{\partial}{\partial t} \left(\sum_{j=1}^{k-1} c_{k,j}\, q_{k-j} \cdot q_j \right) - \sum_{j=1}^{k-1} C_{k,j} \frac{\partial\phi_j}{\partial t} \Delta\phi_{k-j}$$

$$+ (1-\gamma)U \sum_{j=1}^{k-1} C_{k,j} \frac{\partial\phi_{k-j}}{\partial x} \Delta\phi_j + \left(\frac{1-\gamma}{2} \right) k!$$

$$\times \sum_{i=1} \sum_{j=1} \sum_{k=1} \frac{1}{i!} \frac{1}{j!} \frac{1}{m!} q_i \cdot q_j \Delta\phi_m, \qquad i+j+m = k$$

$$+ \sum_{n=1} \sum_{m=1} \sum_{p=1} \frac{k!}{n!m!p!} \left(\frac{\partial\phi_n}{\partial x} \frac{\partial\phi_m}{\partial x\cdot} \frac{\partial^2\phi_p}{\partial x^2} + \frac{\partial\phi_n}{\partial x} \frac{\partial\phi_m}{\partial z} \frac{\partial^2\phi_p}{\partial x \partial z} \right.$$

$$\left. + \frac{\partial\phi_n}{\partial z} \frac{\partial\phi_m}{\partial x} \frac{\partial^2\phi_p}{\partial z\partial x} + \frac{\partial\phi_n}{\partial z} \frac{\partial\phi_m}{\partial z} \frac{\partial^2\phi_p}{\partial z^2} \right)$$

$$- 2U \sum_{p=1}^{k-1} C_{k,p} \left(\frac{\partial\phi_{k-p}}{\partial x} \frac{\partial^2\phi_p}{\partial x^2} + \frac{\partial\phi_{k-p}}{\partial z} \frac{\partial^2\phi_p}{\partial z^2} \right), \qquad n+m+p = k.$$

$$\tag{6.14}$$

Note that g_k depends only on ϕ_j, $j \leq k$.

In particular for $k = 1$,

$$g_1 = -2\frac{\partial}{\partial t}|q_1|^2 - 2\frac{\partial \phi_1}{\partial t}\Delta\phi_1 + 2(1-\gamma)U\frac{\partial \phi_1}{\partial x}\Delta\phi_1$$

$$-4U\left(\frac{\partial \phi_1}{\partial x}\frac{\partial^2 \phi_1}{\partial x^2} + \frac{\partial \phi_1}{\partial z}\frac{\partial^2 \phi_1}{\partial z^2}\right).$$

$\square$

Next we deduce the corresponding boundary conditions.

Boundary Conditions

Flow Tangency

$$\frac{\partial}{\partial z}\phi(\lambda,t,x,y,0+) = -\lambda\left[\frac{\partial}{\partial t}(h(t,y) + (x-a)\theta(t,y))\right.$$

$$\left.+ \frac{\partial \phi(\lambda,t,x,y,0+)}{\partial x}\theta(t,y)\right]\sum_{k=1}^{\infty}\lambda^k\frac{\partial}{\partial z}\phi_k(t,x,0+),$$

$$= -\lambda\left[\frac{\partial}{\partial t}(h(t,y) + (x-a)\theta(t,y))\right]$$

$$-\sum_{k=1}^{\infty}\lambda^k k\frac{\partial}{\partial x}\phi_{k-1}(t,x,y,0+)\theta(t,y).$$

Hence

$$\frac{\partial}{\partial z}\phi_1(t,x,0+) = -\left[\frac{\partial}{\partial t}(h(t,y) + (x-a)\theta(t,y))\right] - U\theta(t,y)\frac{\partial}{\partial z}\phi_k(t,x,y,0+)$$

$$= -k\frac{\partial}{\partial x}\phi_{k-1}(t,x,0+)\theta(t,y) \quad \text{for } k \geq 2,$$

$$\frac{\partial}{\partial z}\phi_1(t,x,y,0-) = -\left[\frac{\partial}{\partial t}(h(t,y) + (x-a)\theta(t,y))\right] - U\theta(t,y)\frac{\partial}{\partial z}\phi_k(t,x,y,0-)$$

$$= -k\frac{\partial \phi_{k-1}(t,x,y,0-)}{\partial x}\theta(t,y) \quad \text{for } k \geq 2.$$

Hence

$$\frac{\partial}{\partial z}\phi_1(t, x, y, 0\pm) = -\left[\frac{\partial}{\partial t}(h(t, y) + (x - a)\theta(t, y))\right] - U\theta(t, y), \quad (6.15)$$

$$\frac{\partial}{\partial z}\phi_k(t, x, y, 0\pm) = -k\frac{\partial\phi_{k-1}(t, x, y, 0\pm)}{\partial x}\theta(t, y) \quad \text{for } k \geq 2. \quad (6.16)$$

Kutta–Joukowsky Conditions

It is important to note that we need the Kutta–Joukowsky conditions in the nonlinear case as well for uniqueness of solution.

For this we need to consider as before the acceleration potential first.
We have

$$\psi(.) = \frac{\partial\phi}{\partial t} + \frac{1}{2}\nabla\phi \cdot \nabla\phi\psi,$$

$$(\lambda, t, x, y, z) = \sum_{k=1}^{\infty}\frac{\lambda^k}{k!}\frac{\partial\phi_k}{\partial t} + \frac{U^2}{2} + U\sum_{k=1}^{\infty}\frac{\lambda^k}{k!}\frac{\partial\phi_k}{\partial x} + \sum_{k=1}^{\infty}\sum_{j=1}^{\infty}\frac{\lambda^j\lambda^k}{j!k!}q_k \cdot q_j.$$

With

$$\psi_k(t, x, z) = \frac{\partial^k\psi(0, t, x, z)}{\partial\lambda^k},$$

we have the expansion

$$\psi(\lambda, t, x, z) = \frac{U^2}{2} + \sum_{k=1}^{\infty}\frac{\lambda^k}{k!}\psi_k(t, x, z), \quad (6.17)$$

$$\psi_k(t, x, z) = \frac{\partial\phi_k}{\partial t} + U\frac{\partial\phi_k}{\partial x} + \sum_{j=1}^{k-1}C_{k,j}q_{k-j} \cdot q_j, \quad (6.18)$$

where $C_{k,j}$ are the binomial coefficients.

$$\delta\psi(\lambda, t, x, 0) = \sum_{k=1}^{\infty}\frac{\lambda^k}{k!}\delta\psi_k(t, x, 0), \quad (6.19)$$

$$\delta\psi(\lambda, t, x, 0) = 0 \quad |x| > b$$

implies that

$$\delta\psi_k(t, x, 0) = 0 \quad \text{for } |x| > b \text{ for every } k.$$

Next define: (Kussner doublet function)

$$A(t, x) = -\delta\psi(t, x, 0)/U \tag{6.20}$$

and for every k:

$$A_k(t, x, y) = -\delta\psi_k(t, x, 0)/U. \tag{6.21}$$

Then

$$(-U)A_1(t, x) = \delta\psi_1(t, x, 0) = \frac{\partial\delta\phi_1}{\partial t} + U\frac{\partial\delta\phi_1}{\partial x}. \tag{6.22}$$

And from the boundary conditions

$$\frac{\partial\delta\phi_1}{\partial t} = 2\frac{\partial\phi_1(t, x, 0+)}{\partial t},$$

$$\frac{\partial\delta\phi_1}{\partial x} = 2\frac{\partial\phi_1(t, x, 0+)}{\partial x}.$$

As in the time invariant case we consider next

Flow Decomposition

The next important step is to take advantage of the fact that (6.14) is a linear nonhomogeneous equation with nonzero boundary conditions and decompose the potentials as

$$\phi_k = \phi_{L,k} + \phi_{0,k},$$

where $\phi_{0,k}$ satisfies the nonhomogeneous equation (6.14),

$$\Xi(\phi_{0,k}) = g_{k-1} \quad -\infty < x, z < \infty \tag{6.23}$$

(where g_{k-1} depends of course on ϕ_j, $j \leq k-1$) and there no discontinuities on $z = 0$. Although $\phi_{L,k}$ satisfies the homogeneous equation,

$$\Xi(\phi_{L,k}) = 0 \tag{6.24}$$

except on $z = 0$ where it satisfies the boundary conditions given above. Note in particular that $\phi_1 = \phi_{L,1}$.

Furthermore, from the field equation

$$\frac{\partial^2 \phi_{L,k}}{\partial t^2} + 2U \frac{\partial^2 \phi_{L,k}}{\partial t \partial x} - \left[a_\infty^2 (1 - M^2) \frac{\partial^2 \phi_{L,k}}{\partial x^2} + a_\infty^2 \frac{\partial^2 \phi_{L,k}}{\partial z^2} \right] = 0,$$

we deduce that

$$\phi_{L,k}(t, x, -z) = -\phi_{L,k}(t, x, z) \qquad z > 0,$$

because we require that

$$\frac{\partial}{\partial z} \phi_{L,k}(t, x, 0-) = \frac{\partial}{\partial z} \phi_{L,k}(t, x, 0+). \tag{6.25}$$

Hence

$$\frac{\partial}{\partial x} \phi_{L,k}(t, x, 0-) = -\frac{\partial}{\partial x} \phi_{L,k}(t, x, 0+). \tag{6.26}$$

Hence from (6.27)

$$\delta \left(\frac{\partial \phi_{L,k}}{\partial x} \right)^2 = 0.$$

In as much as

$$\delta \left(\frac{\partial \phi_{L,k}}{\partial z} \right)^2 = \delta \left[\left(\frac{\partial \phi_{L,k}(., 0+)}{\partial z} + \frac{\partial \phi_{L,k}(., 0-)}{\partial z} \right) \right.$$

$$\left. \cdot \left(\frac{\partial \phi_{L,k}(., 0+)}{\partial z} - \frac{\partial \phi_{L,k}(., 0-)}{\partial z} \right) \right],$$

we have

$$\delta \left(\frac{\partial \phi_{L,k}}{\partial z} \right)^2 = 0 \text{ by (6.26)}.$$

Hence we define

$$v_k(t, x) = \frac{\partial}{\partial z} \phi_{L,k}(t, x, 0).$$

Next we have the crucial result.

Theorem 6.2.

$$(-U)A_k(t,x,y) = \delta\psi_k(t,x,y,0) = \partial\delta\phi_{L,k}/\partial t + U\,\partial\delta\phi_{L,k}/\partial x$$

$$\text{for every } k \geq 1. \qquad (6.27)$$

Proof. From (6.18) we have

$$\delta\psi_k(t,x,y,0) = \partial\delta\phi_k/\partial t + U\,\partial\delta\phi_k/\partial x + \sum_{j=1}^{k-1} C_{k,j}\delta(q_{k-j}\cdot q_j).$$

And hence we need only to prove that

$$\sum_{j=1}^{k-1} C_{k,j}\delta(q_{k-j}\cdot q_j) = 0, \qquad (6.28)$$

or

$$\delta(q_{k-j}\cdot q_j) = 0.$$

Now the left side

$$= q_{k-j}(.,0+)\cdot q_j(.,0+) - q_{k-j}(.,0-)\cdot q_j(.,0-). \qquad (6.29)$$

Let v_k denote $\partial\phi_k/\partial z \gamma_k$ denote $\partial\phi_k/\partial x$.
Then (6.29),

$$= v_{k-j}(.,0+)v_j(.,0+) - v_{k-j}(.,0-)v_j(.,0-)$$

$$+ (\gamma_{k-j}(.,0+)\gamma_j(.,0+) - \gamma_{k-j}(.,0-)\gamma_j(.,0-)).$$

And

$$v_{k-j}(.,0+)v_j(.,0+) - v_{k-j}(.,0-)v_j(.,0-)$$

$$= (v_{k-j}(.,0+) + v_{k-j}(.,0-))v_j(.,0+)$$

$$- (v_j(.,0-) + v_j(.,0+))v_{k-j}(.,0-)$$

$$= 0 \text{ by (6.25),}$$

and

$$(\gamma_{k-j}(.,0+)\gamma_j(.,0+) - \gamma_{k-j}(.,0-)\,\gamma_j(.,0-))$$

$$= (\gamma_{k-j}(.,0+) - \gamma_{k-j}(.,0-))\gamma_j(.,0+) + (\gamma_j(.,0+) - \gamma_j(.,0-))\gamma_{k-j}(.,0-),$$

which by (6.26)= 0.
Hence (6.29) holds, proving the theorem.   □

Remarks: The importance of (6.27) is, among others, that the formula for the nonlinear case is the same as for the linear case. We have seen this already in the time invariant case in Chap. 4 for zero angle of attack. In particular we use it to calculate the pressure jump:

$$\delta p = -\rho_\infty \delta\psi; \delta\psi = \sum_{k=1}^{\infty} \lambda^k / k! (\partial\delta\phi_k/\partial t + U\partial\delta\phi_k\partial x).$$

Hence the boundary conditions satisfied by $\phi_{L,k}$ are:

$$\frac{\partial}{\partial z}\phi_{L,k}(t, x, 0\pm) = -k\frac{\partial\phi_{L,k-1}(t, x, 0\pm)}{\partial x}\theta(t, y) \quad \text{for } k \geq 2, \qquad (6.30)$$

$$A_k = -\delta\psi_{L,k}/U$$

$$= -(1/U)\partial\delta\phi_k/\partial t + \partial\delta\phi_k/\partial x$$

and

$$A = \sum_{k=1}^{\infty} A_k/k! \qquad (6.31)$$

Note in particular that

$$\delta\phi_k(t, x) = 0, \quad |x| > b \text{ for every } k. \qquad (6.32)$$

To proceed further we need to determine the time domain functions $A_k(.,.)$ in $\mathcal{X}$ which we show satisfy the linear time domain Possio equation.

The Linear Time Domain Possio Equation

Let us recall that we began with the field equation for ϕ_1:

$$\partial^2\phi_1/\partial t^2 + 2U\partial^2\phi_1/(\partial t\ \partial x) - [a_\infty^2(1 - M^2)\partial^2\phi_1/\partial x^2 + a_\infty^2\partial^2\phi_1/\partial z^2] = 0$$

with the flow tangency boundary condition and the Kutta–Joukowsky conditions.

The solution technique was to take the Laplace transform in the time domain and the Fourier transform in the space domain and obtain the Laplace transform version of the Possio integral equation. Showing that the solution is the Laplace Transform of a time domain function can be nontrivial—see [10, 41]. Indeed the basic problem of when a function analytic in a right-half-plane is the Laplace transform of a time domain function is still largely an open question. For $M = 0$, for example, we had to invoke a special result due to Sears.

Here we obtain the time domain version by formally inverting the Laplace domain equation. This, it should be noted, would not be possible without the use of the function $a(M, s)$ introduced in Chap. 5. We begin with the version:

$$2\hat{W}_a(\lambda, .) = \sqrt{(1 - M^2)}\mathbb{P}\mathbb{H}(I + \mathbb{B}(k))\hat{A}(\lambda, .), \tag{6.33}$$

where the operator $\mathbb{B}(k)$, it will be recalled, is defined by

$$\mathbb{B}(k)A = (-k\mathbb{R}(k, \mathbb{D})A)/(\sqrt{(1 - M^2)}) - \int_{-\alpha_2}^{\alpha_1} k\mathbb{R}(ks, \mathbb{D})A\, a(M, s)\mathrm{d}s,$$

$$\hat{A}(\lambda, x) = \int_0^\infty e^{-\lambda t} A(t, x)\mathrm{d}t \qquad \text{Re. } \lambda > \sigma_a, \quad |x| < b,$$

where $A(., .)$ is in $\mathcal{X}$. Here

$$R(k, D)\hat{A}(\lambda, .) \text{ is the function in } L_p(R^1):$$

$$\int_{-1}^x e^{-\frac{\lambda}{U}(x-\xi)}\hat{A}(\lambda, \xi)\mathrm{d}\xi, \quad -\infty < x < \infty,$$

whose inverse Laplace transform is

$$\int_{-1}^x \int_0^t \delta(\tau - (x - \xi)/U)A(t - \tau, \xi)\mathrm{d}\tau\mathrm{d}\xi$$

$$= \int_0^t \int_{-1}^x \delta(\tau - (x - \xi)/U)A(t - \tau, \xi)\mathrm{d}\xi\mathrm{d}\tau$$

$$= \int_0^t A(t - \sigma, x - U\sigma)\mathrm{d}\sigma. \tag{6.34}$$

We note the mixing up here between the time and space variables.
    Hence $R(k, D)\hat{A}(\lambda, .)$ is the Laplace transform of

$$\int_0^t S(U\sigma)A(t - \sigma, .)\mathrm{d}\sigma.$$

Next we need to take the inverse transform of $kR(k, D)\hat{A}(\lambda, .)$. By the usual rules the candidate would be

$$\mathrm{d}/\mathrm{d}t \int_0^t S(U\sigma)A(t - \sigma, .)\mathrm{d}\sigma \qquad \text{in } 0 < t, \tag{6.35}$$

which if $A(t,.)$ is in the domain of $\mathbb{D}$ for $t > 0$ and $\mathbb{D}A(t,.)$ is in $\mathcal{X}$ can be defined as:

$$= A(t) + U \int_0^t \mathbb{P}S(Ut - U\sigma)\mathbb{D}A(\sigma,.)d\sigma, \qquad 0 < t$$

and would be ok. But in our case, we simply cannot assume that $A(t,.)$ is in the domain of $\mathbb{D}$. Instead we require that $A(t)$ be absolutely continuous in t with derivative in $\mathcal{X}$.

We can then state the time domain Possio integral (TDP) as an equation in $\mathcal{X}$:

$$2/(\sqrt{(1 - M^2)})w_a(t, x) = 1/\pi \int_{-b}^{b} 1/(x - \xi)\Big[A(t, \xi)$$

$$- \partial/\partial t \int_0^t ((A(t - \sigma, \xi - U\sigma))/(\sqrt{(1 - M^2)})$$

$$+ \int_{-\alpha_2}^{\alpha_1} a(M, s)A(t - \sigma, \xi - U\sigma/s)ds/s)d\sigma\Big]d\xi,$$

$$|x| < b.$$

And the TDP in abstract form is:

$$w_a(t,.) = (\sqrt{(1 - M^2)})/2\mathbb{P}\mathbb{H}\Big[A_1(t,.)$$

$$- d/dt \int_0^t \Big[(\mathbb{P}S(U(t - \sigma))\mathbb{P}A_1(\sigma,.))/(\sqrt{(1 - M^2)})$$

$$+ \int_{-\alpha_2}^{\alpha_1} \mathbb{P}S((U(t - \sigma))/s)\mathbb{P}A_1(\sigma,.)a(M, s)ds/s\Big]\Big]d\sigma, \qquad (6.36)$$

where the sense in which the integral is defined to take care of the singularity at $s = 0$ needs elaboration:

The integral

$$\int_0^t \int_{-\alpha_2}^{\alpha_1} \mathbb{S}((U(t - \sigma))/s)A_1(\sigma,.)a(M, s)ds/s\, d\sigma$$

is interpreted in the Cauchy sense:

$$\text{Lim } \epsilon \to 0$$

$$\int_0^t \int_{-\epsilon}^{\epsilon} \mathbb{S}((U(t-\sigma))/s)A_1(\sigma,.)a(M,s)ds/s\,d\sigma$$

$$+ \frac{M}{\pi\sqrt{(1-M^2)}} \int_0^t \int_{\epsilon}^{\alpha_1} \mathbb{S}(U\sigma/s)A_1(t-\sigma,.)ds/s\,d\sigma$$

$$+ \frac{M}{\pi\sqrt{(1-M^2)}} \int_0^t \int_{-\alpha_2}^{-\epsilon} \mathbb{S}(U\sigma/s)A_1(t-\sigma,.)ds/s\,d\sigma,$$

so that

$$\int_0^t 1/\pi \int_{-\alpha_2}^{\alpha_1} \mathbb{S}(U\sigma/s)A_1(t-\sigma,.)ds/s\,d\sigma$$

$$= 1/\pi \int_0^{\alpha_1} ds \int_0^{t/s} 1/\tau(\mathbb{S}(U\tau)-\mathbb{S}(-U\tau))A_1(t-\tau s,.)d\tau$$

$$= \int_0^{\alpha_1} \left(HA_1(t) - 1/\pi \int_{t/s}^{\infty} 1/\tau(\mathbb{S}(U\tau)-\mathbb{S}(-U\tau))(A_1(t)-A_1(t-\tau s))dt \right) ds$$

$$= \mathbb{PH}A(t,.) - d/dt \int_0^t \mathbb{PH}\mathcal{B}(t-\sigma)A(\sigma)d\sigma,$$

where

$$\mathcal{B}(t)f = (\mathbb{S}(Ut)f)/(\sqrt{(1-M^2)}) + \int_{-\alpha_2}^{\alpha_1} \mathbb{S}(Ut/s)fa(M,s)ds/s.$$

Next let us denote the Possio equation by

$$\Phi(M)A = w$$

corresponding to the linear Possio equation:

$$\left(\sqrt{(1-M^2)}\right)/(2\pi) \int_{-b}^{b} 1/(x-\xi)\left[A(t,\xi) - \partial/\partial t \int_0^t \left((A(t-\sigma,\xi-U\sigma)) \right.\right.$$

$$\left./(\sqrt{(1-M^2)}) + \int_{-\alpha_1}^{\alpha_1} (a(M,s)-a(M,0))A(t-\sigma s,\xi-U\sigma)ds \right.$$

$$\left.\left. + \int_{\alpha_1}^{\alpha_2} a(M,-0)A(t-\sigma s,\xi-U\sigma)ds \right)d\sigma \right]d\xi$$

$$= w_a(t,.) \tag{6.37}$$

with domain in $\mathcal{X}$ and range in $\mathcal{X}$, with the Kutta condition in addition (implicit everywhere, if not specified explicitly otherwise).

If we assume the time domain version has a solution, then of course we may take the Laplace transform which will then satisfy the LDP.

Let us explore next how far we can go in terms of solving the time domain Possio (TDP), inverting the Laplace Transform versions in Chap. 5.

Solving the Linear TDP Equation

We denote the solution of the Linear TDP—when it has a unique solution—by $A = \Psi(M)w_a$ where we have seen that in Sect. 5.5 the Laplace transform of the solution has the form

$$\hat{A} = (I + \mathbb{P}\mathbb{B}(k)\mathbb{P} + W(k))^{-1}2/(\sqrt{(1 - M^2)})T\hat{w}_a. \tag{6.38}$$

Or of the form (5.48)

$$0 < M < 1 : \hat{A} = \left(I + \left(\frac{\sqrt{(1 - M^2)}}{M}\right)\mathbb{P}\mathbb{H}(\mathbb{B}(k) - \mathbb{B}(\infty))\right)^{-1}\frac{2}{M}\hat{w}_a. \tag{6.39}$$

We begin with the version (6.38).

Lemma 6.3. *For A in $L_p[-b,b], 1 < p < 2$: $(I + \mathbb{P}\mathbb{B}(k)\mathbb{P} + W(k))A$ is the Laplace transform:*

$$= \int_0^\infty e^{-kt}d/dt\,\mathcal{P}(\mathcal{B}(t) + \mathcal{W}(t))A\,dt,$$

where $\mathcal{P}\mathcal{W}(t)A$ is the function $1/\pi\sqrt{((b - x)/(b + x))}$

$$\left[1/(\sqrt{(1 - M^2)})\int_0^t A(b - t + \sigma)d\sigma\right.$$

$$+ \int_0^t \int_0^{\alpha_1} a(M, s)A(b - (t - \sigma)/s)\sqrt{(\sigma/(2bs + \sigma))}ds/((b - x)s + \sigma)$$

$$+ a(M, -s)A((t - \sigma)/s - b)\sqrt{((2bs + \sigma)/\sigma)}ds/((b + x)s + \sigma)$$

$$+ \int_{\alpha_1}^{\alpha_2} a(M, -s)A((t - \sigma)/s - b)\sqrt{((2bs + \sigma)/\sigma)}ds/((b + x)s + \sigma)d\sigma\right]$$

$$+ \int_0^{\alpha_2} A(t/s - (2b + x))a(M, -s)ds, \qquad |x| < b$$

and

$$PB(t)PA = -\left[1/(\sqrt{(1-M^2)})PS(t)PA + \int_{-\alpha_2}^{\alpha_1} PS(t/s)PAa(M,s)ds/s\right].$$

Proof. To derive

$$W(k)A = \int_0^\infty e^{-kt}\dot{W}(t)Adt$$

□

we use

$$1/(\sqrt{(1-M^2)})h_-(k)L_-(k)$$

$$= \int_0^\infty e^{-kt}dt\ 1/(\sqrt{(1-M^2)})\left[1/\pi\sqrt{((b-x)/(b+x))}\right.$$

$$\left.\times \int_{-b}^b \sqrt{((t-b+\xi)/(t+b+\xi))}1/(t-(x-\xi))\right]A(\xi)d\xi,$$

$$L_+(ks,A)h_+(ks,x)$$

$$= \int_0^\infty e^{-kt}\int_{-b}^b A(\xi)d\xi\int_0^t \delta(t-\sigma-s(b+\xi))1/\pi$$

$$\times \sqrt{((b-x)/(b+x))}\sqrt{(\sigma/(2bs+\sigma))}d\sigma/((b+x)s+\sigma)dt$$

$$= \int_0^\infty e^{-kt}\int_{-b}^b A(\xi)d\xi\ 1/\pi\sqrt{((b-x)/(b+x))}$$

$$\times \sqrt{((t-s(b+\xi))/(2bs+t-s(b+\xi)))}1/((b+x)s+t-s(b+\xi))$$

$$= \int_0^\infty e^{-kt}\int_{-b}^b A(\xi)d\xi\ 1/\pi\sqrt{((b-x)/(b+x))}$$

$$\times \sqrt{((t-s(b+\xi))/(t+s(b-\xi)))}dt/(t+s(x-\xi)).$$

And following the notation in Chap. 5

$$\int_0^{\alpha_2} L_+(ks,A)h_+(ks,x)a(M,-s)ds$$

$$= \int_0^\infty e^{-kt}dt\int_{-b}^b A(\xi)d\xi\int_0^{\alpha_2} 1/\pi\sqrt{((b-x)/(b+x))}$$

$$\times \sqrt{((t-s(b+\xi))/(t+s(b-\xi)))}(a(M,-s)ds)/(t+s(x-\xi))$$

$$= \int_0^\infty e^{-kt}dt\ 1/\pi\sqrt{((b-x)/(b+x))}\int_{-b}^b A(\xi)d\xi$$

$$\times \int_0^{\alpha_2} \sqrt{((t-s(b+\xi))/(t+s(b-\xi)))}(a(M,-s)ds)/(t+s(x-\xi)),$$

and

$$\int_0^{\alpha 1} L_-(ks, A)h_-(ks, x)a(M, s)ds$$

$$= \int_0^\infty e^{-kt} dt \left[1/\pi \sqrt{((b-x)/(b+x))} \int_{-b}^b \int_0^{\alpha_1} \sqrt{((t-sb+s\xi)} \right.$$

$$\left. \Big/ (t+sb+s\xi))(a(M,s)ds)/(t-s(x-\xi)) \right] A(\xi)d\xi.$$

Using

$$h_-(ks, x) = \int_0^\infty e^{-kt} 1/\pi \sqrt{((b-x)/(b+x))} \sqrt{(t/(2bs+t))} dt/((b-x)s+t),$$

$$L_-(ks, A) = \int_0^\infty e^{-kt} dt \int_{-b}^b \delta(t - s(b-\xi)) A(\xi)d\xi$$

and

$$j(ks)L_+(ks, A) = \int_{-b}^b e^{ks(b+x)} e^{-ks(b+\xi)} A(\xi)d\xi$$

$$= \int_{-b}^b e^{-ks}(\xi - x) A(\xi)d\xi$$

$$= \int_0^\infty e^{-kt} dt \int_{-b}^b \delta(t - s(\xi - x)) A(\xi)d\xi,$$

we have

$$\int_0^{\alpha_2} j(ks)L_+(ks, A)a(M, -s)ds$$

$$= \int_0^\infty e^{-kt} dt \int_{-b}^b \int_0^{\alpha_2} \delta(t - s(\xi - x))a(M, -s)ds \, A(\xi)d\xi.$$

Hence $\mathcal{W}(t)A$ is the function

$$= \int_{-b}^b \int_0^{\alpha_2} \delta(t - s(\xi - x))a(M, -s)ds \, A(\xi)d\xi$$

$$+ 1/\pi \sqrt{((b-x)/(b+x))} \int_{-b}^b A(\xi)d\xi$$

$$\int_0^{\alpha_2} \sqrt{((t - s(b + \xi))/(t + s(b - \xi)))}$$

$$\times (a(M, -s)ds)/(t + s(x - \xi)) + 1/\pi \sqrt{((b - x)/(b + x))}$$

$$\int_{-b}^{b} \int_0^{\alpha_1} \sqrt{((t - sb + s\xi)/(t + sb + s\xi))} (a(M, s)ds)/(t - s(x - \xi)) A(\xi)d\xi$$

$$+1/(\sqrt{(1 - M^2)})1/\pi \sqrt{((b - x)/(b + x))}$$

$$\int_{-b}^{b} \sqrt{((t - b + \xi)/(t + b + \xi))}1/(t - (x - \xi)) A(\xi)d\xi.$$

Next for A in $L_p[-b, b]$,

$$\mathbb{P}\mathcal{B}(k)\mathbb{P}A = \int_0^\infty e^{-kt} \mathbb{P}\mathcal{B}(t) \mathbb{P}A dt,$$

$$\mathbb{P}\mathcal{B}(t)\mathbb{P}A = -1/(\sqrt{(1 - M^2)})\mathbb{P}\mathbb{S}(t)\mathbb{P}A - \int_{-\alpha_2}^{\alpha_1} \mathbb{P}\mathbb{S}(t/s)\mathbb{P}A\, a(M, s)ds/s.$$

Hence $d/dt\, \mathcal{P}(\mathcal{B}(t) + \mathcal{W}(t))A$ is given by

$$d/dt \int_0^t P\mathcal{B}(t - \sigma) PA(\sigma, .)d\sigma$$

$$+ \left(d/dt \int_0^t d\sigma \int_{-b}^{b} \int_0^{\alpha_2} \delta(\sigma - s(\xi - x))a(M, -s)ds A(t - \sigma, \xi)d\xi\right)$$

$$+d/dt \int_{-b}^{b} \int_0^{\alpha_2} a(M, -s)A(t - s(\xi - x), \xi)ds d\xi$$

$$+1/\pi \sqrt{((b - x)/(b + x))}d/dt \int_0^t \int_{-b}^{b} \int_0^{\alpha_2} \sqrt{((\sigma - s(b + \xi))/(\sigma + s(b - \xi)))}$$

$$\times (a(M, -s)ds)/(\sigma + s(x - \xi)) A(t - \sigma, \xi)d\xi$$

$$+1/\pi \sqrt{((b - x)/(b + x))}d/dt \int_0^t d\sigma \int_{-b}^{b} \int_0^{\alpha_1} \sqrt{((\sigma - sb + s\xi)/(\sigma + sb + s\xi))}$$

$$\times (a(M, s)ds)/(\sigma - s(x - \xi)) A(t - \sigma, \xi)d\xi$$

$$+1/(\sqrt{(1 - M^2)})1/\pi \sqrt{((b - x)/(b + x))}d/dt$$

$$\times \int_0^t d\sigma \int_{-b}^{b} \sqrt{((\sigma - b + \xi)/(\sigma + b + \xi))}1/(\sigma - (x - \xi)) A(t - \sigma, \xi)d\xi. \qquad \square$$

$$(6.40)$$

Hence

$$(I + \mathbb{P}\mathbb{B}(k)\mathbb{P} + W(k))\hat{w}(k, .)$$
$$= \int_0^\infty e^{-kt} \left[w(t, .) + d/dt \int_0^t (P\mathcal{B}(\tau)P + W(\tau))w(t - \tau, .)d\tau \right] dt.$$

$$(6.41)$$

Hence the TDP can be expressed:

$$A(t, .) + d/dt \int_0^t (P\mathcal{B}(\tau)P + W(\tau))A(t - \tau, .)d\tau = \frac{2}{\sqrt{1 - M^2}} T w_a(t, .). \quad (6.42)$$

Theorem 6.4. *The TDP (6.42) has at most one solution. Any solution of the TDP (6.42) will satisfy the Kutta condition.*

Proof. The Laplace transform of any solution satisfies the LDP which we have shown has a unique solution for small enough λ (Theorem 5.16).

Furthermore we have shown that the Laplace transform satisfies the Kutta condition. □

Finally, from the form (6.39):

$$\left(I + \frac{\sqrt{(1 - M^2)}}{M} \mathbb{P}\mathbb{H}(\mathbb{B}(k) - \mathbb{B}(\infty)) \right) \hat{A}(\lambda, .) = \frac{2}{M} \hat{w}(\lambda, .),$$

we have (strong) limit Re $\lambda \to \infty (I + (\sqrt{(1 - M^2)})/M \mathbb{P}\mathbb{H}(\mathbb{B}(k) - \mathbb{B}(\infty))) = I$. Hence the time domain transform has a delta function at the origin for nonzero M. (For $M = 0$ we also have the delta function derivative. Thus $M = 0$ is special, not obtained as the 1imit for small M.) Hence, if the TDP has a solution for $M > 0$, it is of the form

$$A(t, .) = \frac{2}{M} [w_a(t) + \int_0^t P(M, t - \sigma)w_a(\sigma, .)d\sigma], \quad (6.43)$$

where the Laplace transform:

$$\int_0^\infty e^{-kt} P(M, t)wdt = (I + (\sqrt{(1 - M^2)})/M \mathbb{P}\mathbb{H}(\mathbb{B}(k) - \mathbb{B}(\infty)))^{-1} - I$$

and $P(M, .)$ does not have a delta function at the origin. This is useful to us below in isolating the "non-circulatory" terms in the aeroelastic structure equations.

It is interesting to consider what happens for $k = 0$, in (6.39). We have:

$$\left(I + \frac{\sqrt{(1 - M^2)}}{M}\mathbb{P}\mathbb{H}(-\mathbb{B}(\infty))\right)^{-1} - I$$

(because $\mathbb{B}(0) = 0$)

$$= M/(\sqrt{(1 - M^2)})\mathcal{T} - I,$$

which checks with the solution in the time invariant case

$$= -k\frac{\partial\phi_{L,k-1}(t, x, y, 0\pm)}{\partial x}\theta(t, y). \tag{6.44}$$

Let us start with $k = 2$.

The potential $\phi_1(.)$ is given by specializing (5.18) to the typical section and

$$\alpha = 0 : \quad \tilde{\tilde{\phi}}(\lambda, i\omega, z) = -1/2\tilde{\tilde{A}}(\lambda, i\omega)e^{rz}/(k + i\omega) \quad z > 0,$$

where

$$r = -\sqrt{(M^2k^2 + 2M^2ki\omega + (1 - M^2)\omega^2)}.$$

Hence the Laplace–Fourier transform of $-(\partial\phi_{L,1}(t, x, y, 0+))/\partial x$ is

$$= 1/2\tilde{\tilde{A}}_1(\lambda, i\omega)i\omega/(k + i\omega)$$

$$= 1/2(1 - k/(k + i\omega))\tilde{\tilde{A}}_1(\lambda, i\omega)$$

$$= 1/2(1 - \lambda/(\lambda + i\omega U))\tilde{\tilde{A}}_1(\lambda, i\omega).$$

Hence

$$-(\partial\phi_{L,1}(t, x, 0+))/\partial x = 1/2\left(A_1(t, x) - d/dt\int_0^t A_1(t - \sigma, x - U\sigma)d\sigma\right),$$

where we have used

$$1/(\lambda + iU\omega) = 1/U\int_0^\infty e^{-\frac{\lambda}{U}x}e^{-i\omega x}dx.$$

Hence

$$\partial\delta\phi_1/\partial x = -A_1(t, x) + d/dt\int_0^t A_1(t - \sigma, x - U\sigma)d\sigma. \tag{6.45}$$

For $k > 1$: we have again:

$$-A_k(t, x) = \frac{1}{U}\delta\phi_{L,k} + \frac{\partial}{\partial x}\delta\phi_L, \quad |x| < b, \tag{6.46}$$

we omit y being now only a fixed parameter.

Or

$$\partial \delta\phi_k / \partial t = -U \partial \delta\phi_k / \partial x - U A_k(t,x), \quad 0 < t, \ |x| < b$$
$$= -U \partial \delta\phi_k / \partial x \qquad |x| > b. \tag{6.47}$$

Or, solving the equation, we have

$$\delta\phi_k = -U \int_0^t \mathbb{S}(U\sigma) A_k(t - \sigma)d\sigma$$

and

$$\partial \delta\phi_k / \partial x = -A_k(t,x) + d/dt \int_0^t A_k(t - \sigma, x - U\sigma)d\sigma. \tag{6.48}$$

In (6.47), because

$$A(t,x) = 0 \quad \text{for } |x| > b,$$

we have that the integral is zero for $Ut > (b + x)$ and hence for $t > (2b)/U$

$$\partial \delta\phi_k / \partial x = -A_k(t,x) + d/dt \int_0^{(b+x)/U} A_k(t - \sigma, x - U\sigma)d\sigma. \tag{6.49}$$

We omit the subscript L from now on. We start with

$$\tilde{\phi}_k(\lambda, i\omega, z) = e^{rz}/r \ \ \tilde{v}(\lambda, i\omega, 0+) \qquad z > 0, \tag{6.50}$$

where

$$r = -\sqrt{(M^2\kappa^2 + 2M^2\kappa i\omega + (1 - M^2)\omega^2)},$$
$$\kappa = \lambda/U$$

and

$$\tilde{\phi}_k(\lambda, i\omega, z) = -e^{-rz}/r \ \ \tilde{v}_k(\lambda, i\omega, 0-) \qquad z < 0, \tag{6.51}$$

where

$$\partial/\partial z \hat{\phi}_k(\lambda, x, z) = \hat{v}_k(\lambda, x, z).$$

Hence with

$$\partial/\partial x \hat{\phi}_k(\lambda, x, z) = \hat{\gamma}_k(\lambda, x, z),$$

we have

$$\tilde{\gamma}_k(\lambda, i\omega, 0+) = i\omega/r \tilde{v}_k(\lambda, i\omega, 0+), \tag{6.52}$$

$$\tilde{\gamma}_k(\lambda, i\omega, 0-) = -i\omega/r \tilde{v}_k(\lambda, i\omega, 0-), \tag{6.53}$$

which generalize (6.61) and (6.62).

Again:

$$-A_k(t,x) = 1/U \, \partial \delta \phi_{L,k}/\partial t + \partial \delta \phi_{L,k}/\partial x \qquad |x| < b$$
$$= 0 \qquad |x| > b.$$

The transform version of which using (6.52) and (6.53) yields:

$$-\check{\tilde{A}}_k(\lambda, i\omega) = (\kappa + i\omega)\left(\check{\tilde{\phi}}_k(\lambda, i\omega, 0+) - \check{\tilde{\phi}}_k(\lambda, i\omega, 0-)\right)$$
$$= (\kappa + i\omega)/r\left(\check{\tilde{v}}_k(\lambda, i\omega, 0+) + \check{\tilde{v}}_k, (\lambda, i\omega, 0-)\right).$$

Or

$$\frac{1}{2}\frac{-r}{\kappa + i\omega}\check{\tilde{A}}_k(\lambda, i\omega) = 1/2\left(\check{\tilde{v}}_k(\lambda, i\omega, 0+) + \check{\tilde{v}}_k(\lambda, i\omega, 0-)\right). \qquad (6.54)$$

For $k > 1$, we have to respect the possible discontinuity in $v_k(t,x,z)$ at $z = 0$. Equation (6.54) is then the version of the linear Possio equation valid for $k \geq 2$, recalling that $(1-r)/(2\kappa + i\omega)$ is the multiplier corresponding to the time domain operator $\Phi(M)$ on χ into χ.

Hence we have the time domain equation in

$$\Phi(M)A_k = g_k,$$
$$g_k(t,x) = 1/2(v_k(t,x,0+) + v_k(t,x,0-)) \qquad |x| < b.$$

The right side by the boundary condition (6.31)

$$= 1/2(-k)\theta(t)(\gamma_{k-1}(t,x,0+) + \gamma_{k-1}(t,x,0+)) \quad \text{for } |x| < b.$$

The presence of the factor $\theta(t)$ which depends on t makes use of the Laplace transform difficult unlike the time invariant case in Chap. 4 where θ is a constant. Hence we need to work in the time domain from now on. But first we have the frequency domain relations.

By (6.52) and (6.53)

$$r/i\omega \check{\tilde{\gamma}}_{k-1}(\lambda, i\omega, 0+) = \check{\tilde{v}}_{k-1}(\lambda, i\omega, 0+). \qquad (6.55)$$

Similarly

$$(-r)/i\omega \check{\tilde{\gamma}}_{k-1}(\lambda, i\omega, 0-) = \check{\tilde{v}}_{k-1}(\lambda, i\omega, 0-). \qquad (6.56)$$

Hence, adding, we have:

$$r/i\omega\left(\check{\tilde{\gamma}}_k(\lambda, i\omega, 0+) - \check{\tilde{\gamma}}_k(\lambda, i\omega, 0-)\right) = \left(\check{\tilde{v}}_k(\lambda, i\omega, 0+) + \check{\tilde{v}}_k(\lambda, i\omega, 0-)\right),$$
$$(6.57)$$

where we note that

$$\delta \gamma_k(t, x) = (\gamma_k(t, x, 0+) - \gamma_k(t, x, 0-))$$
$$= 0 \quad \text{for } |x| > b.$$

And by (6.54)

$$\delta \check{\tilde{\gamma}}_k = \frac{i\omega}{r} \frac{-r}{\kappa + i\omega} \check{\tilde{A}}_k(\lambda, i\omega) = \frac{-i\omega}{\kappa + i\omega} \check{\tilde{A}}_k(\lambda, i\omega). \tag{6.58}$$

Hence

$$\delta \check{\tilde{\gamma}}_k = \left(-1 + \frac{\kappa}{\kappa + i\omega}\right) \check{\tilde{A}}_k(\lambda, i\omega).$$

Or in the time domain:

$$\delta \gamma_k(t, x) = -A_k(t, x) + d/dt \int_0^t A_k(t - \sigma, x - U\sigma) d\sigma, \qquad |x| < b$$
$$= 0 \quad |x| > b,$$

which is simply the solution of the partial differential equation (6.31).
We now define the time domain operator Γ on the subdomain of χ which is absolutely continuous in t, with derivative in χ, into χ by

$$\Gamma f = g; \quad g(t, x) = f(t, x) - d/dt \int_0^t f(t - \sigma, x - U\sigma) d\sigma, \qquad |x| < b.$$

Then we have

$$\delta \gamma_k = -\Gamma A_k. \tag{6.59}$$

Note that

$$\Gamma f = 0 \quad \text{implies } f = 0.$$

Hence we can define the inverse Γ^{-1} by

$$\Gamma^{-1} f = g; \quad g(t, x) = f(t, x) - \frac{1}{U} \frac{d}{dt} \int_{-b}^x f(t, s) ds, \tag{6.60}$$

on the same domain as that of Γ and in particular:

$$A_k(t, x) = \delta \gamma_k(t, x) + 1/U \int_{-b}^x d/dt \delta \gamma_k(t, s) ds, \qquad |x| < b,$$

where the necessary differentiability in t of $\delta \gamma_k(t, s)$ is assumed.

Also we need now to define the linear operator θ with domain and range in $\chi\,\theta f = h; h(t, x) = \theta(t)f(t, x), |x| < b$ where $\theta(t)$ is the given torsion angle that does not depend on the chord variable x.

We only need to consider the case where the torsion angle is bounded (actually < 1) that θ is actually linear bounded.

Then we have

$$2\Phi(M)A_k = -k\Theta\mathbb{P}\overline{\gamma}_{k-1}, \tag{6.61}$$

where we use

$$\overline{\gamma}_k(t, x) = 1/2(\gamma_k(t, x, 0+) + \gamma_k(t, x, 0-))$$

and similarly we use the notation:

$$\overline{v}_k(t, x) = (v_k(t, x, 0+) + v_k(t, x, 0-)).$$

Again, subtracting (6.42), from (6.41) we have:

$$\left(\check{\tilde{\gamma}}_{k-1}(\lambda, i\omega, 0+) + \check{\tilde{\gamma}}_{k-1}(\lambda, i\omega, 0-)\right)$$
$$= i\omega/r\left(\check{\tilde{v}}_{k-1}(\lambda, i\omega, 0+) - \check{\tilde{v}}_{k-1}(\lambda, i\omega, 0-)\right). \tag{6.62}$$

The boundary conditions yield

$$\delta v_{k-1}(t, x) = -(k-1)\theta(t)\delta\gamma_{k-2}(t, x), \qquad |x| < b,$$

$$\overline{v}_{k-1}(t, x) = -(k-1)\theta(t)\overline{\gamma}_{k-2}(t, x), \qquad |x| < b.$$

Now $i\omega/-r$ is a Mikhlin multiplier for each λ. In fact we have the following; see [55].

Lemma 6.5.

$$i\omega/(-r) = \int_{-\infty}^{\infty} e^{-i\omega x}\hat{L}(\kappa, x)dx, \qquad -\infty < \omega < \infty,$$

where

$$\hat{L}(\kappa, -x) = -\hat{L}(\kappa, x),$$

$$\hat{L}(\kappa, x) = \frac{-\kappa M}{\pi(1 - M^2)^{3/2}}K_1\left(\frac{\kappa M}{1 - M^2}x\right) \qquad x > 0$$

and the inverse Laplace transform

$$L(t,x) = \frac{-M}{\pi(1-M^2)^2} \frac{d}{dt}\left[t \Big/ \sqrt{\left(t^2 - \left(\frac{M}{1-M^2}\frac{x}{U}\right)^2\right)}\right],$$

$$t > \frac{M}{1-M^2}\frac{x}{U} = 0 \qquad 0 < t < \frac{M}{1-M^2}\frac{x}{U} \qquad x > 0,$$

$$L(t,-x) = -L(t,x).$$

Let $L(t)$ denote the bounded linear operator on $L_p(R_1)$ into itself defined by

$$L(t)f = g; \qquad g(x) = \int_{-\infty}^{\infty} L(t, x-s)f(s)ds, \qquad |x| < \infty.$$

And let $\mathcal{L}(M)$ be the corresponding operator on χ into χ defined by

$$\mathcal{L}(M)A = g; \qquad g(t,.) = \int_0^t \mathbb{P}L(t-\sigma)A(\sigma, M)d\sigma. \qquad (6.63)$$

Then we have:
$$\overline{\gamma}_{k-1} = -\mathcal{L}(M)(k-1)\Theta\delta\gamma_{k-2}. \qquad (6.64)$$

Next by (6.58) and (6.59),

$$-\frac{i\omega}{\kappa+i\omega}\check{A}_k(\lambda, i\omega) = \frac{i\omega}{r}\left(\check{v}_k(\lambda, i\omega, 0+) + \check{v}_k(\lambda, i\omega, 0-)\right).$$

Or

$$-\frac{r}{\kappa+i\omega}\check{A}_k(\lambda, i\omega) = \left(\check{v}_k(\lambda, i\omega, 0+) + \check{v}_k(\lambda, i\omega, 0-)\right).$$

Hence $2\Phi(M)A_k = \mathbb{P}\overline{v}_k = -k\Theta\mathbb{P}\overline{\gamma}_{k-1}$ which by $(6.64) = k\Theta\mathcal{L}(M)(k-1)\Theta\delta\gamma_{k-2}$. Hence using (6.61)

$$= k(k-1)\Theta\mathcal{L}(M)\Theta\Gamma A_{k-2}.$$

Hence we obtain

$$2\Phi(M)A_k = k(k-1)\Theta\mathcal{L}(M)\Theta\Gamma A_{k-2}.$$

Or under the assumption of unique solution to the Possio equation:

$$A_k = k(k-1)1/2\Psi(M)\Theta\mathcal{L}(M)\Theta\Gamma A_{k-2}, \qquad (6.65)$$

which is our recursive relation.

This generalizes the time invariant case in Chap. 4, where $1/2\Psi(M)\Theta\mathcal{L}(M)$ $\Theta\Gamma$ corresponds to multiplication by

$$-\mathcal{T}\mathbb{P}\mathbb{H}(\theta(y)^2)/(1 - M^2) = -(\theta(y)^2)/(1 - M^2).$$

An immediate inference from (6.65) is that

$$\Gamma A_{2k} = 0$$

and hence

$$A_{2k} = 0. \tag{6.66}$$

Let

$$J = \frac{1}{2}\Psi(M)\Theta\mathcal{L}(M)\Theta\Gamma. \tag{6.67}$$

Then (6.51) yields:

$$A_{2n+1}/(2n + 1)! = J^n A_1$$

and finally

$$A = \sum_{k=0}^{\infty} A_k/k! = \sum_{k=0}^{\infty} J^k A_1. \tag{6.68}$$

Assuming that

$$||J|| < 1. \tag{6.69}$$

Or

$$\text{Sup}_y |\theta(t, y)|$$

is sufficiently small we have

$$A = (I - J)^{-1}A_1. \tag{6.70}$$

Of course $\theta(t, y)$ is determined as a solution of the aeroelastic equation, but in practice it is indeed much less than one radian, so the convergence holds.

We can express (6.70) as the nonlinear version of the time domain Possio equation.

6.3   The Nonlinear Possio Integral Equation

This is an equation in $L_p[-b, b], 1 < p < 2, t > 0$ for determining the pressure doublet function $A(t, x, y)$, given the structure state variables, $h(t, y), \theta(t, y)$. In other words it replaces the aerodynamics insofar as the structure dynamics is concerned in isentropic flow. It is the nonlinear input–output relation for the system, the structure dynamics in air flow.

We begin with (6.70),

$$A - JA = A_1.$$

Hence

$$\Phi(M)A - \Phi(M)JA = w_a,$$

where

$$\Phi(M)J = \Theta \mathcal{L}(M)\Theta\Gamma.$$

Hence

$$\Phi(M)A - 1/2\Theta\mathcal{L}(M)\Theta\Gamma A = w_a. \tag{6.71}$$

Then the nonlinear time domain Possio equation is given by

$$\frac{\sqrt{(1-M^2)}}{2\pi} \int_{-b}^{b} \frac{1}{x-\xi} A(t,\xi) - \frac{\partial}{\partial t} \left[\int_0^t \Big((A(t-\sigma, \xi - U\sigma))/(\sqrt{(1-M^2)}) \right.$$

$$+ \int_{-\alpha_1}^{\alpha_1} (a(M,s) - a(M,0)) A(t - \sigma s, \xi - U\sigma) ds$$

$$+ \int_{\alpha_1}^{\alpha_2} a(M,0) A(t - \sigma s, \xi - U\sigma) ds \Big) d\sigma \bigg] d\xi$$

$$- 1/2\theta(t,y) \int_0^t \int_{-b}^b L(t - \sigma, x - s)\theta(\sigma, y) D(\sigma, s) ds d\sigma$$

$$+ d/dt(h(t,y) + (x - ab)\theta(t,y)) + U\theta(t,y) = 0, \tag{6.72}$$

where

$$D(t,x) = A(t,x) - \frac{d}{dt} \int_0^t A(t - \sigma, x - U\sigma) d\sigma \tag{6.73}$$

and

$$L(t,x) = (-M) \Big/ \Big(\pi(1-M^2)^2 \Big) \frac{d}{dt} \frac{t}{\sqrt{t^2 - \frac{Mx}{\square}}} t \Big/ \Big(\sqrt{t^2 - (M^2 x^2)}$$

$$\Big/ ((1-M^2)^2 U^2))), \quad \text{for } t > \frac{Mx}{U(1-M^2)}, \quad \text{and } = 0$$

$$0 < t < \frac{Mx}{U(1-M^2)} \qquad x > 0$$

$$L(t,-x) = -L(t,x).$$

Note that complicated as it is, this completely bypasses—the aerodynamic flow equation—the nonlinear Euler equation. We are not concerned with the air flow any more. Note also that the nonlinearity in $\theta(.,.)$ is "physically realizable."

Remarks: We have existence and uniqueness for the solution of (6.72) assuming the regular Possio equation has a unique solution and the torsion angle is small enough. The solution is given by (6.53).

We use the qualifier "nonlinear" to indicate that the solution is no longer linear in the structure state variables, $h(t, y), \theta(t, y)$.

6.4 Nonlinear Aeroelastic Dynamics

Having determined the pressure jump, we can now go on to full nonlinear aeroelastic dynamics: specialized to the typical section, zero angle of attack, $0 < M < 1$, and linear structure dynamics. We show first that the nonlinear aeroelastic dynamics formulates as a nonlinear convolution/evolution equation in a Hilbert space, in fact in the energy space we have already discussed in the previous chapters.

We begin with the series (6.68):

$$A = \Psi(M)w_a + \sum_{k=1}^{\infty} J^k \Psi(M)w_a. \qquad (6.74)$$

We have already studied the first term leading to the linear convolution evolution equation in Chap. 5 for nonzero M. Now we have an additional component. We verify first that there are no additional non-circulatory terms.

Lemma 6.6. *There are no non-circulatory terms in*

$$\sum_{k=1}^{\infty} J^k \Psi(M)w_a. \qquad (6.75)$$

Proof. We have only to note in

$$J = 1/2\Psi(M)\Theta\mathcal{L}(M)\Theta\Gamma, \qquad (6.76)$$

the Laplace transform of the kernel in $\mathcal{L}(M)$ given by

$$\hat{L}(k,.) \to 0 \quad \text{as Re.} \ k \to \infty$$

and therefore there are no delta functions in $L(t,.)$ and hence $J\Psi(M)w_a$ has no non-circulatory components and neither does the sum in (6.76). $\square$

Lift and Moment Calculation

We next calculate the lift and moment corresponding to A(.,.) given by (6.75).

Here we need to bring back the dependence on the structure span variable y. Thus we have:

$$w_a(t, x, y) = f_1(x)a_1(t, y) + f_2(x)a_2(t, y). \tag{6.77}$$

Similarly we need to consider the functions of the form: $\Psi(M)w_a$ as a function of x and y.

Hence we need to introduce the space C of functions of the form:

$$f(x, y) \quad |x| < b; \quad 0 < y < \ell$$

with range in the scalar field such that for each y, $0 < y < \ell$,

$$f(., y) \quad \text{is in } L_p(-b, b).$$

And

$$\int_0^\ell (\|f(., y)\|_p)^2 dy < \infty,$$

where $\|.\|_p$ denotes the norm in $L_p(-b, b)$. This is a Banach space with norm:

$$\sqrt{\int_0^\ell (\|f(., y)\|_p)^2 dy}.$$

Let $\mathcal{L}_1 = L_1[(0, \infty); \mathcal{H}_E]$ denote the space of functions such that $f(t, .)$ is in $\mathcal{H}_E$ for each $t \geq 0$ and

1. $\int_0^T \|f(t, .)\|_E dt < \infty;$ for each $T, 0 < T < \infty$
and Laplace transformable

2. $\int_0^\infty e^{-\sigma t} \|f(t, .)\|_E dt < \infty$ for $\sigma > \sigma_a \geq 0$.

We note that the function $w_a(t, .., y)$ defined in (6.4) is in $\mathcal{L}_1$. Define now on $L_p(-b, b)$:

$$\mathcal{L}_L(f) = U\rho_\infty \int_{-b}^b f(x) dx,$$

$$\mathcal{L}_M(f) = U\rho_\infty \int_{-b}^b (x - ab) f(x) dx.$$

Then for $f(.)$ in C, the functions

$$\mathcal{L}_L(f(t, .., y))\mathcal{L}_M(f(t, .., y)) \qquad 0 < t; \, 0 < y < \ell$$

are in $\mathcal{L}_1$.

Recall now the structure state vector

$$Y(t, y) = \begin{pmatrix} x\,(t, y) \\ \dot{x}\,(t, y) \end{pmatrix}, \qquad 0 < y < \ell$$

as an element of the energy space $\mathcal{H}_E$ for each $t \geq 0$, and also in $\mathcal{L}_1$. Then

$$w_a(t, x, y) = f_1 b_1^\star Y(t, y) + f_2 b_2^\star Y(t, y), \tag{6.78}$$

where b_1, b_2 are column vectors:

$$b_1^\star = [0, U, 1, -ab]; \qquad b_2^\star = [0, 0, 0, -1].$$

And hence

$$\Psi(M)w_a = \Psi(M)[(f_1 b_1^\star + f_2 b_2^\star)Y(., y)],$$

which yields a linear bounded operator denoted L_{12} on $\mathcal{L}_1$ into χ:

$$\Psi(M)w_a = L_{12}Y(., .). \tag{6.79}$$

Let

$$g(t, x, y), \qquad |x| < b, \, 0 < y < \ell$$

denote the function $L_{12}Y$. Then $\Gamma\Psi(M)w_a$ is the function

$$h(t, x, y) = g(t, x, y) - \mathrm{d}/\mathrm{d}t \int_0^t g(t - \sigma, x - U\sigma, y)\mathrm{d}\sigma,$$

where

$$\mathrm{d}/\mathrm{d}t \int_0^t g(t - \sigma, x - U\sigma, y)\mathrm{d}\sigma$$

$$= \int_0^{\frac{b+x}{U}} \dot{g}(t - \sigma, x - U\sigma, y)\mathrm{d}\sigma, \qquad t > \frac{b + x}{U}$$

$$= \int_{t - \frac{b+x}{U}}^t \dot{g}(\tau, x - U(t - \tau), y)\mathrm{d}\tau, \qquad t > \frac{b + x}{U}$$

and can be neglected for large t, compared to the first term. And in as much as our interest is primarily in the asymptotic response we omit it, and thus in what follows we set $\Gamma\Psi(M) = \Psi(M)$, so that

$$Jg = \frac{1}{2}\Psi(M)\Theta\mathcal{L}(M)\Theta L_{12}Y(.),$$

where now

$$\theta(t) = \Theta(t, y) = b^\star Y(t, y), \qquad b = \mathrm{col}[0, 1, 0, 0]$$

and

$$= \frac{1}{2}\Psi(M)(\Theta\mathcal{L}(M)\Theta L_{12}Y(.)) \qquad (6.80)$$

from now on. With this qualification let us evaluate Jg. We have: $\Theta\mathcal{L}(M)\Theta g$ is the function

$$h(t, ., y) = \Theta(t, y)\int_0^t L(t - \sigma)\Theta(\sigma, y)g(\sigma, ., y)d\sigma \quad \text{in } L_p(-b, b)$$

for each y which we can express as

$$= \int_0^t B_2(Y(t, y), Y(\sigma, y))L(t - \sigma)g(\sigma, ., y)d\sigma,$$

where $B_2(Y_1, Y_2)$ is the bilinear functional

$$b^\star Y_1 Y_2^\star b.$$

Next we need to consider $\Psi(M)h$ for h in C

We have already used this for $h = w_a$. Recall (6.67) $\Psi(M)h$ is the function:

$$\frac{2}{M}h + \int_0^t P(M, t - \sigma)h(\sigma, .)d\sigma,$$

where the first term is of importance in the first term in (6.75).

Hence we find it convenient now to express it as

$$\int_0^t \varphi(M, t - \sigma)h(\sigma, .y)d\sigma$$

allowing for a delta function in $\varphi(M, t)$. Thus Jg is the function

$$\frac{1}{2}\int_0^t \varphi(M, t - \sigma)\int_0^\sigma B_2(Y(\sigma, y), Y(\tau, y))L(\sigma - \tau)g(\tau, ., y)d\tau d\sigma \qquad (6.81)$$

valid for large t, because we are taking Γ as the identity, and is recognized as a second-degree polynomial in $Y(.,.)$ the structure state.

Also

$$J\Psi(M)w_a = \frac{1}{2}\int_0^t \varphi(M, t - \sigma)\int_0^\sigma B_2(Y(\sigma, y), Y(\tau, y))L(\sigma - \tau)$$

$$\times \int_0^\sigma \varphi(M, \tau - s)w_a(s, ., y)d\tau d\sigma. \tag{6.82}$$

And the corresponding lift

$$\mathcal{L}_L(Jg)\mathcal{L}_L\left(\frac{1}{2}\int_0^t \varphi(M, t - \sigma)B_2(Y(\sigma, y), Y(\tau, y))\int_0^\sigma L(\sigma - \tau)g(\tau, ., y)d\tau d\sigma\right)$$

can be expressed as

$$= \frac{1}{2}\int_0^t\int_0^\sigma \mathcal{L}_L(\varphi(M, t - \sigma)B_2(Y(\sigma, y), Y(\tau, y))L(\sigma - \tau)g(\tau, ., y))d\tau d\sigma. \tag{6.83}$$

And

$$\mathcal{L}_L(J\Psi(M)w_a(., ., y))$$

$$\mathcal{L}_L\left(\frac{1}{2}\Psi(M)\Theta\mathcal{L}(M)\Theta\Gamma\Psi(M)w_a(., ., y)\right),$$

where $\Psi(M)w_a(., ., y)$ is the function (fixing $M, 0 < M < 1$, in what follows in this chapter)

$$g_0(\tau, ., y) = \int_0^\tau \varphi(\tau - s)(f_1 b_1^\star + f_2 b_2^\star)Y(s, y)ds, \tag{6.84}$$

so that (6.83) becomes

$$\int_0^t\int_0^\sigma\int_0^\tau \frac{1}{2}\mathcal{L}_L(\varphi(t - \sigma)B_2(Y(\sigma, y), Y(\tau, y))$$

$$L(\sigma - \tau)\varphi(M, \tau - s)(f_1 b_1^\star + f_2 b_2^\star))Y(s, y)dsd\tau d\sigma.$$

More generally:

$$\mathcal{L}_L(J^k g_0), \quad k \geq 1, \tag{6.85}$$

$$\frac{1}{2} \int_0^t \int_0^\sigma \mathcal{L}_L(\varphi(t - \sigma) B_2(Y(\sigma, y), Y(\tau, y)) L(\sigma - \tau) g_{k-1}(\tau, ., y)) d\tau d\sigma \quad (6.86)$$

with $g_0(., ., .)$ given by (6.84).
Hence we can express:

$$g_{kL}(t, y) = \mathcal{L}_L(J^k \Psi(M) w_a(., ., y))$$

as

$$\frac{1}{2^k} \int_0^t \int_0^{\sigma_1} .. \int_0^{\sigma_{2k}} \mathcal{L}_L \left[\varphi(t - \sigma_1) b^\star Y(\sigma_1, y) b^\star Y(\sigma_2, y) L(\sigma_1 - \sigma_2) \cdot \right.$$

$$\varphi(\sigma_2 - \sigma_3) b^\star Y(\sigma_3, y) b^\star Y(\sigma_4, y) L(\sigma_3 - \sigma_4)$$

$$\cdots\cdots\cdots\cdots\cdots\cdots$$

$$\varphi(\sigma_{2k-2} - \sigma_{2k-1}) b^\star Y(\sigma_{2k-1}, y) b^\star Y(\sigma_{2k}, y) L(\sigma_{2k-1} - \sigma_{2k})$$

$$\left. \varphi(\sigma_{2k} - \sigma_{2k+1})(f_1 b_1^\star + f_2 b_2^\star) Y(\sigma_{2k+1}) \right] d\sigma_1 \ldots . d\sigma_{2k+1}$$

$$\cdots\cdots\cdots\cdots\cdots\cdots \quad (6.87)$$

Similarly we can calculate the moment

$$g_{kM}(t, y) = \mathcal{L}_M(J^k \Psi(M) w_a(., ., y))$$

$$\text{as} = \frac{1}{2^k} \int_0^t \int_0^{\sigma_1} .. \int_0^{\sigma_{2k}} \mathcal{L}_M(\varphi(t - \sigma_1) B_2(Y(\sigma_1, y), Y(\sigma_2, y)) L(\sigma_1 - \sigma_2) \cdot$$

$$\varphi(\sigma_2 - \sigma_3) B_2(Y(\sigma_3, y), Y(\sigma_4, y)) L(\sigma_3 - \sigma_4)$$

$$\cdots\cdots\cdots\cdots\cdots\cdots$$

$$\varphi(\sigma_{2k-2} - \sigma_{2k-1}) B_2(Y(\sigma_{2k-1}, y), Y(\sigma_{2k}, y)) L(\sigma_{2k-l} - \sigma_{2k})$$

$$\varphi(\sigma_{2k} - \sigma_{2k+1})(f_1 b_1^\star + f_2 b_2^\star) Y(\sigma_{2k+1})) d\sigma_1 \ldots d\sigma_{2k+1}. \quad (6.88)$$

This completes the calculation of the lift and moment. So we can proceed to the following.

The Nonlinear Convolution/Evolution Equation

We can now state the nonlinear convolution/evolution equation, generalizing the linear case in Chap. 5, using the same notation therein.
    We state this in the form of a theorem.

Theorem 6.7. *The dynamics of the structure under aerodynamic loading can be expressed as*

$$\dot{Y}(t) = \mathcal{A}Y(t) + \mathcal{N}(t, Y(\cdot)) \qquad t > 0, \tag{6.89}$$

where $Y(t)$ is the structure state at time t, $\mathcal{N}(t, Y(.))$ defines a physically realizable nonlinear convolution, and for $Y(.)$ such that for each $T, 0 < T < \infty$,

$$|\theta(t, y)| = |b^\star, y(t, y)|$$

is

$$\leq m(T) < \infty, \quad 0 \leq t \leq T, \quad 0 < y < \ell \tag{6.90}$$

is defined by the Volterra expansion: valid for $0 < t < T$

$$\mathcal{N}(t, Y(.)) = \int_0^t \mathcal{L}(t - \sigma)Y(\sigma)\mathrm{d}\sigma \tag{6.91}$$

$$+ \sum_{k=1}^{\infty} \int_0^t \int_0^{\sigma_1} \cdot \int_0^{\sigma_{2k}} \mathcal{L}_{2k}(t - \sigma_1, \sigma_1 - \sigma_2, ., \sigma_{2k} - \sigma_{2k+1};$$

$$Y(\sigma_1), .Y(\sigma_{2k+1})\mathrm{d}\sigma_1 \ldots \mathrm{d}\sigma_{2k+1},$$

where

$$\mathcal{L}_{2k}(t_1, t_2, ..t_{2k+1}, Y_1, y_2, ..Y_{2k+1}), \qquad Y_i \text{ in } R^4$$

$$= \begin{pmatrix} 0 \\ \mathcal{M}^{-1}\mathcal{F}_{2k}(t_1, t_2, \cdots t_{2k+1}; Y_1, Y_2, \cdots Y_{2k+1}) \end{pmatrix}, \tag{6.92}$$

where

$$\mathcal{F}_{2k}(t_1, t_2, \cdots .t_{2k+1}; Y_1, Y_2, .Y_{2k+1})$$

$$= \frac{1}{2^k} \begin{bmatrix} \mathcal{L}_L(\varphi(t_1)B_2(Y_1, Y_2)L(t_2).\varphi(t_3)B_2(Y_3, Y_4)L(t_4) \\ \cdots\cdots\cdots\cdots\cdots \\ \varphi(t_{2k-1})B_2(Y_{2k-1}, Y_{2k})L(t_{2k}) \\ \varphi(t_{2k+1})(f_1, b_1^\star + f_2, b_2^\star)Y_{2k+1} \\ \mathcal{L}_M(\varphi(t_1)B_2(Y_1, Y_2)L(t_2) \cdot \varphi(t_3)B_2(Y_3, Y_4)L(t_4) \\ \cdots\cdots\cdots\cdots\cdots \\ \varphi(t_{2k-1})B_2(Y_{2k-1}, Y_{2k})L(t_{2k}) \\ \varphi(t_{2k+1})(f_1, b_1^\star + f_2, b_2^\star)Y_{2k+1} \end{bmatrix}. \tag{6.93}$$

Proof. In what follows we fix T and so abbreviate $m(T)$ to simply m. We start with the linear part of (6.90) given in Sect. 5.6:

$$\dot{Y}(t) = AY(t) + \int_0^t \mathcal{L}(t - \sigma)Y(\sigma)d\sigma, \tag{6.94}$$

where

$$Y(t) = \begin{pmatrix} x\,(t) \\ \dot{x}\,(t) \end{pmatrix},$$

$$\mathcal{M} = \begin{pmatrix} m & S \\ S & I_\theta \end{pmatrix},$$

$$\mathcal{A} = \begin{pmatrix} 0 & I \\ -\mathcal{M}^{-1}(A + \mathcal{K}) & -\mathcal{M}^{-1}\mathcal{D} \end{pmatrix}.$$

$A, \mathcal{K}$, and $\mathcal{D}$ are as defined in Sect. 5.6.

$$\mathcal{L}(t)Y = \begin{pmatrix} 0 \\ \mathcal{M}^{-1}\mathcal{F}(t)B^*Y \end{pmatrix} \quad \text{for } Y \text{ in } R^4.$$

The nonlinear part is determined by the lift and moment calculated for the pressure doublet $A(.,.)$ determined as a solution of the nonlinear Possio equation (6.73).

Under condition (6.90) we can use the expansion (6.75) for the pressure doublet and our calculation of the corresponding lift and moment above.

This yields the kth term for $k \geq 1$ in the Volterra expansion (6.90):

$$\mathcal{N}(t, Y(.)) = \sum_{k=1}^{\infty} g_k(t, Y(.)), \tag{6.95}$$

$$g_k(t, Y(.)) = \int_0^t \int_0^{\sigma_1} \dots \int_0^{\sigma_{2k}} \mathcal{L}_{2k}(t - \sigma_1, \sigma_1 - \sigma_2, \dots \sigma_{2k} - \sigma_{2k+1}; \tag{6.96}$$

$$Y(\sigma_1), \dots (\sigma_{2k+1}))d\sigma_1 \dots d\sigma_{2k+1}, \tag{6.97}$$

where

$$\mathcal{L}_{2k}(t_1, t_2, \dots t_{2k+1}; Y_1, Y_2, \dots Y_{2k+1}), \qquad Y_i \text{ in } R^4$$

$$= \begin{bmatrix} 0 \\ \mathcal{M}^{-1}\mathcal{F}_{2k}(t_1, t_2, \dots t_{2k+1}; Y_1, Y_2 \dots Y_{2k+1}) \end{bmatrix}$$

and

$$\mathcal{F}_{2k}(t_1, t_2, \dots t_{2k+1}; Y_1, Y_2, \dots Y_{2k+1})$$

$$= \frac{1}{2^k} \begin{bmatrix} \mathcal{L}_L(\varphi(t_1)b^*Y_1b^*Y_2L(t_2) \cdot \varphi(t_3)b^*Y_3b^*Y_4)L(t_4) \\ \cdots\cdots\cdots\cdots\cdots\cdots \\ \varphi(t_{2k-1})(b^*Y_{2k-1}b^*Y_{2k})L(t_{2k}) \\ \varphi(t_{2k+1})(f_1b_1^\star + f_2b_2^\star)Y_{2k+1} \\ \mathcal{L}_M(\varphi(t_1)b^*Y_1b^*Y_2L(t_2) \cdot \varphi(t_3)b^*Y_3b^*Y_4L(t_4) \\ \cdots\cdots\cdots\cdots\cdots\cdots \\ \varphi(t_{2k-1})b^*Y_{2k-1}b^*Y_{2k}L(t_{2k}) \\ \varphi(t_{2k+1})(f_1,b_1^\star + f_2,b_2^\star)Y_{2k+1} \end{bmatrix}$$

and

$$g_k(t,y) = \begin{pmatrix} 0 \\ \mathcal{M}^{-1}\mathcal{F}_{2k}(t,y) \end{pmatrix}, \qquad (6.98)$$

where

$$\mathcal{F}_{2k}(t,y) = \begin{pmatrix} F_{2kL}(t,y) \\ F_{2kM}(t,y) \end{pmatrix}$$

$$F_{2kL}(t,y) = \frac{1}{2^k}\int_0^t \int_0^{\sigma_1} .. \int_0^{\sigma_{2k}} \mathcal{L}_L(\varphi(t-\sigma_1)b^*Y(\sigma_1,y)b^*Y(\sigma_2,y)L(\sigma_1-\sigma_2).$$

$$\varphi(\sigma_2-\sigma_3)b^*Y(\sigma_3,y)b^*Y(\sigma_4,y)L(\sigma_3-\sigma_4)$$

$$\cdots\cdots\cdots\cdots\cdots$$

$$\varphi(\sigma_{2k-2}-\sigma_{2k-1})$$

$$B_2(Y(\sigma_{2k-1},y),Y(\sigma_{2k},y))L(\sigma_{2k-1}-\sigma_{2k})$$

$$\varphi(\sigma_{2k}-\sigma_{2k+1})(f_1b_1^* + f_2b_2^*)Y(\sigma_{2k+1}))d\sigma_1\ldots.d\sigma_{2k+1},$$

$$F_{2kM}(t,y) = \frac{1}{2^k}\int_0^t \int_0^{\sigma_1} .. \int_0^{\sigma_{2k}} \ell_M(\varphi(t-\sigma_1)b^*Y(\sigma_1,y)b^*Y(\sigma_2,y)L(\sigma_1-\sigma_2).$$

$$\varphi(\sigma_2-\sigma_3)b^*Y(\sigma_3,y)b^*Y(\sigma_4,y)L(\sigma_3-\sigma_4)$$

$$\cdots\cdots\cdots\cdots$$

$$\varphi(\sigma_{2k-2}-\sigma_{2k-1})$$

$$b^*Y(\sigma_{2k-1},y)b^*Y(\sigma_{2k},y)L(\sigma_{2k-1}-\sigma_{2k})$$

$$\varphi(\sigma_{2k}-\sigma_{2k+1})(f_1b_1^* + f_2b_2^*)Y(\sigma_{2k+1}))d\sigma_1\ldots\ldots d\sigma_{2k+1}.$$

To prove convergence of the series in (6.91), we need to get bounds for the terms. Let us look at the case $k = 1$. We have

$$F_{2L}(t, y) = \ell_L\left[\frac{1}{2}\int_0^t \varphi(t - \sigma_1)\, b^*Y(\sigma_1, y)v_1(\sigma_1, .., y)d\sigma_1\right],$$

where

$$v_1(t, .., y) = \int_0^t L(t - s)b^*Y(s, y)v_0(s, .., y)ds,$$

$$v_0(t, .., y) = \int_0^t \varphi(t - s)(f_1(.)b_1^* + f_2(.)b_2^*)Y(s, y)ds$$

and

$$\|v_1(t, .., y)\| \le m\int_0^t l(t - s)a_0(s, .., y)ds,$$

$$|F_{2L}(t, y)| \le \frac{1}{2}\|\ell_L\|m^2(a(.) * l(.) * v_1(., y))(t), \tag{6.99}$$

where $*$ denotes convolution, and

$$\gamma(t) = \|\varphi(t)\| \qquad t \ge 0,$$

$$l(t) = \|L(t)\| \qquad t \ge 0,$$

$$a_0(t, y) = \int_0^t \gamma(t - s)\|(f_1b_1^* + f_2b_2^*)Y(s, y)\|_p ds, \qquad t \ge 0.$$

And more generally

$$\left|F_{2kL}(t, y)\right| \le \frac{1}{2^k}\|\ell_L\|m^{2k}(\gamma(.) * l(.))^{*k} * a_0(., y)(t)$$

with a similar estimate for the moment.
Hence

$$\sum_{k=1}^{\infty}\left|F_{2kL}(t, y)\right| \le \sum_{k=1}^{\infty}\frac{1}{2^k}\|\ell_L\|m^{2k}(a(.) * l(.))^{*k}(t) * a_0(t, y), \tag{6.100}$$

where super $*k$ denotes the k-fold convolution.

We want to show that the right side is finite. Note that

$$S_N(t) = \sum_{k=1}^{N} \frac{1}{2^k} ||\ell_L|| m^{2k} (a(.) * l(.))^{*k}(t)$$

is a nonnegative monotone increasing sequence in N.   □

Lemma 6.8. *The series*

$$\sum_{k=1}^{\infty} \frac{1}{2^k} m^{2k} (\gamma(.) * l(.))^{*k}(t) \tag{6.101}$$

converges a.e. in $t \geq 0$, uniformly in y, $0 < y < \ell$.

Proof. Let

$$\alpha(t) = (\gamma(.) * l(.))(t),$$

$$\hat{\alpha}(\sigma) = \int_0^\infty e^{-\sigma t} v(t) dt \qquad \sigma \geq 0.$$

The transform goes to zero as $\sigma \to \infty$, thus we see that for σ sufficiently large

$$0 < \frac{1}{2} m^2 \hat{\alpha}(\sigma) < 1.$$

Now

$$\int_0^T e^{-\sigma t} S_N(t) dt \leq \int_0^\infty e^{-\sigma t} S_N(t) dt = \sum_{k=1}^{N} \frac{1}{2^k} m^{2k} (\hat{\alpha}(\sigma))^k,$$

which converges as $N \to \infty$ and as $N \to \infty \to$ the finite limit

$$= \sum_{k=1}^{\infty} \frac{1}{2^k} m^{2k} (\hat{\alpha}(\gamma))^k = \frac{1}{2} m^2 \hat{\alpha}(\sigma) \frac{1}{1 - \frac{1}{2} m^2 \hat{\alpha}(\sigma)} < \infty.$$

We can then apply Fatou's lemma to the sequence

$$e^{-\sigma t} \sum_{k=1}^{N} \frac{1}{2^k} m^{2k} (a(.) * l(.))^{*k}(t)$$

to see that it converges to a finite limit a.e. in t, $0 < t < T$, because the integrals converge.   □

Hence $\sum_{k=1}^{\infty}|F_{2kL}(t,y)|$ converges for every t, rather than a.e, because of the convolution. With similar results for the moment, we see that we have convergence of the series (6.16):

$$\sum_{k=1}^{\infty}g_k(t,Y(.)) = \mathcal{N}(t,Y(.))$$

in the energy space.

However, in this generality there is no implication that the function is bounded in $[0,\infty]$. Indeed one has only to take $\frac{1}{2}m^2v(t) = 1$ which yields the limit: e^t.

Series Solution

Next we develop a constructive solution of the initial value problem for the aeroelastic equation (6.89) under some restrictive conditions that imply, in particular, the torsion angle is small enough.

We can use the state space theory developed in Chap. 5 which would lead to the theory of nonlinear semigroups (see [45]) but to minimize abstraction we take a more direct and constructive route, especially because our main interest is in the steady-state response.

Let $S(t)$ $t \geq 0$ denote the semigroup generated by $\mathcal{A}$. Let $Y(0)$ denote the initial structure state. Then the "method of variation of parameters" yields:

$$Y(t) = S(t)Y(0) + \int_0^t S(t-\sigma)\mathcal{N}(\sigma,Y(.))d\sigma, \qquad t \geq 0, \qquad (6.102)$$

$$Y(t) - \int_0^t S(t-\sigma)\mathcal{N}(\sigma,Y(.))d\sigma = S(t)Y(0) \qquad t \geq 0. \qquad (6.103)$$

The solution then can be expressed as a Volterra series [8], which will then be our constructive solution.

Here, however, we can take advantage of the fact that we have already constructed the solution to the linear equation (6.15). Thus

$$\dot{Y}(t) = \mathcal{A}Y(t) + \int_0^t \mathcal{L}(t-\sigma)Y(\sigma)d\sigma + u(t), \qquad (6.104)$$

where

$$u(t) = \mathcal{N}(t,Y(.)) - \int_0^t \mathcal{L}(t-\sigma)Y(\sigma)d\sigma$$

and we denote the right side by:

$$\mathcal{N}_{nl}(t, Y(.)) = \sum_{k=1}^{\infty} \int_0^t \int_0^{\sigma_1} \cdot \int_0^{\sigma_{2k}} \mathcal{L}_{2k}(t - \sigma_1, \sigma_1 - \sigma_2, ., \sigma_{2k} - \sigma_{2k+1};$$

$$Y(\sigma_1), .Y(\sigma_{2k+1}))d\sigma_1 .. d\sigma_{2k+1}, \tag{6.105}$$

where the subscript nl stands for nonlinear.

The initial value problem for the linear equation

$$\dot{Y}(t) = AY(t) + \int_0^t \mathcal{L}(t - \sigma)Y(\sigma)d\sigma$$

has the solution

$$Y(t) = W(t)Y(0),$$

where $W(.)$ is given by (5.15) and the solution of (6.102) can be expressed:

$$Y(t) = W(t)Y(0) + \int_0^t W(t - s)\mathcal{N}_{nl}(s, Y(.))ds. \tag{6.106}$$

This is again a Volterra equation.

However, to get a constructive solution we need to constrain the initial condition. We first consider the case where the linear system is (strongly) stable.

Theorem 6.9. *The Volterra equation (6.105) has the constructive Volterra series solution for each $T, 0 \le t < T$, given by*

$$Y(t, y) = Y_0(t, y) + \sum_{k=1}^{\infty} Y_k(t, y) \tag{6.107}$$

uniformly in y.

where

$$Y_0(t) = W(t)Y(0) \qquad t \ge 0,$$

$$Y_1(t) = \int_0^t W(t - s)\mathcal{N}_{nl}(s, Y_0(.))ds,$$

$$Y_k(t) = \int_0^t W(t - s)\mathcal{N}_{nl}(s, Y_{k-1}(.))ds \qquad k \ge 1 \tag{6.108}$$

for

$$m_0 = \max|Y(0, y)|, \qquad 0 < y < \ell \tag{6.109}$$

small enough.

Proof. We use the standard technique for solving Volterra equations [8] which yields the series in (6.106) and the problem is to prove convergence.

For this we need to relate the properties of $Y_k(.)$ to those of $Y_{k-1}(.)$. Hence we start with the properties of $Y_1(.)$ in terms of the function

$$Y_0(t, y):$$

$$Y_0(t,.) = W(t)Y(0,.),$$

$$\mathcal{N}_{nl}(t, Y_0(.)) = \sum_{k=1}^{\infty} g_k(t, Y_0(.)). \tag{6.110}$$

We go back and work with (6.98).

$$v_0(t, y) = \int_0^t \varphi(t-s)(f_1 b_1^* + f_2 b_2^*)Y_0(s, y)ds \qquad t \geq 0, 0 < y < \ell,$$

$$v_1(t, y) = \int_0^t \varphi(t-s)b^* Y_0(s, y) \int_0^s L(s-\tau)b^* Y_0(\tau, y)v_0(\tau, y)d\tau,$$

$$v_k(t, y) = \int_0^t \varphi(t-s)b^* Y_0(s, y) \int_0^s L(s-\tau)b^* Y_0(\tau, y)v_{k-1}(\tau, y)d\tau, \qquad k \geq 1. \tag{6.111}$$

Note that in this breakdown

$$F_{2kL}(t, y) = \frac{1}{2^k}\mathcal{L}_L(v_k(t, y)) \qquad k \geq 1,$$

$$F_{2kM}(t, y) = \frac{1}{2^k}\mathcal{L}_M(v_k(t, y)) \qquad k \geq 1.$$

And of course because we are fixing $Y(.)$ to be $Y_0(.)$, we may define

$$g_k(t, y) = \begin{pmatrix} 0 \\ \mathcal{M}^{-1}F_{2k}(t, y) \end{pmatrix}.$$

From $Y_0(t) = W(t)Y_0$ using the Green's function form of solution:

$$Y_0(t, y) = \int_0^t W(t, y, s)Y_0(s)ds,$$

we have that

$$m(Y_0(.), T) = \text{Max}|Y_0(t, y)| \leq W_T m_0, \qquad 0 < y < \ell, 0 < t < T,$$

where

$$W_T = \text{Max}||W(t, y, s)|| \qquad 0 \le T, 0 \le y, s \le \ell,$$

$$m(Y_0(.), T) = \text{Max}|Y_0(t, y)|, \qquad 0 < y < \ell.$$

Denote $m(Y_0(.), T)$ by m.

Then

$$|v_1(t, y)| \le m^2 \int_0^t ||\varphi(t - s)||ds \int_0^s ||L(s - \tau)||d\tau,$$

$$||\int_0^\tau \varphi(\tau - \sigma)(f_1 b_1^* + f_2 b_2^*)Y_0(\sigma, y)d\sigma||, \qquad 0 \le t \le T$$

$$\le m^2 \gamma(.) * l(.) * c(., Y_0(.), y)(t),$$

with $*$ denoting convolution as before and

$$\gamma(t) = ||\varphi(t)||,$$

$$l(t) = ||L(t)||,$$

$$c(t, Y_0(.), y) = \int_0^t (||\varphi(t - s) f_1|| \, |b_1^* Y_0(s, y)|$$

$$+ ||\varphi(t - s) f_2|| \, |b_2^* Y_0(s, y)|)ds.$$

Then for $k \ge 1$, and $t \le T$,

$$|v_k(t, y)| \le m^{2k} [\gamma(.) * l(.)]^{*k} * c(., Y_0(.), y)(t).$$

Hence

$$\left| \sum_{k=1}^N v_k(t, y) \right| \le \sum_{k=1}^N |v_k(t, y)|$$

$$< \left(\left(\sum_{k=1}^N m^{2k} [\gamma(.) * l(.)]^{*k} \right) * \left| c(., Y_0(.), y) \right| \right)(t) \qquad t \le T, \qquad (6.112)$$

where

$$\left| c(t, Y_0(.), .) \right| \le \int_0^t (\gamma(t - s)||f_1|| |b_1^* Y_0(s, y)| + a(t - s)||f_2|| |b_2^* Y_0(s, y)|)ds$$

$$\le f \int_0^t \gamma(t - s)|Y_0(s, y)|ds,$$

$$f = \text{max}(||f_1||, ||f_2||) \text{ (which depends on } U). \qquad (6.113)$$

Let

$$\Omega_N = \sum_{k=1}^{N} m^{2k} [\gamma(.) * l(.)]^{*k}.$$

Then the Laplace transform

$$\int_0^\infty e^{-\lambda t} \Omega_N(t) dt = \sum_{k=1}^{N} m^{2k} \hat{\gamma}(\lambda)^k \hat{l}(\lambda)^k.$$

Assume now that m_0 is small enough so that

$$m^2 \hat{\gamma}(0) \hat{l}(0) = r < 1.$$

Then $\Omega_N(t)$ converges monotonic increasing to a function we denote by $\Omega(t), t \geq 0$, whose Laplace transform is

$$\frac{m^2 \hat{\gamma}(\lambda) \hat{l}(\lambda)}{1 - m^2 \hat{\gamma}(\lambda) \hat{l}(\lambda)} \quad \text{which is bounded by} \quad \frac{r}{1-r}$$

and $\Omega(t)$ is nonnegative and goes to zero as $t \to \infty$, and

$$\int_0^\infty \Omega(t) dt < \frac{r}{1-r}. \tag{6.114}$$

Hence

$$\left| \sum_{k=1}^{\infty} v_k(t, y) \right| \leq f \int_0^t (\Omega * \gamma)(t - \tau) |Y_0(\tau, y)| d\tau \tag{6.115}$$

$$\leq m \frac{r}{1-r} \hat{\gamma}(0) \qquad 0 < t < T.$$

Hence it follows that

$$\left| \mathcal{N}_{nl}(t, Y_0(.), y) \right| < \text{const. } f \frac{r}{1-r} \hat{\gamma}(0) m \qquad 0 < t < T \tag{6.116}$$

and

$$||\mathcal{N}_{nl}(t, Y_0(.))|| \quad < \sqrt{\ell} \text{ const. } f \frac{r}{1-r} \hat{\gamma}(0) m. \tag{6.117}$$

Next

$$Y_1(t) = \int_0^t W(t-s)\mathcal{N}_{nl}(s, Y_0(.))ds \qquad 0 < t, \qquad (6.118)$$

which is recognized as the response of the linear system to the input

$$\mathcal{N}_{nl}(t, Y_0(.)) \qquad 0 < t.$$

To determine $Y_1(t)$ as a function of y we use the Green's function formula

$$Y_1(t, y) = \int_0^\ell \int_0^t W(t-\sigma, y, s)\mathcal{N}_{nl}(\sigma, Y_0(.), s)d\sigma ds, \qquad (6.119)$$

where

$$||W(t, y, s)|| < w < \infty \quad \text{for } 0 < t < T, \quad \text{and} \quad 0 < y, s < \ell.$$

Hence

$$|Y_1(t, y)| < w \int_0^\ell \int_0^t |\mathcal{N}_{nl}(\sigma, Y_0(.), s)|d\sigma ds \qquad (6.120)$$

$$< w \int_0^\ell \int_0^T |\mathcal{N}_{nl}(\sigma, Y_0(.), s)|d\sigma ds \qquad 0 < t < T.$$

Or

$$m(Y_1(0), T) \le w\,\ell\,T\,\text{const.}\,f\frac{r}{1-r}\hat{\gamma}(0)\,m(Y_0(.), T).$$

Then it follows that

$$m(Y_k(0), T) \le \left(w\,\ell\,T\,\text{const.}\,f\frac{r}{1-r}\hat{\gamma}(0)\right)^k m(Y_0(.), T),$$

where the constant does not depend on $Y_0(.)$
    Next we constrain r so that

$$w\,\ell\,T\,\text{const.}\,f\frac{r}{1-r}\hat{\gamma}(0) = \epsilon < 1, \qquad (6.121)$$

so that

$$|Y_1(t, y)| < \epsilon\,m \quad \text{for } t < \infty.$$

And iteratively:

$$|Y_k(t, y)| < \epsilon^k\,m \quad \text{for every } k \text{ for } t < T.$$

And hence

$$\sum_{k=1}^{\infty} Y_k(t, y) \quad \text{converges for } t < T$$

as required. □

6.5 Stability

Our main interest is the stability of the solution (6.106) and its dependence on the far field speed. Assuming zero angle of attack, as we do, the far field flow velocity is $q_\infty = \vec{i} U_\infty$ and stability thus depends on U_∞, which in this section we simply denote U, the "speed" parameter.

This presents an excellent illustration of Hopf bifurcation theory (see [4, 7, 37]), although our presentation is independent, not invoking that theory. In fact a direct application is not possible, in the sense that we have a single mathematical theorem we can quote from which the result follows.

We start with the steady-state solution of (6.97) given in Chap. 4: $Y(t) = 0$ valid for all speeds, noting that $\mathcal{N}(t, Y(.)) = 0$ for $Y(.) = 0$.

Linearizing the equation about the zero structure state we get the linear equation:

$$\dot{Y}(t) = \mathcal{A} Y(t) + \int_0^t \mathcal{L}(t - \sigma) Y(\sigma) d\sigma. \tag{6.122}$$

The stability theory for this equation has been treated in detail in Chap. 5 and is determined in terms of the aeroelastic modes.

Stability of the Linearized System: $0 < U < U_F$

The linear system (6.1) is "modally stable" for all speeds U such that $0 < U < U_F$ where U_F is the flutter velocity. Each mode—except for the zero frequency—decays to zero in time.

There are stronger notions of stability. A linear system is "strongly" stable if every initial state decays to zero; the elastic energy goes to zero.

Lemma 6.10. *For* $U < U_F$ *the linear system (6.120) is strongly stable.*

Proof. Actually we prove that the response can be characterized in terms of semigroup theory.

Semigroup on Riesz Space:

Recall the time domain solution of the linearized equation (6.1) for $M > 0$ given by

$$W(t)Y = \sum e^{\lambda_k t} P_k Y + Q(t)Y, \tag{6.123}$$

where $Q(t)Y$ is superstable and can be omitted from consideration. Because it depends on the speed U, let $\mathcal{R}(U)$ denote the subspace generated by the Riesz vectors $\{Y_k\}$; we call it the Riesz subspace, which is then also the span of the vectors $\{Z_k\}$. Then $W(t)$ maps $\mathcal{R}(U)$ into itself. Call the restriction of $W(t)$ to $\mathcal{R}$ by $W_{\mathcal{R}}(t)$.

$\square$

SubLemma 6.10. $W_{\mathcal{R}}(t)$ $t \geq 0$, defines a C_0 semigroup over $\mathcal{R}$. It is a contraction with a compact resolvent. For $U < U_F$ it is strongly stable. Furthermore, it has the representation

$$W_{\mathcal{R}}(t)Y = \sum e^{\lambda_k t} P_k Y; = W_{\mathcal{R}}(t)^* Y = \sum e^{\bar{\lambda}_k t} P_k^* Y.$$

The generator denoted $A_{\mathcal{R}}$ has the representation

$$A_{\mathcal{R}} Y = \sum \lambda_k P_k Y$$

and the resolvent $R(\lambda, A_{\mathcal{R}})$ has the representation

$$R(\lambda, A_{\mathcal{R}})Y = \sum_k \frac{P_k Y}{\lambda - \lambda_k} \quad \text{for } \lambda \neq \lambda_k.$$

Proof of Sublemma. We only show the semigroup property of $W_{\mathcal{R}}(t)$. And this is immediate.

$$W_{\mathcal{R}}(t)W_{\mathcal{R}}(t)Y = \sum e^{\lambda_k t} P_k \left(\sum e^{\lambda_j s} P_j Y \right)$$

$$= \sum e^{\lambda_k t} P_k e^{\lambda_k s} P_k Y$$

$$= \sum e^{\lambda_k (s+t)} P_k Y$$

$$= W_{\mathcal{R}}(s+t)Y.$$

And the representations are immediate. Furthermore,

$$[W_{\mathcal{R}}(t)Y, W_{\mathcal{R}}(t)Y] = \sum \left[e^{\lambda_k t} P_k Y, \sum e^{\lambda_j t} Q_j Y \right]$$

$$= \sum [e^{2\sigma_k t} P_k Y, Q_k Y]$$

$$\leq \sum [P_k Y, Q_k Y] = [Y, Y]$$

for every $t \geq 0$.

Given $\in > 0$, we can find $N_\in$ such that the tail:

$$\sum [e^{2\sigma_k t} P_k Y, Q_k Y], \qquad k > N_\in \text{ is } < \in.$$

And if all $\sigma_k < 0$, we can find $t_\in$ such that

$$\sum [e^{2\sigma_k t} P_k Y, Q_k Y], \qquad k < N_\in \text{ is } < \in,$$

which is enough to prove strong stability for the case $U < U_F$. ☐

Stability of Nonlinear System

Let us see what we can say about stability in the nonlinear case for (6.104).

We say that the nonlinear system is strongly stable if every initial condition bounded in the norm by some nonzero constant decays to zero.

Theorem 6.11. *For $U < U_F$, the nonlinear system is strongly stable for a small enough initial condition, by which we mean that $|Y(t, y)| \to 0$ for all $\max_y |Y_0(y)|$ small enough.*

Proof. For $U < U_F$ the linear system is strongly stable. In particular

$$||W(t) Y_0|| \to 0 \text{ as } t \to 0.$$

This is a new condition that we exploit in extending Theorem 6.7. Thus choose

$$m = \max |Y_0(y)|, \qquad 0 \leq y \leq \ell, < 1.$$

Then

$$\max |Y_0(t, y)|, 0 \leq y \leq \ell, \leq m < 1 \quad \text{for all } t > 0.$$

Next choose m so that

$$m^2 \left(\int_0^\infty ||\varphi(t)|| dt \int_0^\infty ||L(t)|| dt \right) \quad < r < 1.$$

Note that $|Y_0(t, y)| \leq m$ for all t, and in fact $\to 0$ as $t \to \infty$ even if not exponentially.

Hence we can replace T by infinity in Theorem 6.7, and say more.

Thus in the same notation as there we have

$$|\mathcal{N}_{nl}(t, Y_0(.), y)| < \text{const. } f \frac{r}{1-r} \hat{y}(0) m \quad \text{for all } t, \quad 0 < t < \infty \qquad (6.124)$$

and further goes to zero as $t \to \infty$. Next we choose

$$\text{const. } f \frac{r}{1-r} \hat{y}(0) = \epsilon < 1.$$

And the iteration:

$$|Y_k(t, y)| < \epsilon^k |Y_0(t, y)|.$$

Hence

$$Y(t, y) = \sum_{k=0}^{\infty} Y_k(t, y),$$

where

$$|Y(t, y)| \leq \frac{1}{1-\epsilon} |Y(t, y)| \to 0 \text{ uniformly in } y, \qquad \text{as } t \to \infty.$$

$\square$

Next let us consider $U = U_F$.

6.6   Limit Cycle Oscillation

At $U = U_F$ at least one mode will have zero real part.

$$d(M, \lambda_1, U) = 0; \quad \text{Re } \lambda_1 = 0.$$

However, there can be more than one despite popular belief to the contrary. Remember that when we say the kth mode flutters, the frequency $\text{Im}\lambda_k(U_F)$ is quite different from $\text{Im}\lambda_k(0)$. So $\text{Im}\lambda_k(U_F)$ can be the same for different values of k. But then for how many values? How many modes can flutter at the same flutter speed?

Indeed it may be nonfinite as, for example, for zero speed, if S is zero. Note that these are questions that no computer program can answer.

Theorem 6.12. *The number of modes that flutter at a given speed $U > 0$ is finite.*

Proof. Suppose now that the number of modes that flutter is infinite. Then denoting them $\{i\omega_k\}$, $\omega_k > 0$, we must have that

$$\omega_k \to \infty, \quad k \to \infty; \qquad \lambda_k = i\omega_k,$$

in as much as they cannot have a finite accumulation point. Thus we have that

$$\sigma_k(U_F) = 0 = \sigma_k(0) + \int_0^{U_F} \frac{\partial \sigma_k}{\partial U} dU = 0.$$

Because

$$\sigma_k(0) = 0, \quad \text{we have that}$$

$$\int_0^{U_F} \frac{\partial \sigma_n}{\partial U} dU = 0 \quad \text{for every } n, \text{ and } n \to \infty, \tag{6.125}$$

where we can use the formula:

$$\frac{\partial \lambda_n}{\partial U} = \frac{\sum_{i=1}^4 \dfrac{\partial d}{\partial w_i} \dfrac{\partial w_i}{\partial U}}{\sum_{i=1}^4 \dfrac{\partial d}{\partial w_i} \dfrac{\partial w_i}{\partial \lambda}}. \tag{6.126}$$

Now as $\omega_n \to \infty$, so does κ_n. And taking limits in (6.124), and using our previous estimates for $M = 0$ and M nonzero, we can see that the limit is a constant. But it is negative at $U = 0$, as we have shown in Sect. 5.6, and hence we reach a contradiction showing that the number of fluttering modes for any speed cannot be infinite. □

Remarks: Note the implication of this result: as the mode number increases they eventually stop fluttering!

Let us first consider the case where exactly one mode has zero real part and all others have strictly negative real parts. Hence let $i\omega = i\omega_F$, where ω is the fluttering mode

$$d(M, i\omega, U_F) = 0, \quad \frac{\partial \sigma}{\partial U}\Big|_{\lambda = i\omega} > 0,$$

$Y_{i\omega}(y)$ the mode shape. Let P_1 denote the corresponding projection.
At $U = U_F$, let us denote $W(.)$ by $W_F(.)$. Then

$$Y(t) = Y_0(t) + \sum_{k=1}^\infty Y_k(t), \tag{6.127}$$

$$Y_0(t, y) = e^{i\omega t} Y_{i\omega}(y),$$

$$Y_1(t) = \int_0^t W_F(t - s) \mathcal{N}_{nl}(s, Y_0(.)) ds,$$

$$Y_k(t, y) = \int_0^t W_F(t - s) \mathcal{N}_{nl}(s, Y_{k-1}(.)) ds,$$

$$\mathcal{N}_{nl}(t, Y(.)) = \sum_{k=1}^\infty g_k(t, y),$$

$$g_k(t, y) = \begin{pmatrix} 0 \\ \mathcal{M}^{-1} F_k(t, y) \end{pmatrix},$$

$$F_k(t, y) = \frac{1}{2^k} \begin{pmatrix} \mathcal{L}_L(v_k(t, y)) \\ \mathcal{L}_M(v_k(t, y)) \end{pmatrix},$$

where

$$v_0(t, y) = \int_0^t \varphi(t - s)(f_1 b_1^* + f_2 b_2^*) Y_0(s, y) ds \qquad t \geq 0, \ 0 < y < \ell,$$

$$v_1(t, y) = \int_0^t \varphi(t - s) b^* Y_0(s, y) \int_0^s L(s - \tau) b^* Y_0(\tau, y) v_0(\tau, y) d\tau.$$

Hence $v_1(t, y)$ asymptotically, in steady state,

$$= e^{3i\omega t} \hat{\varphi}(i\omega) \hat{L}(i\omega) \hat{\varphi}(i\omega)(f_1(.)b_1^* + f_2(.)b_2^*) Y_{i\omega}(y)(b^* Y_{i\omega}(y))^2.$$

And more generally:

$$v_k(t, y) = e^{i(2k+1)\omega t} (\hat{\varphi}(i\omega) \hat{L}(i\omega))^k \cdot \hat{\varphi}(i\omega)$$

$$\times (f_1(.)b_1^* + f_2(.)b_2^*) Y_{i\omega}(y)(b^* Y_{i\omega}(y))^{2k}, \qquad (6.128)$$

where

$$\left\| \left(\hat{\varphi}(i\omega) \hat{L}(i\omega) \right)^k (b^* Y_{i\omega}(y))^{2k} \right\|$$

$$\leq \left(\|\hat{\varphi}(i\omega)\| \, \|\hat{L}(i\omega)\| \, |b^* Y_{i\omega}(y)|^2 \right)^k = (\hat{\gamma}(0)\hat{l}(0))^k.$$

Let

$$m = \max 0 < y < \ell \quad |b^* Y_{i\omega}(y)|$$

and we assume

$$r = m^2 \hat{\gamma}(0) \hat{l}(0) \quad \text{is} \ < 1. \qquad (6.129)$$

Hence the series

$$v(t, y) = \sum_{k=1}^{\infty} v_k(t, y)$$

$$= e^{3i\omega t} \hat{\varphi}(i\omega) \hat{L}(i\omega)(I - e^{2i\omega t} (b^* Y_{i\omega}(y))^2 \hat{\varphi}(i\omega) \hat{L}(i\omega))^{-1}$$

$$\times \hat{\varphi}(i\omega)(f_1(.)b_1^* + f_2(.)b_2^*) Y_{i\omega}(y).$$

Note that this is made up of harmonics of the form $(3 + 2n)\omega$.

Approximation

For m small enough we may approximate the sum just by the first term, just the third harmonic. Denote the approximation by $v_a(.)$:

$$v_a(t, y) = e^{3i\omega t}\hat{\varphi}(i\omega)\hat{L}(i\omega)\hat{\varphi}(i\omega)(f_1(.)b_1^*$$

$$+ f_2(.)b_2^*)Y_{i\omega}(y)(b^*Y_{i\omega}(y))^2 \tag{6.130}$$

and hence within this approximation

$$N_{nl}(t, Y_0(.)) = e^{3i\omega t}g_1(y), \tag{6.131}$$

where

$$g_1(y) = \begin{pmatrix} 0 \\ M^{-1}F_l(y) \end{pmatrix}$$

$$F_1(y) = \frac{1}{2}\begin{pmatrix} \ell_L(v_a(0, y)) \\ \ell_M(v_a(0, y)) \end{pmatrix}$$

$$v_a(0, y) = \hat{\varphi}(i\omega)\hat{L}(i\omega)\hat{\varphi}(i\omega)(f_1(.)b_1^* + f_2(.)b_2^*)Y_{i\omega}(y)(b^*Y_{i\omega}(y))^2. \tag{6.132}$$

Note that

$$|g_1(y)| \leq \text{const.} \frac{r}{1-r} \cdot |\hat{\varphi}(i\omega)(f_1(.)b_1^* + f_2(.)b_2^*)Y_{i\omega}(y)|$$

$$\leq \text{const.} \frac{r}{1-r}\hat{\gamma}(0)f|Y_{i\omega}(y)|, \tag{6.133}$$

where f is given by (6.112). We use (6.130) in what follows. Then

$$Y_1(t) = \int_0^t W_F(t - s)N_{nl}(s, Y_0(.))ds$$

and the steady-state response

$$Y_1(t) = e^{3i\omega t}\int_0^\infty W_F(s)e^{-i3\omega s}ds\, g_1(.)$$

$$= e^{3i\omega t}\hat{W}_F(3i\omega)g_1(.)$$

$$= e^{3i\omega t}Y_1(0), \qquad Y_1(0) = \hat{W}_F(3i\omega)g_1(.), \tag{6.134}$$

which is the same form as $Y_0(t, .)$, and

$$\left| Y_1(0, y) \right| \leq \text{const.} \frac{r}{1-r} \hat{\gamma}(0) f \left| Y_0(0, y) \right|.$$

(6.135)

And similarly by iteration we have

$$Y_k(t, y) = e^{3i\omega t} Y_k(0, y),$$

(6.136)

where

$$\left| Y_k(0, y) \right| \leq \left(\text{const.} \frac{r}{1-r} \hat{\gamma}(0) f \right)^k \left| \sum_{k=1}^{\infty} Y_k(0, y) \right|.$$

(6.137)

Hence

$$Y(t, y) = e^{3i\omega t} \sum_{k=1}^{\infty} Y_k(0, y) + e^{i\omega t} \sum_{k=1}^{\infty} Y_k(0, y),$$

(6.138)

where the series converges if r is small enough so that

$$\left| \text{const.} \frac{r}{1-r} \hat{\gamma}(0) f \right| = \delta < 1,$$

and

$$\left| \sum_{k=1}^{\infty} Y_k(0, y) \right| < \frac{\delta}{1-\delta} \left| Y_{i\omega}(y) \right|.$$

(6.139)

We have thus a first approximation to the response at the flutter given by (6.13) which is thus the LCO.

It is periodic with the same period as that of the flutter mode frequency predicted by the linear model, with a predominant third harmonic. It is thus an illustration of the Hopf bifurcation theory.

Higher-order approximation would involve the $(3 + 2n)$ harmonics. If we have more than one mode fluttering, we would then have to consider $(3 + 2n)$ harmonics of each mode and intermodulation, a messy calculation at best!

But the main point is again that we have a response which is periodic in the steady state with harmonics of the fundamental frequency given by the linear model.

The LCO Amplitude

We define the amplitude of the LCO as the (RMS value):

$$\text{Lim } T \to \infty \sqrt{(1/T) \int_0^T \left| |Y(t, y)| \right|^2 dt},$$

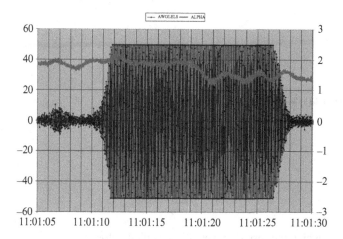

Fig. 6.1 Flutter LCO

which within the approximation (6.12) is given by the Parseval formula for the Fourier series:

$$\text{LCO RMS amplitude} = \sqrt{\left|Y_{i\omega}(y)\right|^2 + \left(\sum_{k=1}^{\infty} Y_k(0, y)\right)^2} \qquad (6.140)$$

$$\leq \left|Y_{i\omega}(y)\right|\sqrt{1 + \left(\frac{\delta}{1-\delta}\right)^2}. \qquad (6.141)$$

In particular we see that for initial amplitudes small enough, the LCO amplitude is proportional to the initial amplitude. Further investigation is required to determine whether this continues to be true for larger initial amplitudes.

Figure 6.1 shows the incidence of flutter LCO encountered in a specially designed Flight Test; see [108] for more details.

Finally we come to the solution to the potential field equation, the air flow.

6.7   The Air Flow Decomposition Theory

As we have noted, our primary concern is the stability of the structure. The air flow per se is of secondary importance. However, as we have seen, we are able to characterize the flow as well. We consider only 2D flow with zero angle of attack.

From Sect. 6.2, we have the decomposition of the kth-order potential

$$\phi_k = \phi_{L,k} + \phi_{0 \cdot k}$$

and

$$\phi = \sum_{k=0}^{\infty} \phi_k / k!$$

recalling (6.134). We define

$$\phi_L = \sum_{k=1}^{\infty} \phi_{L,k} / k!$$

$$\phi_0 = \sum_{k=0}^{\infty} \phi_{0,k} / k!$$

The sense in which the series converge is specified below.
The solution of

$$\Xi(\phi_{0,k}) = g_{k-1}$$

following ([45, p. 692]) is given by

$$\phi_{0,k}(t, x, z) = \int_0^t \int_{-\infty}^{\infty} \int_{-\infty}^{\infty} F(t - \sigma, x - \xi, z - \zeta) g_{k-1}(\sigma, \xi, \eta) d\sigma d\xi d\zeta,$$

$$t > 0; -\infty < x, z < \infty, \tag{6.142}$$

where the kernel

$$F(t, x, z) = 1 / \left(2\pi \sqrt{(1 - M^2)} \right) 1 / \left(\sqrt{ \left((t - Ux/c_1^2)^2 \right.} \right.$$

$$\left. \left. -1 / (1 - M^2) \left(x^2/c_1^2 + z^2/c_2^2 \right) \right) \right),$$

$$c_1^2 = a_\infty^2 (1 - M^2),$$

$$c_2^2 = a_\infty^2.$$

And defining

$$g = \sum_{k=1}^{\infty} g_k / k!$$

where the convergence is in $L_2(R^2)$, we have

$$\phi_0(t, x, z) = \int_0^t \int_{-\infty}^{\infty} \int_{-\infty}^{\infty} F(t - \sigma, x - \xi, z - \zeta) g(\sigma, \xi, \eta) d\sigma d\xi d\zeta,$$

$$t > 0; -\infty < x, z < \infty. \tag{6.143}$$

For $\phi_{L,k}$ we extend Theorem 5.3 to nonzero M. Thus specializing (5.18) to the typical section case and zero angle of attack, we need to take the inverse Laplace/Fourier transform of

$$1/(k + i\omega)e^{-z\sqrt{k^2M^2+2kM^2i\omega+\omega^2(1-M^2)}}\,\tilde{\hat{A}}_k(\lambda, i\omega), \qquad z > 0.$$

Now the inverse Laplace/Fourier transform of

$$1/(k + i\omega)\tilde{\hat{A}}_k(\lambda, i\omega)$$

is

$$\beta_k(t, x) = \int_0^t A_k(t - \sigma, x - U\sigma)\mathrm{d}\sigma, \qquad -\infty < x < \infty.$$

The inverse Fourier transform of

$$e^{-z\sqrt{k^2M^2+2kM^2i\omega+\omega^2(1-M^2)}}$$

is

$$L(t, x, z) = 1/\Big(2\pi\sqrt{(1 - M^2)}\Big)1/\Big(\sqrt{((t - Ux/c_1^2)^2 - r^2)}\Big),$$

where

$$r^2 = (1 - M^2)(x^2/c_1^2 + z^2/c_2^2).$$

Hence we have

$$\phi_{L,k}(t, x, z) = \int_0^t \int_{-b}^{b+Ut} L(t - \sigma, x - \xi, z)\beta_k(\sigma, \xi)\mathrm{d}\sigma\mathrm{d}\xi. \tag{6.144}$$

And hence by (6.143)

$$\phi_L(t, x, z) = \int_0^t \int_{-b}^{b+Ut} L(t - \sigma, x - \xi, z)\beta(\sigma, \xi)\mathrm{d}\sigma\mathrm{d}\xi, \tag{6.145}$$

where

$$\beta(t, x) = \int_0^t A(t - \sigma, x - U\sigma)\mathrm{d}\sigma$$

$$L(t, x, z) = 1/(2\pi\sqrt{(1 - M^2)})1/(\sqrt{((t - Ux/c_{12})^2 - r^2)}), \tag{6.146}$$

$$t > 0, -\infty < x < \infty. \tag{6.147}$$

Thus finally we have for the time domain potential flow decomposition

$$\phi(t, x, z) = \phi_L(t, x, z) + \phi_0(t, x, z), \tag{6.148}$$

where $\phi_L(t, x, z)$ provides the lift, has no discontinuities in the velocity, and can be linearized, and the linearized equation determines the LCO speed and period. We are not interested in $\phi_0(t, x, z)$ because it does not affect the structure stability. It cannot be linearized as observed in [86], and may have discontinuities in the flow velocity. But whether there are "shocks" (see, e.g., [14] for definition of shocks in the flow) is not settled by this theory. Shocks imply a change in entropy [14] and our key assumption is that the flow is isentropic. Also, even if there are shocks, it is not clear [86] that they affect the structural stability, not withstanding the generally held beliefs by aeroelasticians.

Notes and Comments

This chapter illustrates the need for analytical theory. For example, a precise definition of the crucial concept of flutter speed is not possible by numerical computation, which is probably why there is no formal definition in all of the aeroelastic literature.

The continuum formulation and solution of the aeroelastic equations in isentropic flow appears here for the first time. The structure dynamics is assumed to be linear and although the extension to the nonlinear case would be of interest, the extra complications it would bring in even in the "domain" specifications would be too much without proportional gain.

And of course our main interest is in the nonlinear aerodynamics anyhow. We have also limited the treatment to the case of zero angle of attack because the extension to the nonzero case would involve too much effort without justifying the returns. Our main objective here is in showing the Hopf bifurcation and the LCO. We have provided the means, however complex, for calculating the LCO amplitude. This is of course also a tedious process in CFD requiring "time marching."

The existence of shocks in isentropic flow continues to be controversial. It is indeed established in 1D flow (the Riemann problem, see, e.g., [4]). However, there is no mathematical proof (yet) in 2D or higher dimensions [private communication: A. Chorin, 2010]. And furthermore, ours can be termed a singular case because of the finite boundary.

Chapter 7
Viscous Air flow Theory

7.1 Introduction

In this chapter we extend the theory allowing for nonzero viscosity. The field equations then are the Navier–Stokes equations, a subject in fluid dynamics of intense research activity, but still replete with many open problems. Our interest here is again on the effect of viscosity on the wing-structure response rather than the flow itself. The extant aeroelastic literature on this aspect is all computational [82, 83, 93].

7.2 The Field Equation/Conservation Laws

We begin as in Chap. 3, with the governing conservation laws for fluid flow in their differential form, but now including viscosity. We follow the notation there for pressure, density, and other thermodynamic variables, with $q(t, x, y, z)$ in particular denoting the fluid velocity.

1. Conservation of mass

$$\frac{\partial \rho}{\partial t} + \nabla \cdot (\rho q) = 0. \tag{7.1}$$

2. Conservation of momentum (Navier–Stokes equation)

$$\rho \frac{\partial q}{\partial t} + \rho (q. \nabla) q = \mu \Delta q - \nabla p + \left(\zeta + \frac{2\mu}{3} \right) \nabla (\nabla \cdot q), \tag{7.2}$$

where
μ is the coefficient of viscosity. (1.78×10^{-6} (for air)).
ζ is the second coefficient of viscosity; more on this later.
Δq is the 3×1 vector with components.
$\nabla \cdot \nabla q_i \quad i = 1, 2, 3$, and q_i are the components of q.

A V Balakrishnan, *Aeroelasticity: The Continuum Theory*,
DOI 10.1007/978-1-4614-3609-6_7, © Springer Science+Business Media, LLC 2012

Similarly $(q.\nabla)q$ is the 3×1 vector with components

$$q.\nabla q_i \, i = 1, 2, 3,$$

$$q = \vec{i} q_1 + \vec{j} q_2 + \vec{k} q_3,$$

$\vec{i}, \vec{j}, \vec{k}$ being the unit orthonormal vectors.

And finally the law that is of crucial importance in our theory:

3. Energy-Flux Equation

As noted by Meyer [14], this is no more than the first law of thermodynamics in a Eulerian version.

The energy per unit volume

$$E(t) = c_v T = \frac{p}{(\gamma - 1)\rho}$$

and the flux of energy per unit mass

$$\rho \frac{DE}{Dt} = -p\nabla \cdot q + \sum_{i=1}^{3} \sum_{k=1}^{3} \sigma_{ik} e_{ik} + \kappa \Delta T, \qquad (7.3)$$

$$\sigma_{ik} = \delta_{ik} \left(\zeta - \frac{2\mu}{3} \right) \nabla.q + 2\mu e_{ik},$$

$$e_{ik} = \frac{1}{2} \left(\frac{\partial q_k}{\partial x_i} + \frac{\partial q_i}{\partial x_k} \right)$$

with the notation $x_1 = x$, $x_2 = y$, $x_3 = z$, δ_{ik} is the Kronecker delta and

$$\Phi = \sum_{i=1}^{3} \sum_{k=1}^{3} \sigma_{ik} e_{ik}$$

is called the dissipation function and we have the thermodynamic relation [14]

$$\rho T \frac{DS}{dt} = \Phi + \kappa \Delta T, \qquad (7.4)$$

where S is the entropy. κ (assumed constant) is the coefficient of thermal diffusivity $\sim 1,000 \, \mu$, $\gamma = c_p/c_v \sim\sim 1.4$. The flow is incompressible if the flow divergence

$$\nabla.q = 0.$$

These are the field equations. But what distinguishes the aeroelastic problem are the boundary conditions. We consider again the finite plane structure model as in Chap. 3. The big difference from the nonviscous flow is:

1. Zero Slip at the boundary:

$$q(t, x, y, 0+) = \vec{k}\frac{Dz(t, x, y)}{Dt}, \quad |x| < b, \; 0 < y < \ell, \qquad (7.5)$$

where $z(., ., ., .)$ is the structure displacement along the z-axis, as in Chap. 3. In other words, there is no slip between the air and the wing motion, the latter being normal to the wing plane in our beam model. This is of course more restrictive than just flow tangency, as in the nonviscous model.

2. The Kutta–Joukowsky conditions remain the same as described in Chap. 3. However, whether the Kutta condition is needed is examined later. The far field is $\vec{i}U$ where we omit the subscript infinity. Our main interest again is the structure stability as a function of $U > 0$.

The basic relevant references for viscous flow for our needs are: Meyer [14]; Landau–Lifschitz [12], Schlicting–Kersten [15], 2003 edition of the 1965 classic; Chorin–Marsden [4], R. Temam [19]; and Oleinik–Samokhin [13].

The main difficulty here starts with the basic question of existence and uniqueness whatever the function space and sense (weak) of convergence chosen for the flow solution.

Of the huge readily available literature on Navier–Stokes equations, it is fortunate that we are only concerned with aspects of the theory that affect the boundary-structure stability.

It should be noted that if we set $\mu = 0; \quad \zeta = 0$ in the Navier–Stokes equation (7.2), it does formally reduce to the Euler equation (3.2) we start with. But the further condition of isentropy is required to reach the full-potential equation, and the fluid-structure boundary conditions imposed are of course quite different.

Incompressible Flow

Furthermore, we make the "simplifying" assumption that the flow is incompressible in the sense that the flow divergence can be neglected

$$\nabla . q = 0.$$

This in particular allows us to sidestep the question of including the second coefficient of viscosity. As a result, the Navier–Stokes equation (7.2) simplifies to:

$$\rho\frac{\partial q}{\partial t} + \rho(q.\nabla)q - \mu\Delta q + \nabla p = 0 \qquad (7.6)$$

and the energy-flux equation to:

$$\rho\frac{DE}{Dt} = 2\mu\sum_{i=1}^{3}\sum_{k=1}^{3}e_{ik^2} + \kappa\Delta T, \qquad (7.7)$$

$$e_{ik} = \frac{1}{2}\left(\frac{\partial q_k}{\partial x_i} + \frac{\partial q_i}{\partial x_k}\right) \tag{7.8}$$

and we do not need to worry about ζ.

Before we can consider the stability of the structure we need to consider the following.

The Static/Steady-State Equation

The static Navier–Stokes equation simplifies to

$$\rho(q.\nabla)q + \mu\Delta q = 0 \tag{7.9}$$

and the energy-flux equation to

$$2\mu\sum_{i=1}^{3}\sum_{k=1}^{3}e_{ik^2} + \kappa\Delta T = 0. \tag{7.10}$$

But this condition by (7.4) is equivalent to saying that the entropy is constant in time. Hence in the time-invariant case this equation is automatically satisfied because entropy is taken to be time invariant. Also the continuity equation (7.1) using the incompressibility condition yields

$$q.\nabla\rho = 0,$$

so that we may take ρ to be constant. Thus (7.9) holds with ρ a constant. The surprise here is that the static solution for the nonviscous case, where the structure is at rest and the flow vector is a constant, does not hold here because of the no-slip boundary condition that the structure and the air move together so that the air velocity has to be zero over the structure. Hence the flow cannot be a constant if the far field velocity is nonzero.

Typical Section/Zero Angle of Attack

To reduce complexity further we now consider a flexible wing so that we may specialize to the typical section/zero angle of attack, as in the previous chapters.

Thus we have the field equation

$$\rho(q.\nabla)q - \mu\Delta q = 0,$$

where ρ is a constant.

The 2D NS in the xz-plane can now be written out as

$$q_1 \frac{\partial q_1}{\partial x} + q_3 \frac{\partial q_1}{\partial z} + \frac{1}{\rho} \frac{\partial p}{\partial x} = v \left(\frac{\partial^2 q_1}{\partial x^2} + \frac{\partial^2 q_1}{\partial z^2} \right) \quad q_1(x, 0) = 0, \qquad (7.11)$$

$$q_1 + \frac{\partial q_3}{\partial x} + q_3 \frac{\partial q_3}{\partial z} + \frac{1}{\rho} \frac{\partial p}{\partial z} = v \left(\frac{\partial^2 q_3}{\partial x^2} + \frac{\partial^2 q_3}{\partial z^2} \right), \qquad (7.12)$$

$$\nabla . q = \frac{\partial q_1}{\partial x} + q_3 \frac{\partial x}{\partial z} = 0,$$

$$v = \frac{\mu}{\rho}.$$

Next we come to the all-important flow-structure boundary conditions.

Boundary Conditions

$$q_3(x, 0) = 0 = q_1(x, 0) \qquad |x| < b.$$

Far Field

$$q_3(x, \infty) = 0 = q_3(\pm\infty, z),$$
$$q_1(x, \infty) = U = q_1(\pm\infty, z).$$

The entropy S is given by ([14], p. 144)

$$pe^{-(s/c_v)} = \text{constant } \rho^\gamma.$$

So the first task is to determine the static solution. At the present time the only way is to invoke the following celebrated theory.

Prandtl Boundary Layer Theory

Proposed by Prandtl in 1904 [15] for fluids of small viscosity ("vanishingly small" and air qualifies with ($\mu \sim 10)^{-6}$); but more important, of large Reynold's number; see [4]. The concept of the boundary layer is to subdivide the flow region into two parts: a thin-boundary layer at the wing boundary where the viscosity must be taken into account where the no-slip boundary condition holds, and the "outer" layer which is the bulk of the region where we may neglect viscosity and assume isentropic flow. For the 'matching' problems here see [3,4].

A word of caution! It should be noted right away that Oleinik writing in 1999 includes this problem in [13] in the section on "Some Open Problems:" "Is it possible to give a strict mathematical justification of this procedure and find the limits of applicability of the Prandtl equations?"

In other words, there is no mathematical justification for this procedure, yet.

The Prandtl Limit Equations and the Blasius Solution

Here we follow Landau–Lifschitz [12]. The Kutta condition is not invoked. In the usual formulation, as in [12], the boundary is allowed to be infinite. But here we do not have that luxury; the boundary is finite. There is another additional complication.

In reference to (7.2) we need to make additional simplifications in the inner layer:

$$\frac{\partial p}{\partial z} = 0 = \frac{\partial p}{\partial x}$$

or the pressure is a constant, along with the density. We set

$$v = \frac{p}{\rho}.$$

Next we assume:

$$\frac{\partial^2 q_3}{\partial z^2} = 0; \qquad \frac{\partial^2 q_3}{\partial x^2} = 0.$$

Hence the Navier–Stokes equation (7.2) becomes

$$q_1 \frac{\partial q_1}{\partial x} + q_3 \frac{\partial q_1}{\partial z} = v \frac{\partial^2 q_1}{\partial z^2}, \tag{7.13}$$

$$q_1 \frac{\partial q_3}{\partial x} + q_3 \frac{\partial q_3}{\partial z} = 0,$$

which using

$$\frac{\partial q_1}{\partial x} + \frac{\partial q_3}{\partial z} = 0$$

becomes

$$q_1 \frac{\partial q_3}{\partial x} = q_3 \frac{\partial q_1}{\partial x}. \tag{7.14}$$

Here we can state the next theorem.

Theorem 7.1 (Oleinik [13]). *The field equation:*

$$q_1 \frac{\partial q_1}{\partial x} + q_3 \frac{\partial q_1}{\partial z} = v \frac{\partial^2 q_1}{\partial z^2}, \tag{7.15}$$

$$\frac{\partial q_1}{\partial x} + \frac{\partial q_3}{\partial z} = 0 \tag{7.16}$$

in the strip $\Omega = \{ z > 0; \; -b < x < b \}$, *with the boundary conditions*

$$q_1(x, 0+) = 0; \qquad q_3(x, 0+) = 0,$$
$$q_1(x, \infty) = u, \qquad -b < x < b$$

has a unique solution such that

$$\frac{\partial q_1}{\partial z} \quad and \quad \frac{\partial^2 q_1}{\partial z^2}$$

are continuous and bounded in Ω.

$$\frac{\partial q_1}{\partial x}, \; \frac{\partial q_3}{\partial z}$$

are continuous and bounded in any finite part of Ω. □

We do not have a constructive solution technique. We do have the following however.

Theorem 7.2 (Blasius [cf. 12, 14, 15]). *If the chord is infinite, that is, we have in our notation* $-b < x < \infty$, *the solution is unique and the limit value as z goes to infinity*

$$q_3(x, \infty) = 0.43 \sqrt{\frac{vU}{x+b}} \qquad -b < x < \infty. \tag{7.17}$$

Proof. See [14]. □

We use this as the boundary condition to calculate the steady-state potential flow in the outer layer.

Thus we have the steady-state problem as in Chap. 4, (4.1), now specialized to typical section and zero angle of attack. The field equation for the potential $\phi(x, z)$ is:

$$0 = a_\infty^2 \left[1 + \frac{(\gamma - 1)}{2a_\infty^2} \left(U^2 - \left(\frac{\partial \phi}{\partial x}\right)^2 - \left(\frac{\partial \phi}{\partial z}\right)^2 \right) \right] (\Delta \phi)$$
$$- \frac{1}{2} \left[\frac{\partial \phi}{\partial x} \frac{\partial}{\partial x} |\nabla \phi|^2 + \frac{\partial \phi}{\partial z} \frac{\partial}{\partial z} |\nabla \phi|^2 \right], \tag{7.18}$$

$$\Delta\phi = \frac{\partial^2\phi}{\partial x^2} + \frac{\partial^2\phi}{\partial z^2},$$

$$|\nabla\phi|^2 = \left(\frac{\partial\phi}{\partial x}\right)^2 + \left(\frac{\partial\phi}{\partial z}\right)^2$$

with the boundary condition:

$$\frac{\partial\phi(x,\,0)}{\partial z} = w(x), \qquad |x| < b,$$

where $w(\cdot)$ is given by (7.9), and the far field condition that

$$q_\infty = iU$$

and the Kutta–Joukowsky conditions.

This problem has already been treated in Chap. 4. The solution for the potential as calculated there is given by

$$\phi(x,\,z) = xU + \sum_{k=1}^{\infty} \frac{1}{k!}\phi_k(x,\,z), \tag{7.19}$$

where

$$\begin{pmatrix} \dfrac{\partial\phi_I}{\partial x} \\[2mm] \dfrac{\partial\phi_I}{\partial z} \end{pmatrix} = \rho U \left(\begin{array}{c} z\dfrac{\sqrt{1-M^2}}{2}\displaystyle\int_{-b}^{b} \dfrac{A(\xi)d\xi}{z^2(1-M^2)+(x-\xi)^2} \\[4mm] z\dfrac{\sqrt{1-M^2}}{2}\displaystyle\int_{-\infty}^{\infty} \dfrac{B(\xi)}{z^2(1-M^2)+(x-\xi)^2}\,d\xi \end{array} \right), \tag{7.20}$$

$$A(x) = \frac{0.86}{\sqrt{1-M^2}}\sqrt{\frac{b-x}{b+x}}\int_{-b}^{b}\sqrt{\frac{1}{b-\xi}}\frac{1}{\xi-x}\,d\xi, \qquad |x| < b, \tag{7.21}$$

$$B(x) = \frac{1}{\pi}\int_{-b}^{b}\frac{A(\xi)}{x-\xi}\,d\xi, \qquad |x| < \infty. \tag{7.22}$$

For $k \geq 2$:

$$\phi_k(x,\,z) = \int_{-\infty}^{\infty}\int_{-\infty}^{\infty} L(x-\xi,\,z-\eta)g_{k-1}(\xi,\,\eta)d\xi d\eta,$$

$$x,\,z \text{ in } R^2 \quad (4.82),$$

$$g_{k-1} = \frac{Y-1}{2} \sum_{i=1}^{2} \sum_{j=1}^{k-1} c_j^k \Delta\phi_{k-j} \sum_{m=0}^{j} C_m^j \frac{\partial\phi_{j-m}}{\partial x_i} \frac{\partial\phi_m}{\partial x_i}$$

$$+ \sum_{l=1}^{k-2} \sum_{i=1}^{2} \sum_{j=1}^{2} \frac{\partial^2\phi_{k-1}}{\partial x_i \partial x_j} \sum_{m=0}^{1} c_m^l \frac{\partial\phi_{1-m}}{\partial x_i} \frac{\partial\phi_m}{\partial x_j} \,,$$

$$x_1 = x \quad x_2 = z. \tag{7.23}$$

Linearization

Next we linearize the full-potential equation (typical-section theory) about this steady-state solution given by (7.11). Denoting it by $\phi_0(x, z)$ and $\phi(\lambda, t, x, z)$ the solution corresponding to $\lambda\theta(t)$, $\lambda h(t)$ with

$$\phi(0, t, x, z) = \phi_0(x, z).$$

Then we have the expansion:

$$\phi(\lambda, t, x, z) = \phi_0(x, z) + \sum_{k=1}^{\infty} \frac{\lambda^k}{k!} \phi_k(t, x, z),$$

where

$$\phi_k(t, x, z) = \frac{\partial^k \phi(t, x, z)}{\partial\lambda^k}.$$

We are only (!) interested in $\phi_1(t, ., .)$.

Hence differentiating

$$\frac{\partial^2\phi(\lambda)}{\partial t^2} + \frac{\partial}{\partial t}|\nabla\phi(\lambda)|^2 = a_\infty^2 \left(1 + \frac{\gamma-1}{a_\infty^2}\left(U_\infty^2 - |\nabla\phi(\lambda)|^2 - \frac{\partial}{\partial t}\phi(\lambda)\right)\right) \Delta\phi(\lambda)$$

$$-\nabla\phi(\lambda) \cdot \nabla\frac{|\nabla\phi(\lambda)|^2}{2} \tag{7.24}$$

with respect to λ and setting it equal to zero, we obtain:

$$\frac{\partial^2\phi_1}{\partial t^2} + 2\frac{\partial}{\partial t}\nabla\phi_1 \cdot \nabla\phi_0 = a_\infty^2(1 + \frac{\gamma-1}{a_\infty^2}(U_\infty^2 - |\nabla\phi_0|^2))\Delta\phi_1$$

$$- (\gamma-1)\left(2\nabla\phi_1 \cdot \nabla\phi_0 + \frac{\partial}{\partial t}\phi_1\right)\Delta\phi_0$$

$$- \nabla\phi_1 \cdot \nabla\frac{|\nabla\phi_0|^2}{2} - \nabla\phi_0 \cdot \nabla(\nabla\phi_1 \cdot \nabla\phi_0). \tag{7.25}$$

The flow-tangency boundary condition is now

$$\frac{\partial \phi_1(t, x, 0)}{\partial z} = -\left(\dot{h}(t, \ y) + (x - ab)\dot{\theta}(t, \ y) + \frac{\partial \phi_1(x, 0)}{\partial x}\theta(t, \ y) \right) \qquad (7.26)$$

and we have used the fact that

$$\phi_k(x, \ 0) = 0 \qquad k \geq 2.$$

Plus: the Kutta–Joukowsky conditions as in Chap. 3.

To proceed further we have to construct the whole linear theory in Chap. 5 for (7.25), where now we face the new difficulty that the coefficients depend on the space variables. So we stop here in this version of the book, leaving it as part of suggested future work.

Chapter 8
Optimal Control Theory: Flutter Suppression

8.1 Introduction

We have examined the problem of feedback control of structures for stability enhancement in Chap. 2. In this chapter we return to this problem for aeroelasticity where the structure is subject to aerodynamic loading. An important point to be noted here is the dependence of stability properties on the speed parameter. Hence the control action will need to depend on the speed as well.

Our continuum model is as described in Chap. 3. As we have shown in Chap. 6, stability is determined completely by the linearized model linearized about the steady-state solution. Hence it follows that the controller need be based only on this linear model, an important conclusion of the theory.

Hence we consider only the linear model in Chap. 5. In particular we limit ourselves to isentropic flow.

As is now well established, to begin with for control design we need a state-space model. The complexity here is that the state is no longer finite-dimensional. For finite-dimensional approximations see [17, Chap. 12].

Fortunately we have developed a state-space model in Chap. 5 which we now invoke. Thus we have:

$$\dot{Z}(t) = \mathring{A}Z(t),$$

$$Y(t) = \mathcal{P}Z(t).$$

We need to add to this the control actuator model. Whatever it is, however, it has to be on the structure. This involves a sensor that inputs to the actuator, a compensator.

Thus for $M = 0$, we have the control model:

$$\dot{Y}(t) = \mathcal{A}Y(t) + Cx(t) + \mathcal{B}u(t),$$

$$\dot{x}(t) = A_c x(t) + L\dot{Y}(t). \tag{8.1}$$

Or, going back to the convolution/evolution form:

$$\dot{Y}(t) = \mathcal{A}_S Y(t) + TY(t) + \int_0^t \mathcal{L}(t-\sigma)\dot{Y}(\sigma)d\sigma + \mathcal{B}u(t), \qquad (8.2)$$

where

$$\mathcal{B}u(t) = \begin{pmatrix} 0 \\ Bu(t) \end{pmatrix} \qquad (8.3)$$

and thus the control may be force or moment, and

$$u(\cdot) \in L_2[H_u, [0, T]] \qquad 0 < T < \infty,$$

where H_u is a separable Hilbert space, and $\mathcal{B}$ is a linear bounded operator into H, following the notation in Chap. 2.

For $M \neq 0$, we have the modification of the convolution term (cf. 5.4.)

$$\dot{Y}(t) = \mathcal{A}Y(t) + \int_0^t \mathcal{L}(t-\sigma)Y(\sigma)d\sigma.$$

But this can also be subsumed in the state equation in terms of the full state $Z(.)$:

$$\dot{Z}(t) = \dot{\mathcal{A}}Z(t) + \begin{pmatrix} 0 \\ Bu(t) \end{pmatrix}. \qquad (8.4)$$

The first question is whether this class of controls, and in particular the canonical feedback control of Chap. 2: $-\mathcal{B}^*\mathcal{P}Z(t)$ can yield controllability in the Z-state space.

Here we can use the explicit semigroup solution

$$S_z(t)Z = \begin{pmatrix} S_c(t)x + \int_0^t S_c(t-\sigma)L\dot{Y}(\sigma)d\sigma \\ Y(t) \end{pmatrix},$$

$$Y(t) = W(t)Y + \int_0^t W(t-\sigma)(CS_c(\sigma)x - \mathcal{L}(\sigma)Y)d\sigma,$$

$$Z = \begin{pmatrix} x \\ Y \end{pmatrix}$$

for the solution to the control equation for zero initial conditions:

$$Z(t) = \int_0^t S_z(t-\tau)\begin{pmatrix} 0 \\ Bu(\tau) \end{pmatrix}d\tau$$

$$= \begin{pmatrix} \int_0^t d\tau \int_0^{t-\tau} S_c(t-\tau-\sigma)L\dfrac{d}{d\sigma}\left[W(\sigma)Bu(\tau) - \int_0^\sigma W(\sigma-s)\mathcal{L}(s)Bu(\tau)ds\right] \\ \int_0^t W(t-\sigma)Bu(\sigma)d\sigma - \int_0^t \int_0^\tau W(\tau-\sigma)\mathcal{L}(\sigma)d\sigma\,Bu(t-\tau)d\tau \end{pmatrix}.$$

Controllability–Stabilizability

As we have seen in Chap. 2, controllability requires that

$$\Omega = \bigcup_u \bigcup_{t>0} S_z(t) \begin{pmatrix} 0 \\ \mathcal{B}u \end{pmatrix}$$

must be dense in the state-space, where

$$S_z(t) \begin{pmatrix} 0 \\ \mathcal{B}u \end{pmatrix} = \begin{pmatrix} \int_0^t S_c(t-\sigma) L \dfrac{d}{d\sigma} \left[W(\sigma)\mathcal{B}u - \int_0^\sigma W(\sigma-\tau)\mathcal{L}(\tau)\mathcal{B}ud\tau \right] d\sigma \\ W(t)\mathcal{B}u - \int_0^t W(t-\sigma)\mathcal{L}(\sigma)\mathcal{B}ud\sigma \end{pmatrix}.$$

For $M \neq 0$,

$$S_z(t) \begin{pmatrix} 0 \\ \mathcal{B}u \end{pmatrix} = \begin{pmatrix} \int_0^t S_c(t-\sigma)L\Phi(\sigma)d\sigma \\ \Phi(t) \end{pmatrix},$$

$$\Phi(t) = W(t)\mathcal{B}u(t).$$

Let

$$\Phi_i = \begin{pmatrix} X_i \\ Y_i \end{pmatrix}$$

denote the eigenvector of the generator $\underset{\sim}{A}$ corresponding to the eigenvalue λ_i. These are the aeroelastic modes, as we have seen, with Y_i the structure mode shape.

As we show in Chap. 5 they are confined to a bounded real part strip in the complex plane with a finite number with positive real part for each speed.

Let us recall from Chap. 2 our key result in control theory, Theorem 2.1, and our canonical candidate for a robust stability enhancing controller.

Here, however, the state space is no longer a Hilbert space. So it does not apply directly.

Theorem 8.1. *Suppose for some i,*

$$B^*Y_{i,2} = 0 \quad where\ Y_i = \begin{pmatrix} Y_{i,1} \\ Y_{i,2} \end{pmatrix}, \quad and \quad Re\ \lambda_i \geq 0.$$

Then the state space is not controllable and the canonical robust feedback control

$$u(t) = -\mathcal{B}^*\mathcal{P}Z(t)$$

cannot stabilize the system.

Proof. This control yields the state equation:

$$\dot{Z}(t) = \left(\dot{A} - \begin{pmatrix} 0 \\ B \end{pmatrix} B^* \mathcal{P} \right) Z(t)$$

and

$$\mathcal{P} \left(\dot{A} - \begin{pmatrix} 0 \\ B \end{pmatrix} B^* \mathcal{P} \right) \Phi_i = \lambda_i Y_i - BB^* Y_i.$$

But

$$B^* Y_i = B^* Y_{i,2} = 0.$$

Hence

$$\left(\dot{A} - \begin{pmatrix} 0 \\ B \end{pmatrix} B^* \mathcal{P} \right) \Phi_i = \dot{A} \Phi_i = \lambda_i \Phi_i.$$

Or this unstable mode is not stabilized.   □

Remark: This result is not surprising because a mode cannot be affected if the corresponding mode shape leaves no trace on the control input. In fact we have:

$$\mathcal{M} \ddot{x}(t) + \mathcal{D} \dot{x}(t) + \mathcal{K} x(t) + A x(t) = \int_0^t \mathcal{F}(\sigma) x(t - \sigma) d\sigma.$$

Then the question is: Is it possible to find a controller that has a trace of every mode? For which, in other words,

$$B^* \mathcal{P} \Phi_i \neq 0$$

or

$$B^* Y_{i,2} \neq 0. \tag{8.5}$$

Or is there a "u_i" in the control space such that $[Y_{i,2}, Bu_i] \neq 0$ for every i?

This requires that we calculate the mode shapes, and take note of the dependence on the far field speed. Suppose then we can find B and u such that

$$B^* \mathcal{P} \Phi_i \neq 0 \quad \text{for any } i.$$

Theorem 8.2. *Suppose*

$$B^* \mathcal{P} \Phi_i \neq 0 \quad \text{for every } i.$$

Then the feedback control $u(t) = -B^* \mathcal{P} Z(t)$ *will enhance the stability of the structure response, if we can assume that the eigenfunctions are little changed in closed loop.*

Proof. At the simplest level, the argument would be this. In practice, the control effort is not large enough to change the mode shapes. Hence the eigenfunctions Φ_i

will remain the same as without control. Hence the eigenvalues of the generator are given by

$$\left(\dot{A} - \begin{pmatrix} 0 \\ B \end{pmatrix} B^* P \right) \Phi_i = \gamma_i \Phi_i$$

and hence

$$\gamma_i \, ||P\Phi_i||^2 = \left[P \left(\dot{A} - \begin{pmatrix} 0 \\ B \end{pmatrix} B^* P \right) \Phi_i, P\Phi_i \right]$$

$$= [P\dot{A}\Phi_i, P\Phi_i] - ||B^* P\Phi_i||^2,$$

$$\mathrm{Re}\, \gamma_i = \mathrm{Re}\, \lambda_i - \frac{||B^* Y_i||^2}{||Y_i||^2},$$

which is strictly less than $\mathrm{Re}\, \lambda_i$. Hence the damping of every mode is increased. In particular the structure response is more stable.

Note, however, that (cf. Chap. 2)

$$\frac{||B^* Y_i||^2}{||Y_i||^2} \quad \text{goes to zero as } i \to \infty.$$

And in any event we may not be able to make every mode stable: we cannot necessarily make $\mathrm{Re}\, \gamma_i$ negative unless $\mathrm{Re}\, \lambda_i$ is zero.

Now let us consider the more general proof.

Let Ψ_i denote the eigenvectors of the controlled system stability operator so that

$$\dot{A}\Psi_i - \begin{pmatrix} 0 \\ BB^* P\Psi_i \end{pmatrix} = \gamma_i \Psi_i.$$

Then

$$P\dot{A}\Psi_i - BB^* P\Psi_i = \gamma_i P\Psi_i.$$

Suppose

$$BB^* P\Psi_i = 0.$$

Then

$$\dot{A}\Psi_i = \gamma_i \Psi_i,$$

so that Ψ_i is an eigenvector of $\dot{A}$, which would contradict our hypothesis. But then

$$P\dot{A}\Psi_i - BB^* P\Psi_i = \gamma_i P\Psi_i$$

and the assumption about eigenfunctions needed is that

$$[P\dot{A}\Psi_i, P\Psi_i] = \lambda_i. \qquad \square$$

In any event it would appear that

$$u(t) = -\mathcal{B}^*\mathcal{P}Z(t) \tag{8.6}$$

is a good candidate feedback control law for augmenting stability, in particular for increasing flutter speed.

Flutter Suppression

Note that flutter instability cannot be totally avoided; the system will never be stable. The terminology "flutter suppression" is usually used in this sense of increasing the flutter speed.

In particular this means that there is no "optimal" control! Also here is an area where the control based on "discretized" system models may be misleading because here the number of modes is finite and this would make the controller like the one above be able to stabilize all the modes, whereas the continuum model shows it is not possible.

Then the question is: Is it possible to find a controller that has a trace of every mode. For which

$$\mathcal{B}^*\mathcal{P}\Phi_i \neq 0$$

or

$$\mathcal{B}^*Y_{i,2} \neq 0.$$

Or is there a "u_i" in the control space such that

$$[Y_{i,2}, \; Bu_i] \neq 0$$

for every i.

This requires that we calculate the mode shapes and take note of the dependence on the far field speed that the control may have to be dependent on the speed, which is an open problem at present.

Self-straining Actuators: Incompressible Flow

Because of the remarkable stabilizability properties of the self-straining actuators, described in Chap. 2, it is natural to consider them for flutter suppression even if the control effort may be limited. Here following [8] we examine how effective they can be in incompressible flow where we have an explicit solution of the Possio equation. For an account of experimental verification see [25].

We only consider the torsion controller because of its "super stability" capability. As in Sect. 2.5 we have to change the definition of the state to include the end points. Thus following [60], we have:

$$\dot{Y}(t) = \mathcal{A}_S Y(t) + \mathcal{T}Y(t) + \int_0^t \mathcal{L}(t-s)\dot{Y}(s)ds, \qquad (8.7)$$

where $Y(\,.\,)$ is in the energy space

$$\mathcal{H}_E = \mathcal{D}\left(\sqrt{\mathcal{A}_s}\right) \times \mathcal{H},$$

$$\mathcal{H} = H_b \times H_p,$$

$$\mathcal{H}_b = L_2[0,\ell] \times E^1,$$

$$\mathcal{H}_t = L_2[0,\ell] \times E^1,$$

$$\mathcal{A}_S = \begin{pmatrix} A_b & 0 \\ 0 & A_t \end{pmatrix},$$

$$\mathcal{D}(A_h) = \left(\begin{matrix} h \\ h'(\ell) \end{matrix} \text{ in } H_b \middle| h''''\epsilon L_2(0,\ell); h(0) = 0 = h'(0) = h'''(1)\right),$$

$$A_h \begin{pmatrix} h \\ h'(\ell) \end{pmatrix} = \begin{pmatrix} EIh''''(\,.\,) \\ EIh''(\ell) \end{pmatrix},$$

$$\mathcal{D}(A_t) = \left(\begin{matrix} \theta \\ \theta(\ell) \end{matrix} \text{ in } H_t \middle| \theta''\epsilon L_2(0,\ell); \theta(0) = 0\right),$$

$$A_t \begin{pmatrix} \theta \\ \theta(\ell) \end{pmatrix} = \begin{pmatrix} -GJ\theta'' \\ GJ\theta'(\ell) \end{pmatrix}.$$

$\mathcal{H}_E = \mathcal{D}\left(\sqrt{\mathcal{A}_S}\right) \times \mathcal{H}$ with energy inner product$\mathcal{A}_S$ with domain and range in $\mathcal{H}_E$,

$$\mathcal{D}(\mathcal{A}_S) = Y = \begin{pmatrix} x_1 \\ z_1 \end{pmatrix} \quad x_1 = \begin{pmatrix} h_1 \\ h_1'(\ell) \\ \theta_1 \\ \theta_1(\ell) \end{pmatrix} \epsilon \mathcal{D}(\mathcal{A}_S) \quad z_1 = \begin{pmatrix} h_2 \\ \theta_2 \end{pmatrix},$$

and

$$\begin{pmatrix} h_2 \\ -EI\frac{h_1''(\ell)}{g_h} \\ \theta_2 \\ -GJ\frac{\theta_1'(\ell)}{g_\theta} \end{pmatrix} \epsilon \mathcal{D}\left(\sqrt{\mathcal{A}_s}\right),$$

$$h_2'(\ell) = -EI\frac{h_1''(\ell)}{g_h}; \qquad \theta_2(\ell) = \frac{-GJ\,\theta_1'(\ell)}{g_\theta},$$

$$\mathcal{A}_s Y = \begin{pmatrix} h_2 \\ -EI\dfrac{h_1''(\ell)}{g_h} \\ \theta_2 \\ -GJ\dfrac{\theta_1'(\ell)}{g_\theta} \\ -\mathcal{M}^{-1}\begin{pmatrix} EI h_1''''(\ell) \\ -GJ\theta_1''(\ell) \end{pmatrix} \end{pmatrix}.$$

If $g_h = 0$, then $h_1''(\ell) = 0$. If $g_\theta = 0$, then $\theta_1'(\ell) = 0$.

$\mathrm{Re}[\mathcal{A}_s Y, Y] = -g_h|h_1(\ell)|^2 - g_\theta|\theta_1(\ell)|^2$ so that $\mathcal{A}_s$ generates a contraction semigroup.

Next $\mathcal{T}$ and $\mathcal{L}(\,.\,)$ are as defined in Sect. 5.4.

To determine the aeroelastic modes we can continue to use the development in Sect. 5.4. We can continue to use (5.7) except that we must change the matrix P to

$$\begin{pmatrix} 0 & \dfrac{\lambda g_h}{EI} & 1 & 0 & 0 & 0 \\ 0 & 0 & 0 & 0 & 0 & 0 \\ 0 & 0 & 0 & 0 & \dfrac{\lambda g_\theta}{GJ} & 1 \end{pmatrix}.$$

Torsion Control

We only consider the pitch or torsion control because of their exceptional stabilizing properties, as shown in Chap. 2. Thus we set $g_h = 0$.

Our main interest is in the root locus and the stability curve as in Sect. 5.7 and how the stability is enhanced by the self-straining controls. Let P_θ denote the matrix

$$\begin{pmatrix} 0 & 0 & 1 & 0 & 0 & 0 \\ 0 & 0 & 0 & 0 & 0 & 0 \\ 0 & 0 & 0 & 0 & \dfrac{\lambda g_\theta}{GJ} & 1 \end{pmatrix}$$

and let

$$d(0, g_\theta, \lambda, U) = \det P_\theta e^{A(\lambda)\ell} Q \tag{8.8}$$

and following [18], using the notation

$$\{a_{ij}\} \text{ for } e^{A(\lambda)\ell},$$

we have:

$$P_\theta e^{A(\lambda)\ell} Q = \begin{pmatrix} & a_{33} & & a_{34}\ a_{36} \\ & a_{43} & & a_{44}\ a_{46} \\ a_{63} + \dfrac{\lambda g_\theta}{GJ} a_{53} & a_{64} + \dfrac{\lambda g_\theta}{GJ} a_{54} & a_{66} + \dfrac{\lambda g_\theta}{GJ} a_{56} \end{pmatrix}$$

and hence it follows that we can express

$$d(0, g_\theta, \lambda, U) = d(0, 0, \lambda, U) + \frac{\lambda g_\theta}{GJ} d(0, \infty, \lambda, U), \tag{8.9}$$

$$d(0, \infty, \lambda, U) = \det P_3 e^{A(\lambda)\ell} Q$$

$$= \det \begin{pmatrix} a_{33}\ a_{34}\ a_{36} \\ a_{43}\ a_{44}\ a_{46} \\ a_{53}\ a_{54}\ a_{56} \end{pmatrix}, \tag{8.10}$$

where

$$P_3 = \begin{pmatrix} 0\ 0\ 1\ 0\ 0\ 0 \\ 0\ 0\ 0\ 1\ 0\ 0 \\ 0\ 0\ 0\ 0\ 1\ 0 \end{pmatrix}.$$

If $U = 0$, the modes are given by (Sect. 2.3):

$$\sqrt{GJI_\theta} \cosh\lambda v\ell + g_\theta \sinh\lambda v\ell = 0,$$

$$\lambda_k = -\sigma + i\omega_k, \quad \omega_k = \frac{(2k+1)\pi}{2\ell} \sqrt{\frac{GJ}{I_\theta}} \quad \text{for } g_\theta < g_c = \sqrt{GJI_\theta},$$

$$\sigma = \frac{1}{2\ell} \sqrt{\frac{GJ}{I_\theta}} \log \left| \frac{g + g_c}{g - g_c} \right|. \tag{8.11}$$

Hence

$$\sigma_k(0) = -\sigma.$$

Now from Sect. 2.5

$$d(0, g_\theta, \lambda, 0) = \sqrt{GJI_\theta} \cosh\lambda v\ell + g_\theta \sinh\lambda v\ell, \quad v = \sqrt{\frac{I_\theta}{GJ}}$$

and hence we have, comparing with (8.9),

$$d(0,0,\lambda,0) = \sqrt{GJI_\theta} \cosh\lambda v\ell,$$

$$d(0,\infty,\lambda,0) = \frac{GJ}{\lambda} \sinh\lambda v\ell.$$

Hence

$$\frac{\partial d(0, g_\theta, \lambda_k, 0)}{\partial \lambda} = \ell v\left(\sqrt{GJI_\theta} \sinh\lambda_k v\ell + g_\theta \cosh v\ell\lambda_k\right) \qquad (8.12)$$

and it is important to note that the right side is not zero. If $g_\theta = 0$, this is immediate because $\cosh\lambda_k v\ell$ is then zero. If g_θ is not zero, then $\cosh\lambda_k v\ell$ cannot be zero and because

$$\text{Tanh } \lambda_k v\ell = -\frac{\sqrt{GJI_\theta}}{g_\theta},$$

it follows that (8.12) cannot be zero.

To calculate

$$\frac{\partial d(0, g_\theta, \lambda_k, 0)}{\partial U}$$

we note first that by the perturbation formula in Sect. 2.3, we have that

$$d(0,0,\lambda,U) = \frac{1}{2}\left(1 + (\cos \gamma\ell \cosh \gamma\ell)\cosh \mu\ell + w_2 w_3 \Delta(\gamma, \mu)\right)$$

as in Sect. 5.4. And:

$$d(0,\infty,\lambda,U) = \frac{1}{2}w_1(1 - \cos \gamma\ell \cosh \gamma\ell)\Big)\text{Cosh } \mu\ell$$

omitting cross-product terms involving w_i.

Hence

$$\frac{\partial d(0, g_\theta, \lambda_k, 0)}{\partial U} = \frac{1}{2}\frac{\partial}{\partial U}\left((1 + \cos \gamma\ell \cosh \gamma\ell) \cosh \mu\ell\right.$$

$$\left. + \frac{\lambda g_\theta}{GJ}w_1(1 - \cos\gamma\ell \cosh \gamma\ell)\cosh \mu\ell\right).$$

Although we can continue calculating, we see that

$$\frac{\lambda_k g_\theta}{GJ} < \frac{\lambda_k g_c}{GJ} = \lambda_k\sqrt{\frac{I_\theta}{GJ}}$$

is small enough to be neglected, so that the slope of the stability curve is essentially the same as for zero-control gain and is of course negative. Unfortunately we don't have even an approximate formula for the flutter speed; and although the value of the damping is now large and negative and the slope is the same as before, we cannot still claim that the flutter speed is going to be less.

Chapter 9
Aeroelastic Gust Response

9.1 Introduction

A safety issue as well as a material fatigue issue is the effect of wind gust on the aircraft. This is usually examined as part of the rigid-body response to wind turbulence; see [9, 33]. A sinusoidal-gust model is considered in most aeroelastic work starting with [6]. See [17,84].

Here we follow [64] and use, instead, a random-gust model, the random-field model of wind turbulence due to Kolmogorov; see [26,35]. We use a linear-structure model because the theory for a nonlinear structure model would involve nonlinear equations. It is not clear to us that the nonlinearity is any more important here than in the rigid-body case, where the gust is allowed to be random, but the rigid-body model is linear. In any event, we only consider the linear Goland beam model and typical-section aerodynamics, as appropriate for a flexible wing of high-aspect ratio.

9.2 The Turbulence Field

We begin with a description of the Kolmogorov model of wind turbulence. Let $v(x, y, z)$ denote the wind field in 3D. Let k denote the unit normal to the wing plane so that the induced normal-wing velocity is

$$w_g(x, y, z) = v(x, y, z).k. \tag{9.1}$$

According to the Kolmogorov theory, the wind field $v(\cdot)$ is isotropic Gaussian with the 3D spectral density:

$$P_3(f_1, f_2, f_3) = \frac{c_n^2}{\left(\kappa^2 + f_1^2 + f_2^2 + f_3^2\right)^{11/6}}, \qquad -\infty < f_i < \infty, \tag{9.2}$$

A V Balakrishnan, *Aeroelasticity: The Continuum Theory*,
DOI 10.1007/978-1-4614-3609-6_9, © Springer Science+Business Media, LLC 2012

where κ is the scale parameter [35] and c_n^2 is the turbulence strength whose value may be considered arbitrary for our purposes here.

The 2D field given by (9.1) has the spectral density:

$$P_2(f_1, f_2) = \frac{\text{constant}}{\left(\kappa^2 + f_1^2 + f_2^2\right)^{4/3}}, \qquad -\infty < f_i < \infty. \qquad (9.3)$$

Temporal Wind Gust Model

For an aircraft in motion the wind-gust velocity at any given point on the wing is a function of time, obtained by invoking the Taylor "frozen field" hypothesis. Thus we assume that at the given altitude z, say z_0, and the wing velocity—the far-field velocity in our aeroelastic model—is along the x-axis equal to U. We also take the angle of attack to be zero. We assume that the wing displacement normal to the wing is small compared to z_0 and neglect the change in the altitude due to wing motion. Then at any point along the chord (x, s, z_0), where s denotes the span variable, the vertical wind-gust component, as a function of time is given by:

$$W_g(t, x, s) = w_g(x - Ut, s, z_0), \qquad |x| < b, \, 0 < s < \ell. \qquad (9.4)$$

We have thus a 2D space–time random field which is Gaussian and stationary in both space and time. The main thing to note is the dependence on the speed in the temporal variable. The space–time covariance function of the process:

$$E[W_g(t_1, x_1, s_1) W_g(t_2, x_2, s_2)],$$

$E[\cdot]$ denoting expected value, is then given by

$$\int_{-\infty}^{\infty} \int_{-\infty}^{\infty} e^{2\pi i (v_1(x-Ut)+v_2 s)} \frac{1}{\left(k^2 + 4\pi^2 \left(v_1^2 + v_2^2\right)\right)^{4/3}} dv_1 dv_2, \qquad (9.5)$$

where

$$t = t_2 - t_1; \qquad x = x_2 - x_1; \qquad s = s_2 - s_1.$$

In particular, we see that if we fix the point x_0 on the chord, the spectral density of the process $W_g(t, x_0, s)$ is given by

$$P(x_0, f_1, f_2) = \frac{1}{U \left(k^2 + 4\pi^2 \left(\frac{f_1^2}{U^2} + f_2^2\right)\right)^{4/3}}, \qquad -\infty < f_1, f_2 < \infty. \qquad (9.6)$$

Figure 9.1 plot the spectral density in db for various values (FPS units) of speed and scale factor. There is significant low-frequency content for small speeds and small κ, but the curves flatten out resembling white noise for higher values.

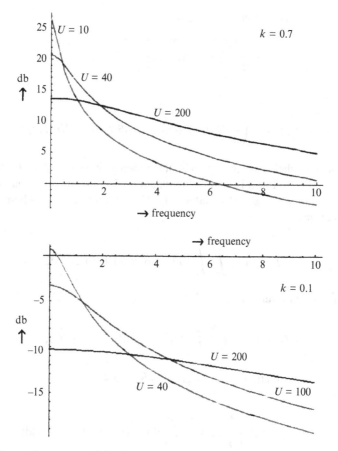

Fig. 9.1 Temporal guest spectral density

Our objective in this chapter is to consider this as the random disturbance input
to the aeroelastic system and calculate the structure response for the Goland model.

9.3   The Gust Forced Aeroelastic Equations

We assume that the structure model is the Goland model with two degrees of
freedom treated in Chap. 2. The aerodynamics equations are as in the typical-section
zero angle of attack linear case treated in Chap. 5. The field equations are as given
in (6.11):

$$\frac{\partial^2 \phi}{\partial t^2} + 2U \frac{\partial^2 \phi}{\partial t \, \partial x} - \left[a_\infty^2 (1 - M^2) \frac{\partial^2 \phi}{\partial x^2} + a_\infty^2 \frac{\partial^2 \phi}{\partial z^2} \right] = 0$$

with the now stochastic boundary condition

$$\frac{\partial \phi(t, x, 0)}{\partial z} = w_a(t, x, s) + W_g(t, x, s), \tag{9.7}$$

where

$$w_a(t, x, s) = -\left(\frac{\mathrm{d}}{\mathrm{d}t}(h(t, s) + (x - ab)\theta(t, s)) + U\theta(t, s)\right).$$

We need to calculate first the lift and the moment corresponding to the input "downwash" $W_g(.)$. For this we need to calculate first the pressure doublet, using the Possio equation. Here we consider only $M = 0$ and $M = 1$, where we have explicit solutions.

Thus

$$A_g(t, x, s) = \int_0^t \int_{-b}^b \Psi(t - \sigma, x, \xi) W_g(\sigma, \xi, s) \mathrm{d}\xi$$

with the steady-state version:

$$A_g(t, x, s) = \int_0^\infty \int_{-b}^b \Psi(\sigma, x, \xi) W_g(t - \sigma, \xi, s) \mathrm{d}\xi,$$

where the "system function" $\Psi(.)$ is given in Sect. 5.4.

The lift

$$L_g(t, s) = \rho U \int_{-b}^b \mathrm{d}x \int_0^\infty \int_{-b}^b \Psi(\sigma, x, \xi) W_g(t - \sigma, \xi, s) \mathrm{d}\xi,$$

where it is convenient to define

$$l(t, \xi) = \int_{-b}^b \rho U \Psi(t, x, \xi) \mathrm{d}x, \tag{9.8}$$

so that the lift due to gust

$$L_g(t, s) = \int_0^\infty \int_{-b}^b l(\sigma, \xi) W_g(t - \sigma, \xi, s) \mathrm{d}\xi \tag{9.9}$$

and the moment

$$M_g(t, s) = \rho U \int_{-b}^b (x - ab) \mathrm{d}x \int_0^\infty \int_{-b}^b \Psi(\sigma, x, \xi) W_g(t - \sigma, \xi, s) \mathrm{d}\xi,$$

and defining

$$m(t, \xi) = \int_{-b}^{b} \rho U(x - ab)\Psi(t, x, \xi)dx, \qquad (9.10)$$

we have

$$M_g(t, s) = \int_{0}^{\infty} \int_{-b}^{b} m(\sigma, \xi) W_g(t - \sigma, \xi, s)d\xi. \qquad (9.11)$$

These are then the random lift and moment, stationary Gaussian random processes in time variable t and space variable s. The corresponding covariance functions are given by

$$R_L(t, s) = E(L_g(t_1, s_1)L_g(t_2, s_2)), \quad t = t_2 - t_1; \quad s = s_2 - s_1$$

$$= \int_{0}^{\infty} \int_{0}^{\infty} d\sigma_1 d\xi_1 \int_{0}^{\infty} \int_{-b}^{b} l(\sigma_1, \xi_1) E[W_g(t_1 - \sigma_1, \xi_1, s_1)$$

$$\times W_g(t_2 - \sigma_2, \xi_2, s_2)] . l(\sigma_2, \xi_2)d\sigma_2 d\xi_2,$$

where

$$E[W_g(t_1 - \sigma_1, \xi_1, s_1) W_g(t_2 - \sigma_2, \xi_2, s_2)]$$

$$= \int_{-\infty}^{\infty} \int_{-\infty}^{\infty} e^{2\pi i (v_1(\xi_2 - \xi_1 - U(t - (\sigma_2 - \sigma_1))) + v_2 s)}$$

$$\times \frac{1}{(k^2 + 4\pi^2 (v_1^2 + v_2^2))^{4/3}} dv_1 dv_2$$

and hence:

$$R_L(t, s) = \int_{0}^{\infty} \int_{0}^{\infty} d\sigma_1 d\xi_1 \int_{0}^{\infty} \int_{-b}^{b} l(\sigma_1, \xi_1)$$

$$\times \int_{-\infty}^{\infty} \int_{-\infty}^{\infty} \frac{e^{2\pi i (v_1(\xi_2 - \xi_1 - U(t - (\sigma_2 - \sigma_1))) + v_2 s)}}{(k^2 + 4\pi^2 (v_1^2 + v_2^2))^{4/3}} dv_1 dv_2 l(\sigma_2, \xi_2)d\sigma_2 d\xi_2$$

$$= \int_{-\infty}^{\infty} \int_{-\infty}^{\infty} e^{2\pi i v_1 U t + 2\pi i v_2 s} \left| \int_{0}^{\infty} \int_{-b}^{b} e^{-2\pi i v_1 \xi_2 - 2\pi i v_1 U \sigma_2} \right.$$

$$\times \left. l(\sigma_2, \xi_2)d\sigma_2 d\xi_2 \right|^2 \cdot \frac{1}{(k^2 + 4\pi^2 (v_1^2 + v_2^2))^{4/3}} dv_1 dv_2$$

$$= \int_{-\infty}^{\infty} \int_{-\infty}^{\infty} e^{-2\pi i f_1 t + 2\pi i f_2 s} \left| \int_0^\infty \int_{-b}^b e^{-2\pi i (f_1/\underline{U})\xi_2 - 2\pi i f_1 \sigma_2} \right.$$

$$\left. \times \, l(\sigma_2, \xi_2) d\sigma_2, d\xi_2 \right|^2 \cdot \frac{1}{U\left(k^2 + 4\pi^2 \left(\left(\frac{f_1}{U}\right)^2 + f_2^2\right)\right)^{4/3}} \, df_1 df_2$$

$$= \int_{-\infty}^{\infty} \int_{-\infty}^{\infty} e^{2\pi i f_1 t + 2\pi i f_2 s} \left| H_L\left(f_1, \frac{f_1}{U}\right) \right|^2$$

$$\times \frac{1}{U\left(k^2 + 4\pi^2 \left(\left(\frac{f_1}{U}\right)^2 + f_2^2\right)\right)^{4/3}} \, df_1 df_2$$

$$H_L(f_1, f_2) = \int_0^\infty \int_{-b}^b e^{-2\pi i f_1 t - 2\pi i f_2 \xi} l(t, \xi) dt d\xi \qquad (9.12)$$

and the spectral density

$$= P_L(f_1, f_2) = \left| H_L\left(f_1, \frac{f_1}{U}\right) \right|^2 \frac{1}{U\left(k^2 + 4\pi^2 \left(\left(\frac{f_1}{U}\right)^2 + f_2^2\right)\right)^{4/3}}. \qquad (9.13)$$

Similarly

$$R_M(t, s) = \int_{-\infty}^{\infty} \int_{-\infty}^{\infty} e^{2\pi i \nu_1 U t + 2\pi i \nu_2 s} |H_M(U\nu_1, \nu_1)|^2$$

$$\times \frac{1}{\left(k^2 + 4\pi^2 \left(\nu_1^2 + \nu_2^2\right)\right)^{4/3}}$$

$$= \int_{-\infty}^{\infty} \int_{-\infty}^{\infty} e^{2\pi i f_1 t + 2\pi i f_2 s} \left| H_M\left(f_1, \frac{f_1}{U}\right) \right|^2$$

$$\times \frac{1}{U\left(k^2 + 4\pi^2 \left(\left(\frac{f_1}{U}\right)^2 + f_2^2\right)\right)^{4/3}} df_1 df_2,$$

where

$$H_M(f_1, f_2) = \int_{-b}^b \int_0^\infty e^{-2\pi i f_1 t - 2\pi i f_2 \xi} m(t, \xi) dt d\xi \qquad (9.14)$$

and the spectral density

$$P_M(f_1, f_2) = \left| H_M\left(f_1, \frac{f_1}{U}\right) \right|^2 \frac{1}{U\left(k^2 + 4\pi^2\left(\left(\frac{f_1}{U}\right)^2 + f_2^2\right)\right)^{4/3}}.$$

To evaluate H_L, H_M we need to specify M. We begin with $M = 0$.

The Case $M = 0$

Lemma 9.1. *See [6, 64].*

$$H_L\left(v_1, \frac{v_1}{U}\right) = \rho U\left[2\pi T\left(2\pi i \frac{v_1}{U}b\right)\left(J_0\left(2\pi b\frac{v_1}{U}\right)\right.\right.$$
$$\left.\left. -i J_1\left(2\pi b\frac{v_1}{U}\right)\right) + 2\pi i J_1\left(2\pi b\frac{v_1}{U}\right)\right]. \qquad (9.15)$$

This is called the Kussner Function in [6]. $T(\cdot)$ is the Theodorsen function; see Sect. 5.4.

Proof. Our derivation is a variant on both [6] and [64] and follows our version of the solution to the Possio equation for $M = 0$ given in Sect. 5.4. It is mostly a tedious exercise in Fourier transforms and evaluating various integrals, as in [64].

$$H_L\left(v_1, \frac{v_1}{U}\right) = \int_0^\infty e^{-2\pi i v_1 t}\, dt \int_{-b}^b l(t, \xi) e^{-2\pi i(v_1 \xi/U)}\, d\xi$$

$$= \rho U \int_{-b}^b \int_{-b}^b \hat{\Psi}(2\pi i v_1, x, \xi) e^{-2\pi i(v/1\xi U)}\, d\xi dx$$

and from Sect. 5.4:

$$\int_{-b}^b \hat{\Psi}(2\pi i v_1, x, \xi) e^{-2\pi i(v_1/U)\xi}\, d\xi$$

is the solution to the Possio equation for $\hat{w}_a(., \xi) = e^{-2\pi i(v_1/U)\xi}$.

Hence using (5.46) let

$$q(\cdot) = (I + k\mathcal{L}(0))\left[u + h\frac{L(k, (I + k\mathcal{L}(0))u))}{be^{kb}(K_0(kb) + K_1(kb))}\right],$$

where

$$u = 2\tau \hat{W}_a(\cdot),$$

$$k = 2\pi i \frac{v_1 b}{U},$$

$$u(x) = \frac{2}{\pi} \sqrt{\frac{b-x}{b+x}} \int_{-b}^{b} \sqrt{\frac{b+\xi}{b-\xi}} \frac{e^{-2\pi i (v_1/U)\xi}}{\xi - x} d\xi.$$

Hence we have

$$H_L\left(v_1, \frac{v_1}{U}\right) = \int_{-b}^{b} (1 + k(b-x))u(x)dx$$

$$+ \int_{-b}^{b} (1 + k(b-x))h(x)dx \left(\frac{L(k, (I + k\mathcal{L}(0)u))}{be^{kb}(K_0(kb) + K_1(kb))}\right),$$

where

$$h(x) = \frac{1}{\pi} \sqrt{\frac{b-x}{b+x}} \int_{0}^{\infty} e^{-k\sigma} \sqrt{\frac{2b+\sigma}{\sigma}} \frac{1}{x-b-\sigma} d\sigma.$$

Hence

$$\int_{-b}^{b} (1 + k(b-x))h(x)dx$$

$$= \int_{-b}^{b} \frac{1}{\pi} \sqrt{\frac{b-x}{b+x}} \frac{1-k(x-b)}{x-b-\sigma} dx \int_{0}^{\infty} e^{-k\sigma} \sqrt{\frac{2b+\sigma}{\sigma}} d\sigma$$

$$= b \int_{-1}^{1} \frac{(1-kb\sigma)dx}{\pi(x-1-\sigma)} \sqrt{\frac{1-x}{1+x}} \int_{0}^{\infty} e^{-kb\sigma} \sqrt{\frac{2+\sigma}{\sigma}} d\sigma$$

$$- \int_{-1}^{1} \frac{kb}{\pi} \sqrt{\frac{1-x}{1+x}} dx \int_{0}^{\infty} e^{-kb\sigma} \sqrt{\frac{2+\sigma}{\sigma}} d\sigma$$

$$= b \int_{0}^{\infty} (1-kb\sigma) \left(1 - \sqrt{\frac{2+\sigma}{\sigma}}\right) e^{-kb\sigma} d\sigma - kb \int_{0}^{\infty} e^{-kb\sigma} \sqrt{\frac{2+\sigma}{\sigma}} d\sigma$$

$$= b \left[\int_{0}^{\infty} (1-kb\sigma)e^{-kb\sigma} d\sigma - \int_{0}^{\infty} \sqrt{\frac{2+\sigma}{\sigma}} e^{-kb\sigma} d\sigma \right.$$

$$\left. + kb \int_{0}^{\infty} \sigma \sqrt{\frac{2+\sigma}{\sigma}} e^{-kb\sigma} d\sigma \right] - kb \int_{0}^{\infty} e^{-kb\sigma} \sqrt{\frac{2+\sigma}{\sigma}} d\sigma$$

$$= -b \int_0^\infty \sqrt{\frac{2+\sigma}{\sigma}} e^{-kb\sigma} d\sigma - k \frac{d}{dk} \int_0^\infty \sqrt{\frac{2+\sigma}{\sigma}} e^{-kb\sigma} d\sigma - kb^2$$

$$\times \int_0^\infty e^{-kb\sigma} \sqrt{\frac{2+\sigma}{\sigma}} d\sigma$$

$$= -be^{kb}(K) - k(be^{kb}(K) + e^{kb}(K)') - kb^2 e^{kb}(K)$$

$$= -be^{kb}(K) - kbe^{kb}(K) + kbe^{kb}\left(K_1 + \frac{K_1}{kb} + K_0\right) - kbe^{kb}(K)$$

$$= -be^{kb}(K) + e^{kb}K_1 - kb(K)e^{kb},$$

where

$$\int_0^\infty e^{-k\sigma} \sqrt{\frac{2b+\sigma}{\sigma}} d\sigma = be^{kb} K.$$

Next:

$$\int_{-b}^b u(x)dx = \int_{-b}^b dx \frac{2}{\pi} \sqrt{\frac{b-x}{b+x}} \int_{-b}^b \sqrt{\frac{b+\xi}{b-\xi}} \frac{e^{-2\pi i \frac{v}{U}\xi}}{\xi - x} d\xi$$

$$= 2 \int_{-b}^b \sqrt{\frac{b+\xi}{b-\xi}} e^{-2\pi i (v/U)\xi} d\xi$$

$$= 2\pi b J$$

$$J = J_0\left(2\pi \frac{bv}{U}\right) - iJ_1\left(2\pi \frac{bv}{U}\right) \int_{-b}^b (b-x)u(x)dx$$

$$= b \int_{-1}^1 \frac{2}{\pi} \sqrt{\frac{1-x}{1+x}} \frac{1-s+s-x}{s-x} dx \int_{-1}^1 \sqrt{\frac{1+s}{1-s}} e^{-2\pi i (bv/U)s} ds$$

$$= \int_{-1}^1 2b(1 + (1-s)) \sqrt{1+s/1-s} e^{-2\pi i (bv/U)s} ds$$

$$= 2\pi b J + \frac{J_1\left(2\pi \frac{bv}{U}\right) U}{v}.$$

Hence

$$\int_{-b}^b (1 + k(b-x))u(x)dx = 2\pi b J + k2\pi b J + 2\pi i J_1\left(2\pi \frac{bv}{U}\right).$$

And

$$L(k, (I + k\mathcal{L}(0)u)) = \int_{-b}^{b} e^{-k(b-x)} \left(u(x) + k \int_{-b}^{x} u(s)\mathrm{d}s \right) \mathrm{d}x$$

$$= \int_{-b}^{b} u(x)\mathrm{d}x = 2\pi b J.$$

Hence

$$\frac{L(k, (I + k\mathcal{L}(0)u))}{be^{kb}(K_0(kb) + K_1(kb))} = \frac{2\pi(J_0(kb) - iJ_1(kb))}{(K_0(kb) + K_1(kb))e^{kb}} = \frac{2\pi J}{e^{kb} K}.$$

And finally

$$\int_{-b}^{b} (1 + k(b - x))h(x)\mathrm{d}x \left(\frac{L(k, (I + k\mathcal{L}(0)u))}{be^{kb}(K_0(kb) + K_1(kb))} \right)$$

$$= \left(-be^{kb}(K) + e^{kb} K_1 - kb(K)e^{kb} \right) \frac{2\pi J}{e^{kb} K} + \int_{-b}^{b} (1 + k(b - x))u(x)\mathrm{d}x$$

$$= 2\pi b J + k2\pi b J + 2\pi i J_1 \left(2\pi \frac{bv}{U} \right) + (-b + T(kb) - kb)2\pi J$$

$$= 2\pi \left(T(kb)J + iJ_1 \left(2\pi \frac{bv}{U} \right) \right),$$

which yields (9.15) upon multiplication by ρU. $\square$

Lemma 9.2.

$$H_M \left(v_1, \frac{v_1}{U} \right) = \rho U \left[\frac{1 + 2(1 + kb)^2}{kb^2} - \frac{(1 + kb)}{k} e^{kb} K_1(kb) \right.$$

$$- \frac{U}{v}((1 + kb)J_1 + ikbJ_2)$$

$$\left. + \left(2kb + 5k^2b^2 + \frac{1 + kb}{ke^{kb} K} - \frac{(3 + 2kb)}{k} T \right) 2\pi J \right]$$

$$- ab\rho U \left[2\pi T \left(J_0 \left(2\pi b\frac{v_1}{U} \right) - iJ_1 \left(2\pi b\frac{v_1}{U} \right) \right) + 2\pi i J_1 \left(2\pi b\frac{v_1}{U} \right) \right],$$

where

$$T = T(kb); \quad J = J_0 \left(2\pi b\frac{v_1}{U} \right) + iJ_1 \left(2\pi b\frac{v_1}{U} \right),$$

$$K = K_0(kb) + K_1(kb); \quad k = \frac{2\pi i v}{b},$$

$$J_1 = J_1\left(2\pi b \frac{v_1}{U}\right); \quad J_2 = J_2\left(2\pi b \frac{v_1}{U}\right),$$

which is clearly more complicated than H_L and there is no name attached to it yet.

Proof.

$$H_M\left(f_1, \frac{f_1}{U}\right) = \int_{-b}^{b} \int_{0}^{\infty} e^{-2\pi i f_1 t - 2\pi i(f_1/U)\xi} m(t, \xi) dt \, d\xi$$

$$= \int_{-b}^{b} \int_{0}^{\infty} e^{-2\pi i f_1 t - 2\pi i(f_1/U)\xi} \int_{-b}^{b} \rho U(x - ab)\Psi(t, x, \xi) dx \, dt \, d\xi,$$

$$= \int_{-b}^{b} e^{-2\pi i(f_1/U)\xi} \int_{-b}^{b} \rho U(x - ab)\hat{\Psi}(2\pi i f_1, x, \xi) dx \, d\xi$$

$$= \rho U \int_{-b}^{b} e^{-2\pi i(f_1/U)\xi} \int_{-b}^{b} \rho U(x - ab)\hat{\Psi}(2\pi i f_1, x, \xi) dx \, d\xi - abH\left(f_1, \frac{f_1}{U}\right).$$

So we need only to calculate the first term. Now

$$\int_{-b}^{b} e^{-2\pi i(f_1/U)\xi} \int_{-b}^{b} x\hat{\Psi}(2\pi i f_1, x, \xi) dx \, d\xi = \int_{-b}^{b} xq(x) dx,$$

$$q(\cdot) = (I + k\mathcal{L}(0))\left[u + h\frac{2\pi J}{e^{kb} K}\right].$$

Now

$$\int_{-b}^{b} x(1 + k(b - x))u(x) dx = \int_{-b}^{b} (1 + kb)xu(x) dx - k\int_{-b}^{b} x^2 u(x) dx$$

$$\int_{-b}^{b} x^2 u(x) dx = \int_{-b}^{b} x^2 dx \frac{2}{\pi} \sqrt{\frac{b - x}{b + x}} \int_{-b}^{b} \sqrt{\frac{b + \xi}{b - \xi}} \frac{e^{-2\pi i(v/U)\xi}}{\xi - x} d\xi$$

$$= \int_{-b}^{b} \frac{x^2}{\xi - x} dx \frac{2}{\pi} \sqrt{\frac{b - x}{b + x}} \int_{-b}^{b} \sqrt{\frac{b + \xi}{b - \xi}} e^{-2\pi i(v/U)\xi} d\xi$$

$$= -b^2 \int_{-1}^{1} \left(x + \xi + \frac{\xi^2}{x - \xi}\right) dx \frac{2}{\pi} \sqrt{\frac{1 - x}{1 + x}} \int_{-1}^{1} \sqrt{\frac{1 + \xi}{1 - \xi}} e^{-2\pi i(vb/U)\xi} d\xi$$

$$= 2b^2 \int_{-1}^{1} \left(\frac{1}{2} + \xi^2 - \xi \right) \sqrt{\frac{1+\xi}{1-\xi}} e^{-2\pi i (vb/U)\xi} d\xi$$

$$= b^2 J - i \frac{bU}{v} J_2 \left(2\pi \frac{v}{b} \right).$$

Hence

$$\int_{-b}^{b} x(1 + k(b - x))u(x)dx$$

$$= -(1 + kb)J_1 \left(2\pi \frac{v}{b} \right) \frac{U}{v} - kb^2 J + i \frac{kbU}{v} J_2 \left(2\pi \frac{v}{b} \right).$$

Next

$$\int_{-b}^{b} x^2 h(x)dx$$

$$= b^2 \int_{-1}^{1} \frac{1}{\pi} \left(x + 1 + \sigma - \frac{(1+\sigma)^2}{x - 1 - \sigma} \right) \sqrt{\frac{1-x}{1+x}} dx \cdot \int_{0}^{\infty} e^{-kb\sigma} \sqrt{\frac{2+\sigma}{\sigma}} d\sigma,$$

$$= \int_{0}^{\infty} b^2 \left(\frac{1}{2} + 1 + \sigma - (1+\sigma)^2 \left(\sqrt{\frac{\sigma}{2+\sigma}} - 1 \right) \right) e^{-kb\sigma} \sqrt{\frac{2+\sigma}{\sigma}} d\sigma$$

$$= \int_{0}^{\infty} b^2 \left(\frac{1}{2} + 1 + \sigma \right) e^{-kb\sigma} \sqrt{\frac{2+\sigma}{\sigma}} d\sigma$$

$$- \int_{0}^{\infty} b^2 (1+\sigma)^2 \left(1 - \sqrt{\frac{2+\sigma}{\sigma}} \right) e^{-kb\sigma} d\sigma$$

$$= -b^2 \int_{0}^{\infty} (1+\sigma)^2 e^{-kb\sigma} d\sigma + b^2 \int_{0}^{\infty} \left(\frac{1}{2} + 2 + 3\sigma + \sigma^2 \right) e^{-kb\sigma} \sqrt{\frac{2+\sigma}{\sigma}} d\sigma$$

$$= -\frac{2 + 2bk + b^2 k^2}{bk^3} + \frac{1}{2k^2} e^{bk} ((4 + 2kb)K_1(kb) + (2kb + 5k^2 b^2)(K_0 + K_1)).$$

Hence

$$\int_{-b}^{b} (1 + kb)xh(x)dx - k \int_{-b}^{b} x^2 h(x)dx$$

$$= \frac{(1 + kb)}{k} \left(1 + \frac{1}{kb} - e^{kb} K_1(kb) \right) + \frac{2 + 2bk + b^2 k^2}{bk^2}$$

$$- \frac{1}{2k} e^{bk} ((4 + 2kb)K_1(kb) + (2kb + 5k^2 b^2)K).$$

Hence

$$\int_{-b}^{b} q(x)dx - (1 + kb)J_1\left(2\pi\frac{v}{b}\right)\frac{U}{v} - kb^2 J + i\frac{kbU}{v}J_2\left(2\pi\frac{v}{b}\right)$$
$$+ \left[\frac{(1 + kb)}{k}\left(-T + \frac{(1 + kb)}{k}\frac{1}{e^{kb}K}\right)\right.$$
$$\left. - \frac{1}{2k}(4 + 2kb)T + (2kb + 5k^2b^2)\right]2\pi J,$$

which, collecting like terms

$$= \frac{1 + 2(1 + kb)^2}{kb^2} - \frac{(1 + kb)}{k}e^{kb}K_1 - \frac{U}{v}((1 + kb)J_1 + ikbJ_2)$$
$$+ \left(2kb + 5k^2b^2 + \frac{1 + kb}{ke^{kb}K} - \frac{(3 + 2kb)}{k}T\right)2\pi J.$$

Hence

$$H_M\left(f_1, \frac{f_1}{U}\right) = \rho U\left[\frac{1 + 2(1 + kb)^2}{kb^2} - \frac{(1 + kb)}{k}e^{kb}K_1 - \frac{U}{v}((1 + kb)J_1 + ikbJ_2)\right.$$
$$\left. + \left(2kb + 5k^2b^2 + \frac{1 + kb}{ke^{kb}K} - \frac{(3 + 2kb)}{k}T(kb)\right)2\pi J\right]$$
$$- ab\rho U\left[2\pi T(kb)J + 2\pi i J_1\left(2\pi b\frac{v_1}{U}\right)\right],$$

which is clearly more complicated than H_L and there is no name attached to it yet. □

The dependence on the speed does appear to be complicated. However, as $U \to \infty$, we can see that both spectral densities go to zero rapidly. Turbulence is not significant at transonic speeds, and so $M = 0$ is a good assumption.

Finally if we define:

$$N_g(t, s) = \begin{pmatrix} L_g(t, s) \\ M_g(t, s) \end{pmatrix}$$

the corresponding spectral density (matrix) is defined as

$$P_g(f_1, f_2) = \begin{pmatrix} P_L & P_{LM} \\ P_{ML} & P_M \end{pmatrix},$$

where

$$P_{LM}(f_1, f_2) = H_L\left(f_1, \frac{f_1}{U}\right)\frac{conj \cdot H_M\left(f_1, \frac{f_1}{U}\right)}{U\left(k^2 + 4\pi^2\left(\left(\frac{f_1}{U}\right)^2 + f_2^2\right)\right)^{4/3}}$$

$$P_{ML}(f_1, f_2) = H_M\left(f_1, \frac{f_1}{U}\right) \frac{\text{conj} \cdot H_L\left(f_1, \frac{f_1}{U}\right)}{U\left(k^2 + 4\pi^2\left(\left(\frac{f_1}{U}\right)^2 + f_2^2\right)\right)^{4/3}}$$

and P_L, P_M as already defined.

9.4 Structure Response to Turbulence

The structure being the Goland beam, the dynamics are now given by the stochastic differential equation forced by the stochastic lift and moment:

$$m\ddot{h}(t, s) + S\ddot{\theta}(t, s) + EI\frac{\partial^4 h(t, s)}{\partial s^4} - \int_0^t 1(\sigma, \xi)w_a(t - \sigma, \xi, s)d\sigma d\xi = L_g(t, s) \tag{9.16}$$

$$I_\theta\ddot{\theta}(t, s) + S\ddot{h}(t, s) - GJ\frac{\partial^2 \theta(t, s)}{\partial s^2} - \int_0^t m(\sigma, \xi)w_a(t - \sigma, \xi, s)d\sigma d\xi$$

$$= M_g(t, s) \qquad 0 < t; 0 < s < \ell. \tag{9.17}$$

With CF or FF end conditions.

At this point we may use the state representation in Chap. 5 for an abstract formulation; but here it suffices to work the Laplace Transforms to find the transfer function of the system represented by the left-hand side of the equation. Thus we have

$$\lambda^2 \mathcal{M}\hat{x}(\lambda) + \lambda \mathcal{D}\hat{x}(\lambda) + \mathcal{K}\hat{x}(\lambda) + A\hat{x}(\lambda) - \hat{\mathcal{F}}(\lambda)\hat{x}(\lambda) = 0, \tag{9.18}$$

where

$$\mathcal{M} = \begin{pmatrix} m + \pi\rho b^2 & S - \pi\rho b^3 a \\ S - \rho b^3 a\pi & I_\theta + \pi\rho b^4\left(a^2 + \frac{1}{8}\right) \end{pmatrix},$$

$$\mathcal{D} = \begin{pmatrix} 0 & \pi\rho b^2 U_\infty \\ -\pi\rho b^2 U_\infty & 0 \end{pmatrix},$$

$$\mathcal{K} = \begin{pmatrix} 0 & 0 \\ 0 & -\pi\rho b^2 U_\infty^2 \end{pmatrix},$$

$$\hat{\mathcal{F}}(\lambda) = \begin{pmatrix} \frac{T(k)}{k}\rho b U_\infty \\ \frac{T(k)}{k}\rho U_\infty b^2\left(\frac{1}{2} + a\right) - \frac{\rho b^2 U_\infty}{2} \end{pmatrix}$$

and $\hat{x}(\lambda,.)$ is the Laplace transform of

$$x(t,.) = \begin{pmatrix} h(t,.) \\ \theta(t,.) \end{pmatrix}.$$

Thus the transfer function is the operator

$$(\lambda^2 \mathcal{M} + \lambda \mathcal{D} + \mathcal{K} + A - \hat{\mathcal{F}}(\lambda))^{-1}$$

defined for λ in a right half-plane. To determine this we solve:

$$(\lambda^2 \mathcal{M} + \lambda \mathcal{D} + \mathcal{K} + A - \hat{\mathcal{F}}(\lambda))\hat{x}(\lambda,.) = g.$$

Or, following Sect. 5.4:

$$\hat{h}(\lambda,s)'''' - w_1\hat{h}(\lambda,s) - w_2\hat{\theta}(\lambda,s) = \frac{1}{EI}g_1(s),$$

$$\hat{\theta}(\lambda,s)'' - w_3\hat{h}(\lambda,s) - w_4\hat{\theta}(\lambda,s) = \frac{-1}{GJ}g_2(s),$$

$$g = \begin{pmatrix} g_1 \\ g_2 \end{pmatrix}$$

with CF end conditions, to be specific. Then with $y(\lambda,s)$ defined as in Sect. 5.4, we have:

$$y(\lambda,s)' = A(\lambda)y(\lambda,s) + Bg(s) \qquad 0 < s < \ell,$$

where

$$B = \begin{pmatrix} 0 & 0 \\ 0 & 0 \\ 0 & 0 \\ \frac{1}{EI} & 0 \\ 0 & 0 \\ 0 & -\frac{1}{GJ} \end{pmatrix}.$$

This is a two-point boundary-value problem and continuing as in Sect. 5.4, we have

$$\hat{x}(\lambda,s) = \int_0^\ell G(\lambda,s,\sigma)g(\sigma)d\sigma$$

excepting the aeroelastic modes, where $G(\lambda,.,.)$ is the Green's function (see [64]):

$$G(\lambda,s,\sigma) = Ce^{A(\lambda)(s-\sigma)}B - Ce^{A(\lambda)s}QD(\lambda)^{-1}Pe^{A(\lambda)(\ell-\sigma)}B \qquad 0 < \sigma < s$$

$$- Ce^{A(\lambda)s}QD(\lambda)^{-1}Pe^{A(\lambda)(\ell-\sigma)}B \qquad s < \sigma < \ell,$$

where C is the matrix

$$\begin{pmatrix} 1 & 0 & 0 & 0 & 0 & 0 \\ 0 & 0 & 0 & 0 & 1 & 0 \end{pmatrix}.$$

With $W(t, s, \sigma)$ denoting the inverse Laplace transform of $G(\lambda, s, \sigma)$, we have the steady-state structure response to gust

$$x(t, s) = \int_0^\infty \int_0^\ell W(\tau, s, \sigma) N_g(t - \tau, \sigma) d\sigma d\tau$$

at any point s along the span, where the gust load is

$$N_g(t, s) = \begin{pmatrix} L_g(t, s) \\ M_g(t, s) \end{pmatrix} \qquad 0 < s < \ell.$$

Lemma 9.3. *The spectral density of the response at any points along the span is given by*

$$P_g(s, f) = \int_{-\infty}^\infty \hat{G}(2\pi i f, s, 2\pi f_2) P_g(f, f_2) \hat{G}(2\pi i f, s, 2\pi f_2)^* d f_2,$$

where

$$\hat{G}(2\pi i f, s, 2\pi f_2) = \int_0^\ell e^{-2\pi i f_2 \xi} G(2\pi i f, s, \xi) d\xi.$$

Proof. Straightforward calculation.   □

We note that the wing displacement at any point (x, s) is given by:

$$z(t, s) = h(t, s) + (x - ab)\theta(t, s), \qquad |x| < b, 0 < s < \ell.$$

Hence the corresponding gust response spectral density is given by:

$$P_g(f, x, s) = L P_g(s, f) L^*,$$

where

$$L = \text{Row}[1 \ (x - ab)].$$

9.5   Illustrative Example

By way of an illustrative numerical example we consider (see [12]): a model used by Lin [25] in his road-runner ground vehicle experiments. This is typical of a forward-swept wing [6], where the divergence speed is less than the flutter speed. This is a torsion-mode flutter at 1.4 H flutter speed 10.7 fps and divergence Speed 9.59 fps. The parameters are:

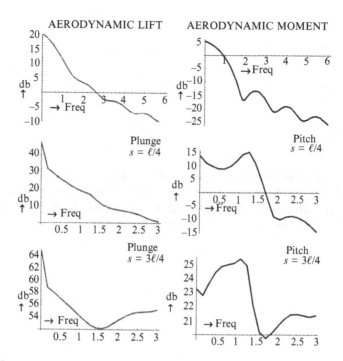

Fig. 9.2 Structure response spectral density at divergence speed: $U = U_d$ Lin model

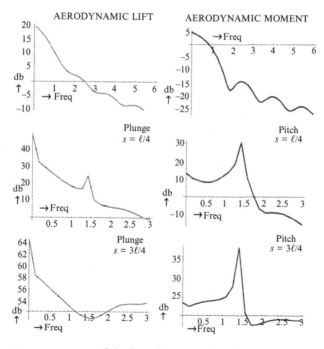

Fig. 9.3 Structure response spectral density at flutter speed: $U = U_F$ Lin model

FPS Units
$m = 0.009937$
$I_\theta = 0.0004403$
$GJ = 3.8383$
$EI = 1.7542$
$l = 10.19$
$a = 0; \quad b = 0.364593$
$S = -0.0003623$
$IT2.3H$

The normalized frequency is larger than for the Goland model. The plots are given in Figs. 9.2. and 9.3. As expected, the spectral density peaks at zero corresponding to the divergence speed and also of course at the flutter mode. Most of the turbulence energy is in the low frequencies, and of course low speeds.

Chapter 10
Addendum: Axial Air flow Theory—Continuum Models

10.1 Introduction

In contrast to the previous chapters the present chapter has nothing to do with aircraft! It is presented more as an addendum devoted to the currently emerging non-aircraft areas of application of aeroelastic flutter such as for a biomedical application "palatal flutter" [102, 105] where the air flow is axial in contrast to normal. And piezoelectric power harvesting from wind as alternate energy source [106, 109] where the orientation of the structure with respect to the flow whether normal or axial is left open as part of system optimization. Both involve the dynamics of a structure modeled as a plate or beam in an air stream and the main interest is again in the aeroelastic flutter phenomenon. But this time our objective is not to suppress it but rather to maintain it, as in the power harvesting application.

The corresponding continuum aeroelastic theory for axial flow turns out to be much more complicated than that for normal flow though the starting point is of course the same, and the basic ideas are the same, which is the reason for presenting it here as a natural continuation of the previous theory enabling us to call attention in particular to the differences from normal flow even as an addendum and limited largely to problem formulation. It serves also as a harbinger of a succeeding volume devoted to a detailed treatment including full-scale experimental results that can be run on low speed wind tunnels and do not require hangars or expensive flight tests.

10.2 The Aeroelastic Equations

There is no longer of course the notion of the rigid-body aircraft axis. Instead we go with the structure beam axis which can be oriented as desired with respect to the air flow.

A V Balakrishnan, *Aeroelasticity: The Continuum Theory*,
DOI 10.1007/978-1-4614-3609-6_10, © Springer Science+Business Media, LLC 2012

Here we consider only axial flow. Thus with $\vec{j}$ as before denoting the unit vector along the structure axis, the far field flow is specified by

$$q_\infty = \vec{j}\,U_\infty. \tag{10.1}$$

Or, more generally, $q_\infty = U_\infty(\vec{j}\cos\alpha + \vec{k}\sin\alpha)$ so that α is the angle of attack. Here, however, we consider only $\alpha = 0$. The right-hand screw convention would make $\vec{i}\times\vec{j}$ the "plunge" unit vector. Hence the unit z direction is given by $-\vec{k}$. The structure model generally depends on the application; for example, for the application to palatal flutter [51] it is a thin plate and torsion per se is not considered.

But here we continue with the linear Goland beam model with two degrees of freedom—bending and torsion—as in Chap. 2 underscoring in particular the profound difference from normal flow. With no association to an aircaft wing we change the span variable to $-\ell < y < \ell$ so that the structure dimensions are:

$$\Gamma = [-b, b] \times [-\ell,\ \ell].$$

The structure equations continue to be

$$m\ddot{h}(t,\ y) + S\ddot{\theta}(t,\ y) + EIh''''(t,\ y) = \int_{-b}^{b} \delta p(t,\ x,\ y)\mathrm{d}x, \quad -\ell < y < \ell, \tag{10.2}$$

$$I_\theta\ddot{\theta}(t,\ y) + S\ddot{h}(t,\ y) - GJ\theta''(t,\ y) = \int_{-b}^{b}(x - ab)\delta p(t,\ x,\ y)\mathrm{d}x,$$
$$- \ell < y < \ell, \tag{10.3}$$

where the pressure jump $\delta p(.)$ is to be calculated from the aerodynamics.

We consider only CF end conditions

$$h, h', \theta = 0 \quad \text{at } y = -\ell,$$
$$h'', h''', \theta' = 0 \quad \text{at } y = \ell.$$

The flow is assumed to be inviscid and isentropic and hence described again as in Chap. 3 by a potential function $\phi(t,\ x,\ y,\ z)$.

The axial flow makes the theory considerably more complicated. Fortunately for the applications indicated above we have "low speed" flow so that we may take $M = 0$, incompressible flow.

This simplifies the field equation for the potential $\phi(t,\ x,\ y,\ z)$, and its Laplace transform

$$\hat{\phi}(\lambda,\ x,\ y,\ z) = \int_0^\infty e^{-\lambda t}\phi(t,\ x,\ y,\ z)\mathrm{d}t, \quad \mathrm{Re}\lambda > \sigma_a \geq 0$$

to the 3D Laplace equation

$$\frac{\partial^2 \phi(t,.)}{\partial x^2} + \frac{\partial^2 \phi(t,.)}{\partial y^2} + \frac{\partial^2 \phi(t,.)}{\partial z^2} = 0, \qquad -\infty < x, \, y, \, z < \infty,$$

$$\frac{\partial^2 \hat{\phi}(\lambda,.)}{\partial x^2} + \frac{\partial^2 \hat{\phi}(\lambda,.)}{\partial y^2} + \frac{\partial^2 \hat{\phi}(\lambda,.)}{\partial z^2} = 0, \qquad -\infty < x, \, y, \, z < \infty, \qquad (10.4)$$

omitting of course the structure boundary.

We recall the definitions of the pressure $p(.)$ and δp; and acceleration potential $\psi(.)$ and $\delta \psi$, where by the isentropy condition

$$\delta p(t, \, x, \, y) = -\rho \delta \psi(t, \, x, y) \quad x, \, y \text{ in } \Gamma = 0 \text{ otherwise,}$$

where the acceleration potential

$$\psi = \frac{\partial \phi}{\partial t} + \frac{1}{2} \nabla \phi \cdot \nabla \phi.$$

And the Kushner doublet function:

$$A(t, \, x, \, y) = -\frac{\delta \psi}{U} = -\left(\frac{1}{U} \frac{\partial \delta \phi}{\partial t} + \frac{1}{2} \frac{\delta(\nabla \phi \cdot \nabla \phi)}{U} \right), \quad x, \, y \text{ in } \Gamma,$$

so that

$$\delta p(t, x, y) = \rho \, UA(t, x, y).$$

Plus the Kutta condition $A(t, \, x, \, \ell-) = 0$.

And

$$v(t, \, x, \, y, \, z) = \frac{\partial \phi(t, \, x, \, y, \, z)}{\partial z}.$$

As before we use

$$q(.) \text{ for } \nabla \phi.$$

Fluid-Structure Boundary Conditions

Continuing with the same notation as in Chap. 3, we have for the far field:

$$q_\infty = \vec{j} U_\infty. \qquad (10.5)$$

The flow tangency condition is again

$$\frac{\partial \phi(t, x, y, 0\pm)}{\partial z} = \frac{\partial \phi_\infty}{\partial z} + \frac{Dz(t)}{Dt},$$

where

$$z(t, x, y, 0+) = -1(h(t, \ y) + (x - a)\theta(t, y))$$

and

$$\frac{Dz(t)}{D(t)} = \frac{\partial z(t)}{\partial t} + q.\nabla z.$$

We have again (3.20) and (3.21), where now $\partial \phi_\infty / \partial z = 0$.
   Thus

$$v(t, x, y, 0\pm) = (-1)\left[\dot{h}(t, y) + (x - a)\dot{\theta}(t, y) + (q(t, x, y, 0\pm) \cdot \vec{i})\theta(t, y) \right.$$

$$\left. + (q(t, x, y, 0\pm) \cdot \vec{j}) \left(\frac{\partial}{\partial y}(h(t, y) + (x - ab)\theta(t, y)) \right) \right], \qquad x, y \ e \ \Gamma.$$

We begin with the steady state solution.

10.3   Steady-State Solution

We begin with the steady-state solution about which we need to linearize the system
equations. Following Chap. 4, this is given by

$$h(y) = 0 = \theta(y) \qquad |y| < \ell \phi(x, y, z) = yU_\infty,$$

where we may note that the steady-state potential is different from that for normal
flow.

10.4   Power Series Expansion

As in Chap. 5, (5.1), and in the same notation, we invoke the power series expansion
for the potential in terms of the scalar multiplicative parameter λ (to be distinguished
from the Laplace transform variable).

$$\phi(\lambda, t, x, y, z) = yU_\infty + \sum_{k=1}^{\infty} \frac{\lambda^k}{k!} \phi_k(t, x, y, z), \qquad (10.6)$$

where

$$\phi_k(t, x, y, z) = \frac{\partial^k \phi(\lambda, t, x, y, z)}{\partial \lambda^k}\bigg|\lambda = 0,$$

where for each $k \geq 1$ we have the field equation

$$\frac{\partial^2 \phi_k(t,.)}{\partial x^2} + \frac{\partial^2 \phi_k(t,.)}{\partial y^2} + \frac{\partial^2 \phi_k(t,.)}{\partial z^2} = 0, \quad -\infty < x, y, z < \infty.$$

And correspondingly

$$\nabla \phi(\lambda, t, x, y, z) = U_\infty \vec{j} + \sum_{k=1}^{\infty} \frac{\lambda^k}{k!} q_k(t, x, y, z),$$

where $q_k(t, x, y, z) = \nabla \phi_k(t, x, y, z)$ with the boundary conditions

$$\frac{\partial \phi(\lambda, t, x, y, 0\pm)}{\partial z} = \sum_{k=1}^{\infty} \frac{\lambda^k}{k!} \frac{\partial}{\partial z} \phi_k(t, x, y, 0\pm)$$

$$= (-1)\left[\lambda \dot{h}(t, y) + (x - a)\lambda \dot{\theta}(t, y) \right.$$

$$+ \left(\sum_{k=2}^{\infty} \frac{\lambda^k}{(k-1)!} \frac{\partial}{\partial x} \phi_{k-1}(t, x, y, 0\pm) \right) \theta(t, y)$$

$$+ \left(\lambda U_\infty + \sum_{k=2}^{\infty} \frac{\lambda^k}{(k-1)!} \frac{\partial}{\partial y} \phi_k(t, x, y, 0\pm) \right)$$

$$\left. \times \left(\frac{\partial}{\partial y}(h(t, y) + (x - ab)\theta(t, y)) \right) \right], \quad x, y \in \Gamma.$$

Hence

$$\frac{\partial}{\partial z} \phi_1(t, x, y, 0\pm) = (-1)\left[[\dot{h}(t, y) + (x - a)\dot{\theta}(t, y)] \right.$$

$$\left. + U_\infty \left(\frac{\partial}{\partial y}(h(t, y) + (x - a)\theta(t, y)) \right) \right], \quad x, y \in \Gamma$$

and hence in particular $\delta v_1(t, x, y) = 0$.
  For $k \geq 2$ we have:

$$v_k(t, x, y, 0\pm) = \left(\frac{\partial}{\partial x} \phi_{k-1}(t, x, y, 0\pm) \theta(t, y) + \frac{\partial}{\partial y} \phi_{k-1}(t, x, y, 0\pm) \right.$$

$$\left. \times \left(\frac{\partial}{\partial y}(h(t, y) + (x - a)\theta(t, y)) \right) \right). \tag{10.7}$$

Note the difference from the corresponding formula for normal flow in Chap. 6, where the second term is absent. Next for the acceleration potential, we have: cf. (5.7):

$$\psi(\lambda,.) = \frac{\partial\phi(\lambda,.)}{\partial t} + \frac{1}{2}\nabla\phi(\lambda,.)\cdot\nabla\phi(\lambda,.) = \sum_{k=1}^{\infty}\frac{\lambda^k}{k!}\frac{\partial}{\partial t}\phi_k(t,x,y,z)$$

$$+ \frac{1}{2}\left(U_\infty\vec{j} + \sum_{k=1}^{\infty}\frac{\lambda^k}{k!}\nabla\phi_k(t,x,y,z)\right)\cdot\left(U_\infty\vec{j} + \sum_{k=1}^{\infty}\frac{\lambda^k}{k!}\nabla\phi_k(t,x,y,z)\right),$$

$$\psi(\lambda,t,x,y,z) = \frac{U^2}{2} + \sum_{k=1}^{\infty}\frac{\lambda^k}{k!}\psi_k(t,x,y,z),$$

$$\psi_1(t,x,y,z) = \frac{\partial\psi(0,t,x,y,z)}{\partial\lambda}$$

$$= \frac{\partial\phi_1}{\partial t} + U\frac{\partial}{\partial y}\phi_1(t,x,y,z).$$

And following (6.18) we have:

$$\psi_k(t,x,y,z) = \frac{\partial\phi_k}{\partial t} + U\frac{\partial\phi_k}{\partial y}$$

$$+ \sum_{j=1}^{k-1}C_{k,j}q_j(t,x,y,z)\cdot q_{k-j}(t,x,y,z), \qquad k \geq 1. \qquad (10.8)$$

And hence for the Kushner doublet:

$$A(\lambda,t,x,y) = -\frac{\delta\psi(\lambda,t,x,y)}{U} = -\frac{1}{U}\sum_{k=1}^{\infty}\frac{\lambda^k}{k!}\delta\psi_k(t,x,y)$$

$$= \sum_{k=1}^{\infty}\frac{\lambda^k}{k!}A_k(t,x,y).$$

Hence

$$A_k(t,x,y) = -\frac{1}{U}\delta\psi_k(t,x,y)$$

$$= -\frac{1}{U}\left[\frac{\partial\delta\phi_k}{\partial t} + U\frac{\partial\delta\phi_k}{\partial y}\right.$$

$$\left. + \sum_{j=1}^{k-1}C_{k,j}\delta(q_j(t,x,y,z)\cdot q_{k-j}(t,x,y,z))\right], \qquad x,y \in \Gamma.$$

$$(10.9)$$

Now for each k we have from (10.2)

$$\frac{\partial^2 \phi_k(t,.)}{\partial x^2} + \frac{\partial^2 \phi_k(t,.)}{\partial y^2} + \frac{\partial^2 \phi_k(t,.)}{\partial z^2} = 0, \qquad -\infty < x, y, z < \infty.$$

As before we rewrite this as

$$\frac{\partial^2 \phi_k(t,.)}{\partial z^2} = -\left(\frac{\partial^2 \phi_k(t,.)}{\partial x^2} + \frac{\partial^2 \phi_k(t,.)}{\partial y^2} \right) \qquad x, \ y \ \text{in} \ R^2.$$

Again as we did in the case of normal flow, we take $L_p - L_q$ Fourier transforms in the spatial domain, in addition to Laplace transforms in the time domain.

$$\tilde{\hat{\phi}}_k(\lambda, \omega_1, \omega_2) = \int_{-\infty}^{\infty} \int_{-\infty}^{\infty} e^{-i\omega_1 x - i\omega_2 y} \hat{\phi}_k(\lambda, x, y) dx dy$$

and then (10.2) becomes

$$\frac{\partial^2 \tilde{\hat{\phi}}_k}{\partial z^2} = (\omega_1^2 + \omega_2^2) \tilde{\hat{\phi}}_k, \tag{10.10}$$

which to satisfy the far field vanishing conditions leads to the unique solution

$$\tilde{\hat{\phi}}_k(\lambda, \omega_1, \omega_2, z) = e^{-z\sqrt{(\omega_1^2 + \omega_2^2)}} \tilde{\hat{\phi}}_k(\lambda, \omega_1, \omega_2, 0+) \qquad z > 0$$

$$= e^{z\sqrt{(\omega_1^2 + \omega_2^2)}} \ \tilde{\hat{\phi}}_k(\lambda, \omega_1, \omega_2, 0-) \qquad z < 0,$$

which we can express also in the time domain as

$$\tilde{\phi}_k(t, \omega_1, \omega_2, z) = e^{-z\sqrt{(\omega_1^2 + \omega_2^2)}} \tilde{\phi}_k(t, \omega_1, \omega_2, 0+) \qquad z > 0$$

$$= e^{z\sqrt{(\omega_1^2 + \omega_2^2)}} (\tilde{\phi})_k(t, \omega_1, \omega_2, 0 - 1), \qquad z < 0$$

with the tilde denoting the Fourier transform.
    Hence

$$-\frac{\tilde{v}_k(t, \omega_1, \omega_2, 0+)}{\sqrt{(\omega_1{}^2 + \omega_2{}^2)}} = \tilde{\phi}_k(t, \omega_1, \omega_2, 0+),$$

$$\frac{\tilde{v}_k(t, \omega_1, \omega_2, 0-)}{\sqrt{(\omega_1{}^2 + \omega_2{}^2)}} = \tilde{\phi}_k(t, \omega_1, \omega_2, 0-). \tag{10.11}$$

The flow tangency boundary condition is

$$v(\lambda, t, x, y, 0\pm) = (-1)\Big[\lambda \dot{h}(t, y) + (x - a)\lambda \dot{\theta}(t, y)$$

$$+ \left(\frac{\partial}{\partial x}\phi(\lambda, t, x, y, 0\pm)\right)\lambda\theta(t, y) + \left(\frac{\partial}{\partial y}\phi(\lambda, t, x, y, 0\pm)\right)$$

$$\times \left(\frac{\partial}{\partial y}\lambda(h(t, y) + (x - a)\theta(t, y))\right)\Big], \qquad x, y \in \Gamma.$$

Hence

$$\sum_{k=1}^{\infty}\frac{\lambda^k}{k!}v_k(t, x, y, 0\pm)$$

$$= (-1)\Big[\lambda\dot{h}(t, y) + (x - a)\lambda\dot{\theta}(t, y)$$

$$+ \left(\sum_{k=1}^{\infty}\frac{\lambda^k}{k-1!}\frac{\partial\phi_{k-1}}{\partial x}(t, x, y, 0\pm)\right)\theta(t, y)$$

$$+ \left(\sum_{k=1}^{\infty}\frac{\lambda^k}{k-1!}\frac{\partial\phi_{k-1}}{\partial y}(t, x, y, 0\pm)\right)\left(\frac{\partial}{\partial y}(h(t, y) + (x - a)\theta(t, y))\right)\Big].$$

For $k = 1$,

$$v_1(t, x, y, 0\pm) = (-1)\Big[\dot{h}(t, y) + (x - ab)\dot{\theta}(t, y)$$

$$+ U\left(\frac{\partial}{\partial y}(h(t, y) + (x - ab)\theta(t, y))\right)\Big]$$

$$\to v_1(t, x, y, 0+) = v_1(t, x, y, 0-).$$

Hence by (10.11),

$$-\frac{\tilde{v}_1(t, \omega_1, \omega_2, 0)}{\sqrt{(\omega_1{}^2 + \omega_2{}^2)}} = \tilde{\phi}_1(t, \omega_1, \omega_2, 0+),$$

$$\frac{\tilde{v}_1(t, \omega_1, \omega_2, 0)}{\sqrt{(\omega_1^2 + \omega_2{}^2)}} = \tilde{\phi}_1(t, \omega_1, \omega_2, 0-)$$

and hence

$$\phi_1(t, \omega_1, \omega_2, 0-) = -\phi_1(t, \omega_1, \omega_2, 0+),$$

$$\delta\tilde{\phi}_1(t, \omega_1, \omega_2) = 2\tilde{v}_1(t, \omega_1, \omega_2, 0)\Big/\left(\sqrt{(\omega_1^2 + \omega_2^2)}\right). \qquad (10.12)$$

For $k \geq 2$:

$$
v_k(t, x, y, 0\pm) = k(-1) \left[\frac{\partial \phi_{k-1}}{\partial x}(t, x, y, 0\pm)\theta(t, y) \right.
$$

$$
\left. + \frac{\partial \phi_{k-1}}{\partial y}(t, x, y, 0\pm) \left(\frac{\partial}{\partial y}(h(t, y) + (x - ab)\theta(t, y)) \right) \right].
$$

$$(10.13)$$

For $k = 2$:

$$
v_2(t, x, y, 0+) = (-2) \left[\frac{\partial \phi_1}{\partial x}(t, x, y, 0+)\theta(t, y) \right.
$$

$$
\left. + \frac{\partial \phi_1}{\partial y}(t, x, y, 0\pm) \left(\frac{\partial}{\partial y}(h(t, y) + (x - ab\theta(t, y))) \right) \right],
$$

$$
v_2(t, x, y, 0-) = (-2) \left[\frac{\partial \phi_1}{\partial x}(t, x, y, 0-)\theta(t, y) \right.
$$

$$
\left. + \frac{\partial \phi_1}{\partial y}(t, x, y, 0-) \left(\frac{\partial}{\partial y}(h(t, y) + (x - ab\theta(t, y))) \right) \right]
$$

$$
= -v_2(t, x, y, 0+)
$$

$\rightarrow$ by (10.11)

$$
\phi_2(t, x, y, 0-) = -\phi_2(t, x, y, 0+).
$$

More generally then:

$$
\phi_k(t, x, y, 0-) = \phi_k(t, x, y, 0+) \text{ for } k \text{ even,}
$$

$$
\tilde{\phi}_k(t, \omega_1, \omega_2, 0-) = -\tilde{\phi}_k(t, \omega_1, \omega_2, 0+) \text{ for } k \text{ odd,}
$$

$$
\tilde{v}_k(t, \omega_1, \omega_2, 0+) = \tilde{v}_k(t, \omega_1, \omega_2, 0-) \text{ for } k \text{ odd,}
$$

$$
\tilde{v}_k(t, \omega_1, \omega_2, 0+) = -\tilde{v}_k(t, \omega_1, \omega_2, 0) \text{ for k even,} \quad (10.14)
$$

$$
A_k(t, x, y) = -\frac{1}{U}\delta\psi_k(t, x, y)
$$

$$
= -\frac{1}{U} \left[\frac{\partial \delta\phi_k}{\partial t} + U\frac{\partial \delta\phi_k}{\partial y} + \sum_{j=1}^{k-1} C_{k,j} \left(q_j(t, x, y, 0+) \cdot q_{k-j}(t, x, y, 0+) \right. \right.
$$

$$
\left. \left. - q_j(t, x, y, 0-) \cdot q_{k-j}(t, x, y, 0-) \right) \right], \quad x, y \in \Gamma. \quad (10.15)
$$

For normal flow we showed in Chap. 6 that

$$\sum_{j=1}^{k-1} c_{k,j} (\nabla \phi_j (t, x, y, 0+) \cdot \nabla \phi_{k-j} (t, x, y, 0+)$$

$$- \nabla \phi_j (t, x, y, 0-) \cdot \nabla \phi_{k-j} (t, x, y, 0-)) = 0.$$

However, as we show below, this won't hold in the present case because obviously if it did, the whole problem would reduce to linear because the field equation is now linear. As before we start with the linear theory.

10.5   Linear Theory

We study first the linear case: $k = 1$.
   Define the Laplace transforms

$$\hat{A}_1(\lambda, x, y) = \int_0^\infty e^{-\lambda t} A(t, x, y) dt \qquad \text{Re } \lambda > \sigma_a \geq 0,$$

$$\hat{\phi}_1(\lambda, x, y) = \int_0^\infty e^{-\lambda t} \phi_1(t, x, y) dt \qquad \text{Re } \lambda > \sigma_a \geq 0,$$

$$\hat{\psi}_1(\lambda, x, y) = \int_0^\infty e^{-\lambda t} \psi_1(t, x, y) dt \qquad \text{Re } \lambda > \sigma_a \geq 0$$

and others similarly below.
   Then we have:

$$\hat{A}_1(\lambda, x, y) = -\frac{\delta \hat{\psi}_1}{U} = k \delta \hat{\phi}_1 + \frac{\partial \delta \hat{\phi}_1}{\partial y}, \qquad \tau = \frac{\lambda}{U}$$

and with the tilde again denoting the Fourier transform, we have:

$$\tilde{\hat{A}}(\lambda, \omega_1, \omega_2) = 2 \frac{\kappa + i \omega_2}{\sqrt{\omega_1^2 + \omega_2^2}} \tilde{\hat{v}}(\lambda, \omega_1, \omega_2).$$

Or

$$\tilde{\hat{v}}(\lambda, \omega_1, \omega_2) = \frac{1}{2} \frac{\sqrt{\omega_1^2 + \omega_2^2}}{\kappa + i \omega_2} \tilde{\hat{A}}(\lambda, \omega_1, \omega_2). \qquad (10.16)$$

This is similar to what we had for the finite plane case in normal flow (cf. (5.18)) with $M = 0$ and $\alpha = 0$ where now of course the x- and y-axes are switched. But there we specialized to the typical section case. Here we need to work with

$$\hat{w}_{a,1}(\lambda, x, y) = (-1)\left[U\left(\hat{h}'(\check{},y) + (x-ab)\hat{\theta}'(\lambda,y)\right)\right.$$

$$\left. + \lambda\hat{h}(\lambda,y) + \lambda(x-ab)\hat{\theta}(\lambda,y)\right], \quad x, y \text{ in } \Gamma.$$

As we have noted, what is different from (5.5) is the appearance of derivatives with respect to y; in (5.5) there are no derivatives with respect to x.

The analogue of the Possio equation is:

$$\hat{w}_{a,1}(\lambda, x, y) = \frac{1}{4\pi^2}\int_{-\infty}^{\infty}\int_{-\infty}^{\infty} e^{ix\omega_1 + iy\omega_2}$$

$$\times \frac{1}{2}\frac{\sqrt{(\omega_1^2 + \omega_2^2)}}{(\kappa + i\omega_2)}\tilde{\hat{A}}_1(\lambda, \omega_1, \omega_2)d\omega_1 d\omega_2, \quad \text{for } x, y \text{ in } \Gamma.$$

$$(10.17)$$

However we cannot quite follow the Possio procedure as in Chap. 5 because $\sqrt{(\omega_1^2 + \omega_2^2)}/(\kappa + i\omega_2)$ is not a Mikhlin multiplier on $L_p\left(R^2\right)$; it is not even bounded in ω_1. Moreover the Fourier transform is not defined. At this point for normal flow we simplified the problem by specializing to the typical section case. We need to determine whether there is a parallel here. We take advantage of the decomposition

$$\hat{w}_{a,1}(\lambda, x, y) = xf_1(\lambda, y) + f_2(\lambda, y) \quad x, y \text{ in } \Gamma,$$

where

$$f_1(\lambda, y) = (-1)\left[U\hat{\theta}'(\lambda, y) + \lambda U\hat{\theta}(\lambda, y)\right],$$

$$f_2(\lambda, y) = (-1)\left[U\left(\hat{h}'(\lambda, y) + (-ab)\hat{\theta}'(\lambda, y)\right)\right.$$

$$\left. + \lambda\hat{h}(\lambda, y) + \lambda(-ab)\hat{\theta}(\lambda, y)\right].$$

Hence for x, y in Γ

$$xf_1(\lambda, y) + f_2(\lambda, y) = \frac{1}{4\pi^2}\int_{-\infty}^{\infty}\int_{-\infty}^{\infty} e^{ix\omega_1 + iy\omega_2}$$

$$\times \frac{1}{2}\frac{\sqrt{(\omega_1^2 + \omega_2^2)}}{(\kappa + i\omega_2)}\tilde{\hat{A}}_1(\lambda, \omega_1, \omega_2)d\omega_1 d\omega_2, \quad (10.18)$$

where, in our usual tilde notation for the Fourier transform:

$$\tilde{\hat{A}}_1(\lambda, \omega_1, \omega_2) = \int_{-b}^{b}\int_{-\ell}^{\ell} e^{-ix_1\omega_1 - iy_1\omega_2}\hat{A}_1(\lambda, x_1, y_1)dx_1 dy_1$$

and we require:

$$\hat{A}_1(\lambda, x, y) \to 0 \quad \text{as } y \to \ell - .$$

This is then the version of the Possio equation valid for axial flow. We can obtain the solution as follows.

We determine $\hat{\tilde{A}}_{11}(\lambda, .)$ from

$$xf_1(\lambda, y) = \frac{1}{4\pi^2} \int_{-\infty}^{\infty} \int_{-\infty}^{\infty} \delta'(\omega_1) e^{i\omega_1 x} e^{+iy\omega_2}$$

$$\times \frac{1}{2} \frac{\sqrt{(\omega_1{}^2 + \omega_2{}^2)}}{(\kappa + i\omega_2)} \tilde{A}_{11}(\lambda, \omega_2) d\omega_1 d\omega_2.$$

Or,

$$xf_1(\lambda, y) = \frac{1}{4\pi^2} \int_{-\infty}^{\infty} e^{+iy\omega_2} \frac{1}{2} \frac{i|\omega_2|}{(\kappa + i\omega_2)} x\tilde{A}_{11}(\lambda, \omega_2) d\omega_2, \qquad |y| < \ell,$$

$$f_1(\lambda, y) = \frac{1}{4\pi^2} \int_{-\infty}^{\infty} e^{+iy\omega_2} \frac{1}{2} \frac{|\omega_2|}{(\kappa + i\omega_2)} \left(i\tilde{A}_{11}(\lambda, \omega_2) \right) d\omega_2, \qquad |y| < \ell,$$

$$(10.19)$$

and similarly $\hat{\tilde{A}}_{12}(\lambda, .)$ from

$$f_2(\lambda, y) = \frac{1}{4\pi^2} \int_{-\infty}^{\infty} \int_{-\infty}^{\infty} \delta(\omega_1) e^{i\omega_1 x} e^{+iy\omega_2} \frac{1}{2} \frac{\sqrt{(\omega_1{}^2 + \omega_2{}^2)}}{(\kappa + i\omega_2)} \tilde{A}_{12}(\lambda, \omega_2) d\omega_1 d\omega_2.$$

Or

$$f_2(\lambda, y) = \frac{1}{4\pi^2} \int_{-\infty}^{\infty} e^{+iy\omega_2} \frac{1}{2} \frac{|\omega_2|}{(\kappa + i\omega_2)} \tilde{A}_{12}(\lambda, \omega_2) d\omega_2, \qquad |y| < \ell. \quad (10.20)$$

And

$$\hat{A}_1(\lambda, x, y) = ix\hat{A}_{11}(\lambda, y) + \hat{A}_{12}(\lambda, y) \qquad x, y \text{ in } \Gamma. \qquad (10.21)$$

We are fortunate that we can use the results from Chap. 5 to solve (10.19) and (10.20).

Thus we have, noting that the functions $f_1(.), f_2(.)$ are in $C_1[-\ell, \ell]$:

$$\left(i\hat{A}_{11}(\lambda, .) \right) = (I + \kappa \mathcal{L}(0))$$

$$\times \left[\hat{v}_1(\lambda, .) + h(\kappa, .) \frac{L\left(\kappa, (I + \kappa \mathcal{L}(0))\hat{v}_1(\lambda, .) \right)}{\ell\kappa (K_0(\ell\kappa) + K_1(\ell\kappa)) e^{\ell\kappa}} \right], \qquad (10.22)$$

where

$$\hat{v}_1(\lambda, .) = 2\mathcal{T} f_1(.)$$

$$\hat{A}_{12}(\lambda, .) = (I + \kappa\mathcal{L}(0)) \left[\hat{v}_2(\lambda, .) + h(\kappa, .) \frac{L\left(\kappa, (I + \kappa\mathcal{L}(0))\hat{v}_2(\lambda, .)\right)}{\ell\kappa(K_0(\ell\kappa) + K_1(\ell\kappa))e^{\ell\kappa}} \right],$$

$$(10.23)$$

where

$$\hat{v}_2(\lambda, .) = 2\mathcal{T} f_2(.), \qquad (10.24)$$

where $\mathcal{T}$ is the linear bounded operator on $C_1[-\ell, \ell]$ into $L_p[-\ell, \ell]$ defined by (as before except for change of axis)

$$\mathcal{T} f = g; \qquad g(y) = \frac{1}{\pi} \sqrt{\frac{\ell - y}{\ell + y}} \int_{-\ell}^{\ell} \sqrt{\frac{\ell + \zeta}{\ell - \zeta}} \frac{f(\zeta)}{\zeta - y} d\zeta, \qquad |y| < \ell.$$

$$\mathcal{L}(0) f = g; \qquad g(y) = \int_{-\ell}^{y} f(s) ds, \qquad |y| < \ell,$$

$$L(\kappa, f) = \int_{-\ell}^{\ell} e^{-\kappa(\ell - s)} f(s) ds$$

$$h(\kappa, y) = \frac{1}{\pi} \sqrt{\frac{\ell - y}{\ell + y}} \int_0^{\infty} e^{-\kappa\sigma} \sqrt{\frac{2\ell + \sigma}{\sigma}} \frac{1}{y - \ell - \sigma} d\sigma, \qquad |y| < \ell.$$

10.6 Stability: Aeroelastic Modes

Our primary concern is of course the stability of the structure under aerodynamic loading. We begin as before with the aeroelastic dynamic equation–linearized about the steady-state solution—in terms of the Laplace transforms with all structure initial conditions set to zero.

We have:

$$m\lambda^2\hat{h}(\lambda, y) + \lambda^2 S\hat{\theta}(\lambda, y) + EI\hat{h}''''(\lambda, y)$$

$$= \rho U \int_{-b}^{b} \hat{A}(\lambda, x, y) dx, \qquad (10.25)$$

$$I_\theta\lambda^2\hat{\theta}(\lambda, y) + S\lambda^2\hat{h}(\lambda, y) - GJ\hat{\theta}''(\lambda, y)$$

$$= \rho l U \int_{-b}^{b} (x - ab)\hat{A}(\lambda, x, y) dx. \qquad (10.26)$$

And we have the "two-point boundary value" conditions:

$$h, h', \theta = 0 \quad \text{at } y = -\ell,$$
$$h''h''', \theta' = 0 \quad \text{at } y = \ell \tag{10.27}$$

that the solution must satisfy.

An important feature of axial flow is that if we set

$$a = 0; \qquad S = 0, \tag{10.28}$$

which certainly holds in the applications of interest, the aeroelastic equations above become uncoupled. In fact we have:

$$m\lambda^2 \hat{h}(\lambda, y) + EI\hat{h}''''(\lambda, y) = \rho U \int_{-b}^{b} \hat{A}(\lambda, x, y)dx, \tag{10.29}$$

$$I_\theta \lambda^2 \hat{\theta}(\lambda, y) - GJ\hat{\theta}''(\lambda, y) = \rho U \int_{-b}^{b} x\hat{A}(\lambda, x, y)dx, \tag{10.30}$$

where the main point is that the right side of (10.26) does not contain the pitching mode and the right side of (10.28) does not contain the bending mode. This does not happen for normal flow.

From now on we assume (10.26). As in Chap. 5, we first look at the $\lambda = 0$.

Divergence Speed

We have defined divergence speed for normal flow in Chap. 5. We follow the same definition here for axial flow. Thus we consider the case where the state variables do not depend on time; all time derivatives are set to zero. And look for values of U for which these time invariant equations have a nonzero solution as in an eigenvalue problem.

Equivalently, we set $\lambda = 0$ in (10.25) and (10.26).

Then we have:

$$f_1(y) = (-1)U\theta'(y),$$
$$f_2(y) = (-1)Uh'(y).$$

And the Possio equation: for x, y in Γ (cf. (10.21))

$$xf_1(y) + f_2(y) = \frac{1}{4\pi^2} \int_{-\infty}^{\infty} \int_{-\infty}^{\infty} e^{ix\omega_1 + iy\omega_2}$$

$$\times \frac{1}{2} \frac{\sqrt{(\omega_1^2 + \omega_2^2)}}{(i\omega_2)} \tilde{A}_1(\omega_1, \omega_2)d\omega_1 d\omega_2,$$

where

$$\tilde{A}_1(\omega_1, \omega_2) = \int_{-b}^{b} \int_{-\ell}^{\ell} e^{-ix_1\omega_1 - iy_1\omega_2} A(x_1, y_1) dx_1 dy_1$$

$$A_1(x, y) \to 0 \text{ as } y \to \ell-$$

The solution is obtained by specializing (10.22) and (10.23) to $\lambda = 0$, yielding

$$A_1(x, .) = x2T f_1 + 2T f_2.$$

And the aeroelastic equations are:

$$EI h''''(.) = \rho U \int_{-b}^{b} A_1(x, .) dx = -4b\rho U^2 T h'$$

with the CF boundary conditions. Note that

$$h' = I_3 h'''',$$

where

$$I_3 f = g; \qquad g(y) = \int_{-\ell}^{y} \int_{s}^{\ell} \int_{t}^{\ell} f(\zeta) d\zeta dt ds.$$

Thus we have the eigenvalue problem in $L_p[-\ell, \ell]$, $1 < p < 2$:

$$f = \mu T I_3 f, \qquad \mu = -\frac{4b\rho U^2}{EI}.$$

Similarly

$$\theta'' = \frac{4\rho b U^2}{3GJ} T \theta'.$$

Or we have the eigenvalue problem in $L_p[-\ell, \ell]$, $1 < p < 2$:

$$f = \mu T I_1 f, \qquad \mu = \frac{4\rho b U^2}{3GJ},$$

$$I_1 f = g; \qquad g(y) = -\int_{y}^{\ell} f(s) ds$$

and we note that the operators $T I_3$ and $T I_1$ are compact.

We have thus at most a sequence of values of the speed and if any sequence of speeds is nonempty, the smallest speed defines the divergence speed, if any. Note that in contrast to the normal flow case we have set the eigenvalue problem in the L_p space, $1 < p < 2$ rather than $p = 2$.

Flutter Speed/Stability Curve

Let us now go on to consider the general case. The first question concerns the space of functions $\theta(.)$ and $h(.)$ in which to seek solutions of (10.28) and (10.29). This depends on the space in which we can locate the right-hand side functions. Here we consider only the bending modes. Note that they are indeed pure bending modes in contrast to the case of normal wind Thus for (10.28) we have:

$$m\lambda^2\hat{h}(\lambda, y) + EI\hat{h}''''(\lambda, y) = \rho U \int_{-b}^{b} \hat{A}(\lambda, x, y)dx$$

$$= 2b\rho U \hat{A}_{12}(\lambda, y), \qquad |y| < \ell,$$

where

$$\hat{A}_{12}(\lambda, .) = (I + \kappa\mathcal{L}(0)) \left[\hat{v}_2(\lambda, .) + h(\kappa, .)\left(L\left(\kappa, (I + \kappa\mathcal{L}(0))\hat{V}_2(\lambda, .)\right)\right)\right.$$

$$\left. \bigg/ \left(\ell\kappa(K_0(\ell\kappa) + K_1(\ell\kappa))e^{\ell\kappa}\right)\right]. \tag{10.31}$$

Note that

$$\hat{v}_2(\lambda, .) = 2\mathcal{T} f_2(\lambda, .),$$

$$f_2(\lambda, y) = (-1)\left[U\hat{h}'(\lambda, y) + \lambda\hat{h}(\lambda, y)\right]$$

and the range of $\mathcal{T}$ is only in $L_p[-\ell, \ell]$ for $1 < p < 2$, even for $f_2(.)$ in $C_1[-\ell, \ell]$.

Hence we can no longer use a Hilbert space formulation as we did for the case of normal flow. In addition we have a more profound difference from the normal flow case. Thus, if we go to the formulation as in Chap. 5, but now consider $x(.)$ as an element in the space $L_p[-\ell, \ell]^4 : x(.) = \text{col}[x_1(.), \ldots x_4(.)]$.

Then we can write

$$\mathcal{D}x(.) = A(\lambda, U)x(.), \tag{10.32}$$

where

$$A(\lambda, U) = \begin{pmatrix} 0 & I & 0 & 0 \\ 0 & 0 & I & 0 \\ 0 & 0 & 0 & I \\ W_1 & W_2 & 0 & 0 \end{pmatrix},$$

$$\mathcal{D}x(.) = z(.); \quad z(s) = x'(s), \quad |s| < \ell$$

defining a closed linear operator on a dense domain in $L_p[-\ell, \ell]^4$, I is the identity operator on $L_p[-\ell, \ell]$ into itself, and w_1, w_2 are no longer scalar multipliers (as in the case of normal flow) but are linear bounded operators on $L_p[-\ell, \ell]$ into itself.

$$W_1h = -\frac{1}{EI}\Big[-m\lambda^2 h + 2b\rho U(I + \kappa\mathcal{L}(0))\Big[2\lambda Th + h(\kappa,.)$$

$$\times(L(\kappa,(I + \kappa\mathcal{L}(0))2\lambda Th))\Big/\big(\ell\kappa(K_0(\ell\kappa) + K_1(\ell\kappa))e^{\ell\kappa}\big)\Big]\Big],$$

$$W_2h = -\frac{1}{EI}\Big[2b\rho U(I + \kappa\mathcal{L}(0))\Big[2UTh + h(\kappa,.)$$

$$\times(L(\kappa,(I + \kappa\mathcal{L}(0))2UTh))\Big/\big(\ell\kappa(K_0(\ell\kappa) + K_1(\ell\kappa))e^{\ell\kappa}\big)\Big]\Big].$$

We go on to formulate the two-point boundary value problem for (10.30).
   We can integrate (10.30) as

$$x(s) = x(-\ell) + g(s), \qquad |s| < \ell,$$
$$g = I_4 A(\lambda, U)x(.),$$

where I_4 is the integral operator

$$I_4 x = z; \quad z(s) = \int_{-\ell}^{s} x(\sigma)d\sigma, \quad |s| < \ell.$$

For $x = \text{col}[x1, x2, x3, x4]$, let

$$L_{24}x = \begin{bmatrix} x_3 \\ x_4 \end{bmatrix}.$$

We choose

$$x(-\ell) = \text{col}[0, 0, x_3, x_4].$$

Then we can write

$$L_{24}x(\ell) = D(\lambda, U)\begin{bmatrix} x_3 \\ x_4 \end{bmatrix},$$

where $D(\lambda, U)$ is 2 by 2 and the two-point boundary value problem can be stated:

$$D(\lambda, U)\begin{pmatrix} x_3 \\ x_4 \end{pmatrix} = 0.$$

Or, we have as before, the aeroelastic modes are the zeros:

$$d(\lambda, U) = \det D(\lambda, U) = 0.$$

We note that the region of analyticity of $d(\lambda, U)$ just as in the normal flow case in Chap. 5 is determined by the Bessel K functions appearing in the denominator in (10.23). In other words the aeroelastic modes are in the same region as before.
   We note also one technique for solving (10.31). Define the function

$$x_0(y) = x(-\ell), \qquad |y| < \ell.$$

Then we can express (10.31) as an equation in $L_p[-\ell, \ell]^4$: $x - I_4 A(\lambda, U)x = x_0$ and for small enough U (as in the applications we are considering) we may use a Neumann expansion

$$x = \sum_{n=0}^{\infty} (I_4 A(\lambda, U))^n x_0.$$

We stop here because this is only intended to be an addendum limited to problem formulation as already noted.

Next we go on to the linear time domain formulation.

10.7   Linear Time Domain Theory: The Convolution/ Evolution Equation

As before, to obtain the time domain equations we take inverse Laplace transforms in (10.22) and (10.23). Again we only consider the case where $a = 0$ and $S = 0$.

This yields

$$m\ddot{h}(t, y) + E I h''''(t, y) = \rho U \int_{-b}^{b} A(t, x, y)\mathrm{d}x,$$

$$I_\theta \ddot{\theta}(t, y) - GJ\theta''(t, y) = \rho U \int_{-b}^{b} (x - ab)A(t, x, y)\mathrm{d}x.$$

With $A_1(t, y)$ denoting the inverse Laplace transform of $i\hat{A}_{11}(\lambda, y)$ and $A_2(t, y)$ denoting the inverse Laplace transform of $\hat{A}(\lambda, y)$,

we have

$$A(t, x, y) = xA_1(t, y) + A_2(t, y).$$

Next the inverse Laplace transform of $\hat{v}_1(\lambda, .)$ is given by

$$v_1(t, .) = \mathcal{T}\left[-4Ubh'(t, .) - 4b\dot{h}(t, .) \right].$$

And similarly that of $\hat{v}_2(\lambda, .)$

$$v_2(t, .) = \mathcal{T}\left[-\frac{4b^3 U}{3}\theta'(t, .) - \frac{4b^3}{3}\dot{\theta}(t, .) \right].$$

And we face the same problem of the choice of spaces as we saw for the Laplace domain. Note that there is no analogue for this in the case of normal flow. There is no longer the distinction of noncircular and circular terms either.

Proceeding further we invert

$$\rho U(I + \kappa \mathcal{L}(0)) \left[\hat{v}_i(\lambda, .) + h(\kappa, .) \left(L \left(\kappa, (I + \kappa \mathcal{L}(0)) \hat{v}_i(\lambda, .) \right) \right) \right]$$

$$\bigg/ \left(\ell \kappa (K_0(\ell \kappa) + K_1(\ell \kappa)) e^{\ell \kappa} \right) \quad i = 1, 2$$

to the time domain. Following Sect. 5.5, this yields

$$\rho U \left[\gamma i(t, y) + \frac{1}{U} \int_{-\ell}^{y} \dot{\gamma}_i(t, s) ds \right], \quad |y| < \ell, \ i = 1, 2,$$

where

$$\gamma_i(t, y) = v_i(t, y) - \int_0^t q(t - \sigma, y) \int_{-\ell}^{y} \dot{v}_i(\sigma, s) \, ds d\sigma,$$

$$q(t, y) = U \int_0^{Ut} r(t - \sigma, y) c(Ut - \sigma) \, d\sigma,$$

$$r(t, y) = \frac{1}{\pi} \sqrt{\frac{\ell - y}{\ell + y}} \frac{1}{\ell - y + t} \sqrt{\frac{2\ell + t}{t}},$$

$$\int_0^{\infty} e^{-\lambda t} c(t) dt = \frac{1}{\int_0^{\infty} e^{-t} \sqrt{\frac{2\ell \lambda + t}{t}} dt}.$$

Thus we have for the time domain equations

$$m \ddot{h}(t, y) + EI h''''(t, y) = \rho U \left[\gamma_1(t, y) + \frac{1}{U} \int_{-\ell}^{y} \dot{\gamma}_1(t, s) ds \right], \quad |y| < \ell,$$

$$I_\theta \ddot{\theta}(t, y) - GJ \theta''(t, y) = \rho U \left[\gamma_2(t, y) + \frac{1}{U} \int_{-\ell}^{y} \dot{\gamma}_2(t, s) ds \right], \quad |y| < \ell.$$

We can again formulate this as a convolution/evolution equation similar to (5.1) for the incompressible normal flow case, except that it will be an equation in a Banach space.

Finally we go on to formulate the stability problem for the nonlinear system continuing with the power series expansion (10.4) following the development in Chap. 6.

10.8   Nonlinear Stability Theory: Hopf Bifurcation/Flutter LCO

It should be noted that we did not consider the incompressible case $M = 0$ for normal flow in Chap. 6 on nonlinear stability. Here we do it for axial flow while still following pretty much the material therein in principle.

Without apology, we specialize to the case: $a = 0$ and $S = 0$. As we have seen, this uncouples the structure equations for the linear case.

For $k > 1$, we have the field equation

$$\frac{\partial^2 \phi_k(t, .)}{\partial x^2} + \frac{\partial^2 \phi_k(t, .)}{\partial y^2} + \frac{\partial^2 \phi_k(t, .)}{\partial z^2} = 0, \quad -\infty < x, y, z < \infty''''$$

with the boundary condition:

$$\frac{\partial \phi_k(x, y, 0)}{\partial z} = (-1) \left[\frac{\partial}{\partial y} \phi_{k-1}(t, x, y, 0) \left(\frac{\partial}{\partial y} h(t, y) + (x - ab) \frac{\partial}{\partial y} \theta(t, y) \right) \right.$$
$$\left. + \left(\frac{\partial}{\partial x} \phi_{k-1}(t, x, y, 0) \right) \theta(t, y) \right] \quad x, \ y \text{ in } \Gamma.$$

The aeroelastic equation is

$$m\ddot{h}(t, y) + EI h''''(t, y) = \rho U \int_{-b}^{b} A(t, x, y) \mathrm{d}x, \quad -\ell < y < \ell,$$

$$I_\theta \ddot{\theta}(t, y) - GJ \theta''(t, y) = \rho U \int_{-b}^{b} x A(t, x, y) \mathrm{d}x \quad -\ell < y < \ell,$$

$$A(t, x, y) = \sum_{n=1}^{\infty} \frac{A_n(t, x, y)}{n!},$$

where A_n is given by (10.15) and again is the solution of a Possio equation for each n, and the expansion leads to a Volterra series.

We end the addendum here, deferring more details to Volume. 2.

References

List of Books

1. Mikhlin, S.G.: Mathematical Physics, an Advanced Course. North-Holland, Amsterdam (1970)
2. Thompson, P.A.: Compressible Fluid Dynamics. McGraw-Hill, New York (1972)
3. Van Dyke, M.: Perturbation Methods in Fluid Mechanics. The Parabolic Press, California (1975)
4. Chorin, A.I., Marsden, J.E.: A Mathematical Introduction to Fluid Mechanics. Springer, New York (1993)
5. Hodges, D.H., Pierce, G.A.: Introduction to Structural Dynamics and Aeroelasticity. Cambridge University Press, Cambridge (2007)
6. Bisplinghoff, R.L., Ashley, H., Halfman, R.L.: Aeroelasticity. Addison-Wesley, Cambridge (1955)
7. Marsden, J.E., M. McCracken: The Hopf Bifurcation and its Applications. Springer, New York (1976)
8. Miller, R.K.: NonLinear Volterra Integral Equations. Benjamin, W.A. (1971)
9. Etkin, B., Reid, L.D.: Dynamics of Flight. John Wiley, New york (1996)
10. Hille, E., Phillips, R.S.: Functional Analysis and Semigroups. American Mathematical Society, Providence (1957)
11. Tricomi, F.G.: Integral Equations. Dover Publications, New York (1955)
12. Landau, L.D., Lifshitz, E.M.: Fluid Mechanics. Elsevier, Amsterdam (2004)
13. Oleinik, O.A., Samokhin, V.N.: Mathematical Models in Boundary Layer Theory. Chapman & Hall, London (1999)
14. Meyer, R.E.: Introduction to Mathematical Fluid Dynamics. Wiley Interscience, New York (1971)
15. Schlichting, H., Gersten, K.: Boundary Layer Theory. Springer, Berlin (2003)
16. Balakrishnan, A.V.: Applied Functional Analysis. Springer, New York (1980)
17. Dowell, E. et al: A Modern Course in Aeroelasticity. Kluwer, New York (2004)
18. Kuethe, A.M., C-Y. Chow: Foundations of Aerodynamics. John Wiley, New York (1976)
19. Temam, R.: Navier Stokes Equations. RMS Chelsey, Providence (1984)
20. Mikhlin, S.G.: Multidimensional singular integrals and integral equations. Pergammon, New York (1965)
21. Cole, J.D., Cook, L.P.: Transonic Aerodynamics. North Holland, Amsterdam (1986)
22. Butzer, P.L., Nessel, R.J.: Fourier Analysis and Approximation, vol. 1. Academic Press, New York (1971)
23. Knopp, K.: Theory of Functions, Part II. Dover, New York (1947)

A V Balakrishnan, *Aeroelasticity: The Continuum Theory*,
DOI 10.1007/978-1-4614-3609-6, © Springer Science+Business Media, LLC 2012

24. Dowell, E.H., Tang, D.: Dynamics of Very High Dimensional Systems. World Scientific, Singapore (2003)
25. Lin, J.: Suppression of Bending Torsion Wing Flutter Using Self Straining Controllers. Dissertation, Engineering UCLA (2003)
26. Balakrishnan, A.V.: Introduction to Random Processes in Engineering. Wiley, New York (2005)
27. Liepmann, H.W., Roshko, A.: Elements of Gas Dynamics. Dover, New York (1985)
28. Carracedo, C.M., Alix, M.S.: The Theory of Fractional Powers of Operators. Elsevier, Amsterdam (2001)
29. Watson, G.N.: A Treatise on Bessel Functions. Cambridge University Press, Cambridge (1955)
30. Anderson, J.D.: Modern Compressible Flow, 3rd edn. McGrawHill, New York (2003)
31. Gohberg, I., Krupnik, N.: One Dimensional Singular Integral Equations, vol. 1, Introduction. Birkhauser Verlag, Boston (1992)
32. Timoshenko, S.: Vibration Problems in Engineering, 2nd edn. Van Nostrand, New York (1937)
33. Schmidt, L.V.: Introduction to Aircraft Flight Dynamics. AIAA Education Series, CA (1998)
34. Timoshenko, S., Young, D.H.: Vibration Problems in Engineering. Van Nostrand Reinhold, Dordrecht (1955)
35. Tatarski, V.I.: The Effects of the Turbulent Atmosphere on Wave Propgation. US Dept. of Commerce, Springfield (1971)
36. Meyer, R.E. (ed.): Transonic Shock and Multidimensional Flows. Academic Press, New York (1982)
37. Kielhofer, H.: Bifurcation Theory An Intrudction with Application to pde. Springer, New York (2004)
38. Nixon, D. (ed.): Unsteady Transonic Aerodynamics, Progress in Astronautics and Aeronautics. AIAA, Washington (1989)
39. Landhal, M.T.: Unsteady Transonic Flow. Pergammon, New York (1961)
40. Liepmann, H.W., Puckett, A.E.: Introduction to Aerodynamics of a Compressible Fluid. Wiley, New York (1947)
41. Arendt, W., Batty, C.J.K., Hieber, M., Neubrander, F.: Vector valued Laplace Transforms and Cauchy Problems. Birkhauser Verlag, Basel (2000)
42. Pontryagin, L.S.: Ordinary Differential Equations. Addison Wesley, London (1962)
43. Mikhlin, S.G.: Multidimensional Singular Integrals and Integral Equations. Pergammon, New York (1965)
44. Courant, R., Hilbert, D.: Methods of Mathematical Physics, vols. 1–2. Interscience, New York (1966)
45. Barbu, V.: Nonlinear Semigroups and Differential Equations in Banach Spaces. Nordhoff, Leyden (1976)
46. Hochstadt, H.: Integral Equations. Wiley, New York (1973)
47. Fung, Y.C.: An Introduction to the Theory of Aeroelasticty. Dover Publications, New York (1983)
48. Goldberg, I.C., Krein, M.G.: Introduction to the Theory of Linear Nonselfadjoint Operators, vol. 18. AMS Translations, Providence (1969)
49. Anderson, J.D.: Modern Compressible Flow. McGrawHill, New York (2003)
50. Paidoussis, M.P.: Fluid Structure Interactions, Slender Structures and Axial Flow, vol. 12. Elsevier Academic Press, London (2004)
51. Rudin, W.: Real Complex Analysis. McGraw–Hill, New York (1966)
52. Lovitt, W.V.: Linear Integral Equations. Dover, New York (1950)

List of Papers

53. Balakrishnan, A.V.: Vibrating Systems with singular mass-inertia matrices. In: Sivasundaram, S. (ed.) Proceedings ICNPAA, Embry-Riddle Aeronautical University Press (1996)
54. Balakrishnan, A.V.: Dynamics and Control of Articulated Anisotropic Timoshenko beams In: Tzou, H.S., Bergman, L.A. (ed.) Dynamics and Control of Distributed Systems. Cambridge University Press, Cambridge (1998)
55. Balakrishnan, A.V.: On the nonnumeric mathematical foundations of linear aeroelasticity. In: Sivasundaram, S. (ed.) Proceedings ICNPAA, August 2002
56. Balakrishnan, A.V.: Possio Integral Equation of Aeroelasticity. J. Aero. Eng. 16(4), 138–154 October (2003)
57. Balakrishnan, A.V.: An Integral Equation in Aeroelasticity. In: Aman, H. et al. (ed.) Functional Analysis and Evolution Equations. Birkhauser, New York (2007)
58. Balakrishnan, A.V.: On superstability of Semigroups. In: Polis, M.P. et al. (ed.) Proceedings, IFIP TC7 System Modelling and optimization 12–19 (1999)
59. Balakrishnan, A.V.: NonLinear Aeroelasticity Theory: Continuum Models, Contemporary Mathematics, vol. 420. American Mathematical Society, Providence (2007)
60. Balakrishnan, A.V.: Subsonic Flutter Suppression using self-straining actuators. J. Franklin Inst. 338, 149–170 (2001)
61. Balakrishnan, A.V.: Control of flexible flight structures. In: Analyse Mathematique et Applications, Gauthier-Villars (1988)
62. Balakrishnan, A.V.: Theoretical Limits of Damping Attainable by Smart beams with rate feedback. In: Proceedings of the SPIE, vol. 3039 (1997)
63. Balakrishnan, A.V.: Damping operators in continuum models of flexible flight structures. Explicit models for proportional damping in Beam Torsion. J. Differ. Integr. Equat. 13 (1990)
64. Balakrishnan, A.V.: Modelling Response of flexible high aspect ratio wings to wind turbulence. J. Aero. Eng. 19(2), 121–132 (2006)
65. Balakishnan, A.V.: Transonic small disturbance potential equation. AIAA J. 16(4), 139–154 (2004)
66. Balakrishnan, A.V.: NonLinear Possio Integral Equation and aeroelastic flutter limit cycle oscillation, NonLinear Studies, vol. 16. I & S. Publishers (2009)
67. Balakrishnan, A.V.: NonLinear aeroelasticity, the steady state theory. AIAA J. 44, 1006–1012 (2007)
68. Balakrishnan, A.V.: State space representation of flexible sturcture dynamics in air flow. J. Franklin Inst. 347(4), 17–29 (2010)
69. Balakrishnan, A.V.: Damping Performance of strain actuated beams. Comput. Appl. Math. 18 (1999)
70. Balakrishnan, A.V.: Control of structures with selfstraining actuators: coupled Euler/Timoshenko model; I. In: Sivasundaram, S. (ed.) NonLinear Problems in Aviation and Aerospace, Gordon and Breach, Amsterdam (2000)
71. Balakrishnan, A.V.: Analytical solution of the Possio Integral Equation for the sonic case. J. Aerosp. Eng. 434 (2009)
72. Balakrishnan, A.V.: In: Ceragioloi, F. et al. (eds.) The Possio Integral Equation of Aeroelasticity: A Modern View, System Modelling and optimization. Springer, Berlin (2005)
73. Balakrishnan, A.V., Iliff, K.W.: Continuum aeroelastic model for inviscid subsonic bending–torsion Wing flutter. J. Aerosp. Eng. 20(3), 152–164 (2007)
74. Balakrishnan, A.V., Shubov, M.A.: Asymptotic behaviour of the aeroelastic modes for an aircraft wing model in a subsonic air flow. Proc. R. Soc. 27(4), 329–362 (2004)
75. Beran, P.S., Straganac, T.W., Kim, K.: Studies of store induced Limit Cycle Oscillations Using a model with full system nonlinearities. AIAA 1730, 2003–1730.
76. Dowell, E.H., Traybar, J., Hodges, H.: An experimental theoretical study of nonlinear bending and torsion deformations of a cantilever bar. J. Sound Vib. 533–544 (1977).
77. Goland, M., Luke, M.L.: The flutter of a uniform wing with tip weights. J. Appl. Math. 15(1) (1948).

78. Goland, M.: The flutter of a uniform cantilever wing. J. Appl. Mech. **12**, 197–208 (1945)
79. Alvarez–Salazar, O.S., Iliff, K.W.: Destabilizing effects of rate feedback on strain actuated beams. J. Sound Vib. **221**(2), 289–307 (1999)
80. Majda, A.: Disappearing solutions for the dissipative wave equation. Indiana J. Math. **24**(12), 132–164 (1975)
81. Patil, M.J., Hodges, D.H.: Flight dynamics of highly flexible flying wings. AIAA J. Aircr. **43**(6), 1790–1799 (2006)
82. Su, W., Cesnick, C.E.S.: Dynamic response of highly flexible flying wings. AIAA J. **49**, 324–339 (2011)
83. Tang, D., Dowell, E.: Experimental and Theoretical study on Aeroelastic response of High Aspect Ratio wings. AIAA J. **39**(8), 1430–1441 (2001)
84. Tang, D., Dowell, E.: Experimental and theoretical study of gust response for high–aspect ratio wings. AIAA J. **40**(3), 419–429 (2002)
85. Edwards, J.W., Ashley, H., Breakwell, J.V.: Unsteady Aerodynamic modelling for arbitrary motions. AIAA J. **15**(4), 593–595 (1977)
86. Williams, M.H.: Linearisation of unsteady transonic flows containing shocks. AIAA J. **17**(4), 394–397 (1978)
87. Jameson, A.: Steady state solution of Euler equations in transonic flow. In: Meyer, R.E. (ed.) Transonic Shock and Multidimensional Flows: Advances in Scientific Computing. Academic, New York (1981)
88. Albano, S., Rodden, R.P.: A double–lattice method of calculating lift distributions on oscillating surfaces in subsonic flows. AIAA J. **7**, 279–285 (1969)
89. Possio, C.: L'azzione aerodynamica sul profilo oscillante in un fluido compressibile a velocite iposonara. L'Aerotecnica **18**(4) (1938)
90. Garrick, I.E.: Bending Torsion Flutter calculations modified by subsonic compressibility corrections. NACA TN 1034, May (1946)
91. Balakrishnan, A.V., Triggiani, R.: Lack of generation of Strongly continuous semigroups by the damped wave operator on H×H. Appl. Math. Lett. **6**, 33–37 (1993)
92. Moretti, G.: Toward a closer cooperation between theoretical and numerical analysis in gas dynamics. In: Meyer, R.E. (ed.) Transonic, Shock, and Multidimensional Flows: Advances in Scientific Computing, pp. 241–257. Academic, New York (1981)
93. Holmes, P.J.: Bifurcations to divergence and flutter in flow-induced oscillations–finite dimensional analysis. J. Sound Vib. **36**(4), 1730–1754 (1977)
94. Grenier, E.: Non derivation des equations de Prandtl, Seminaire Séminaire Équations aux dérivées partielles (Polytechnique) (1997–1998), Exp. No. 18, p. 12.
95. Serre, D.: Von Neumanns comments about existence and uniqueness for the Initial–Boundary value problems. AMS Bull. **47**(1), (2010)
96. Kemp, N.H., Homicz, G.: Approximate unsteady thin airfoil theory for subsonic flow. AIAA J. **14**, 139–144 (1998)
97. Balakrishnan, A.V., Edwards, J.W.: Calculation of the transient motion of elastic aorfoils forced by control surface motion and gust. NASA TM 81351, August (1980)
98. Glimm, J.: Reflections and prospecives. AMS Bull. **47**(1), 127–136 (2010)
99. Bendiksen, O.O.: Transonic flutter. AIAA paper 2002–1488. (2002)
100. Wang, D.M., Amomoto, H.Y., Dowell, E.H.: Flutter and limit cycle oscillation of two dimensional panels in three dimensional axial flow. J. Fluids Struct. (2003)
101. Samko, S.: Singular integral equations: solutions with variable integrability. Unpublished Manuscript
102. Huang, L.: Flutter of cantilevered plates in axial flow. J. Fluids Struct **9**, 127–147 (1995).
103. Balakrishan, A.V.: Explicit LQG optimised control laws for flexible structures/collocated rate sensors. AIAA 94–1657. Adaptive Structures Forum, La Jolla (1993)
104. Friedman, P.P.: Renaissance of Aeroelasticity and its futuré. J. Aircr. **36**, 105–121 (1999)
105. Ellis, P.D.M., Williams, J.E.F., Shneerson, J.W.: Surgical Relief of Snoring due to Palatal Flutter: a preliminary report. Ann. R. Coll. Sugeons Engl. 286–290 (1995).

106. Dunnmon, J.A., Stanton, S.C., Mann, B.P., Dowell, E.H.: Power Extraction from aeroelastic limit cycle oscillations. J. Fluids Struct. **27**, 1182–1194 (2011)
107. Tuffaha, A.: Flutter stability analysis of a wedge shaped airfoil with nonzero thickness in nonviscous air flow. NonLinear Stud. I S publishers **16**(3), 235–257 (2009)
108. O.S. Alvarez Salazar, Balakrishnan, A.V., Iliff, K.W.: Aeroelastic Flight Test NASA–UCLA, Flight Systems Research Center, UCLA, May (2000)
109. Wang, D.A., Kuo, H.H.: Piezoelectric energy harvesing from flow induced vibration. J. Micromech. Microeng. **20**, 1–9 (2010)
110. Sears, W.R.: Operational methods in the theory of air foils in non-uniform motion. J. Franklin Inst. **230**, 95–111 (1940)

Index